Recent Advances and Emerging Challenges in Functional Coatings

Recent Advances and Emerging Challenges in Functional Coatings

Guest Editors

Alessia Serena Perna
Antonio Viscusi

MDPI

Basel • Beijing • Wuhan • Barcelona • Belgrade • Novi Sad • Cluj • Manchester

Guest Editors

Alessia Serena Perna
Department of Chemical, Materials and Production Engineering
University of Naples Federico II
Naples
Italy

Antonio Viscusi
Department of Chemical, Materials and Industrial Production Engineering
University of Naples Federico II
Naples
Italy

Editorial Office
MDPI AG
Grosspeteranlage 5
4052 Basel, Switzerland

This is a reprint of the Special Issue, published open access by the journal *Materials* (ISSN 1996-1944), freely accessible at: https://www.mdpi.com/journal/materials/special_issues/4PA06352F8.

For citation purposes, cite each article independently as indicated on the article page online and as indicated below:

Lastname, A.A.; Lastname, B.B. Article Title. *Journal Name* **Year**, *Volume Number*, Page Range.

ISBN 978-3-7258-5697-8 (Hbk)
ISBN 978-3-7258-5698-5 (PDF)
https://doi.org/10.3390/books978-3-7258-5698-5

Contents

About the Editors

Alessia Serena Perna

Alessia Serena Perna is currently an Assistant Professor at the Department of Chemical, Materials and Industrial Production Engineering, University of Naples Federico II, where she has been involved in research and teaching since October 2023. She earned her master's degree, cum laude, in Mechanical Engineering from the University of Naples Federico II in 2017, and completed her Ph.D. in Technology, Innovation and Management in 2021 through a joint program between the University of Bergamo and the University of Naples Federico II, during which she was also a visiting researcher at École des Mines in Paris. From 2021 to 2023, she worked as a postdoctoral researcher at the University of Naples Federico II. Her research focuses on additive manufacturing technologies, particularly powder-based processes, the mechanical and surface characterization of materials and products, and the development of innovative technological solutions in the context of Industry 4.0. She also investigates the theoretical modeling of additive manufacturing processes, topological and mechanical optimization of components using generative design, and the optimization of technological processes through machine learning methodologies. Her work combines fundamental understanding with applied engineering to advance the design and manufacturing of high-performance materials and functional coatings, contributing to both scientific knowledge and industrial innovation.

Antonio Viscusi

Antonio Viscusi is currently an Associate Professor in Technologies and Manufacturing Systems at the Faculty of Technological Sciences and Innovation, Universitas Mercatorum, a position he has held since April 2024. He earned his master's degree with highest honors in Aerospace Engineering from the University of Naples Federico II in 2014 and completed his Doctorate in Industrial Products and Processes Engineering (Doctor Europaeus) in 2018, with part of his research conducted at the University of Cambridge. His research focuses on advanced additive manufacturing technologies, including cold spray deposition and the production of metallic foam structural components, supported by finite element modeling and detailed mechanical and surface characterization. He is involved in the design and optimization of innovative manufacturing processes, the development of multifunctional materials, and the integration of experimental and numerical approaches for process understanding. Antonio Viscusi has authored over seventy peer-reviewed articles, contributed to national and international research projects, and collaborates with institutions including Politecnico di Milano, CNR, Mines ParisTech, and the University of Cambridge. His work aims to bridge fundamental research with practical engineering applications, advancing sustainable and high-performance manufacturing solutions.

Preface

The present Reprint, "Recent Advances and Emerging Challenges in Functional Coatings", brings together a collection of contributions that were originally published in the Materials Special Issue of the same title. Its scope reflects the rapidly expanding role of functional coatings in modern engineering, where surface properties often determine the performance, durability, and reliability of components in service. The aim of this compilation is to provide a unified reference that captures the state of the art across different approaches including theoretical, numerical, and experimental studies as well as practical applications.

The motivation for assembling this Reprint arises from the recognition that coatings are no longer considered secondary additions to structural materials but are now regarded as essential elements that can impart new functionalities and extend service lifetimes in demanding environments. By presenting diverse perspectives, the Reprint illustrates how innovative coating technologies, advanced characterization methods, and new material design strategies converge to address pressing scientific and industrial challenges.

This collection is intended for a broad audience of researchers, engineers, and practitioners working in the fields of materials science, surface engineering, and manufacturing technologies. It also serves as a resource for graduate students and early career scientists who wish to gain an overview of recent developments and future directions in the area of functional coatings.

We wish to express our gratitude to the authors whose valuable contributions made this Reprint possible, and to the reviewers who provided constructive feedback that ensured the high quality of the published work. Appreciation is also extended to the editorial team of Materials for their professional support throughout the publication process.

Alessia Serena Perna and Antonio Viscusi
Guest Editors

materials

MDPI

Review

Overview of Recent Developments in Composite Epoxy Resin in Organic Coating on Steel (2020–2024)

Jianghua Hao [1,2], Kun Yang [1,3], Jiaye Wu [1], Mingzhu Wu [1,*] and Ying Li [1,*]

1 Chemical Pollution Control Chongqing Applied Technology Extension Center of Higher Vocational Colleges, Chongqing Industry Polytechnic College, Chongqing 401120, China; haojianghua6625@163.com (J.H.); hahayangkun@163.com (K.Y.); wujy@cqipc.edu.cn (J.W.)
2 Avic Xi'an Aircraft Industry Group Co., Ltd., Xi'an 710089, China
3 Avic Shenyang Aircraft Corporation, Shenyang 110086, China
* Correspondence: wumz@cqipc.edu.cn (M.W.); liying@cqipc.edu.cn (Y.L.)

Abstract: Epoxy resin, widely recognized for its excellent performance, is extensively applied in the anti-corrosion field of steel. Continuous enhancement of the anti-corrosion performance of epoxy resins to satisfy more stringent requirements has become a current hot topic of interest in both scientific and industrial circles. This review focuses on recent advancements in composite epoxy resin coatings for steel from 2020 to 2024, emphasizing improvements in anti-corrosion performance through various additive modifications. Modification methods are categorized into metal-based compounds, organic compounds, organometallic compounds, and carbon-based materials. To assist scholars in understanding the latest research advancements, key findings from electrochemical tests, mechanical assessments, and structural characterizations are summarized, highlighting their influence on corrosion resistance, adhesion, mechanical properties, and self-healing capabilities.

Keywords: epoxy coating; modification; steel; anti-corrosion performance

1. Introduction

As a popular structural material, steel is frequently exposed to aggressive environments that accelerate its corrosion. High humidity is one such condition, where water vapor can condense on the steel surface, triggering electrochemical corrosion reactions. Additionally, steel structures located near coastal areas or in chemical industrial zones are continuously subjected to salt spray or corrosive chemicals, further intensifying the corrosion process. The consequences of steel corrosion are twofold: it weakens the structural integrity, and it leads to significant economic losses from repair and replacement costs.

The use of organic coatings to protect metals stands as one of the most prevalent anti-corrosion strategies. These coatings act as a shield, effectively isolating the metal from its surrounding environment and thus preventing damage to the substrate [1]. Epoxy resins, recognized as top-tier anti-corrosion materials, are widely used to protect metals in industry. This is largely attributable to their exceptional mechanical robustness, thermal stability, and chemical resistance [2]. Traditional epoxy coatings have been widely used for their good mechanical properties and corrosion resistance. However, recent modifications have significantly enhanced their performance. For example, metal-based compounds like ZnO and TiO_2 improve barrier properties and active corrosion protection. Organic compounds such as polyaniline enhance adhesion and self-healing capabilities. Organometallic compounds like MOFs offer tunable structures for improved corrosion resistance, while carbon-based materials like graphene provide excellent barrier properties and mechanical strength. These modifications address the limitations of traditional epoxy coatings, offering superior protection in harsh environments. To continuously improve their performance under various application environments, a variety of methods and numerous novel composites have been employed to engineer the structure of epoxy coating.

In this review, we summarize selected typical examples of composite epoxy resin coatings for steel, drawn from peer-reviewed scientific papers from 2020 to 2024. According to the characteristics of the modified coatings, the additives are classified into four categories for in-depth analysis and discussion. These categories are metals and their compounds, organic compounds, organometallic compounds, and carbon materials. Within the metals and their compounds category, inorganic/organic hybrid complexes and silicon compounds are included in this review. It should be noted, however, that an unambiguous classification of certain substances can be quite challenging. Moreover, our discussion does not encompass activation mechanisms and specific testing methodologies.

2. Epoxy Coatings Modified by Metals and Their Compounds

The anti-corrosion performance of epoxy resins modified by metals and their compounds, particularly metal oxides, has been significantly enhanced [3,4]. A wide variety of metal oxides and modified metal oxides, such as ZnO [5], TiO_2 [6], CeO_2 [7], NiO [8], Fe_3O_4 [9], among others, have been added into epoxy resins to optimize the properties of organic coatings on the surface of steel. Through an analysis of XRD, FT-IR, TEM, XPS, and EIS, Kumari et al. [4] revealed that metal oxide (ZnO, TiO_2) and graphene oxide could enhance the interaction of epoxy resin film and the metal surface via the formation of metal–oxygen bonds. This enhanced interaction provided improved corrosion protection for steel in a harsh environment.

2.1. *Modification of Zinc-Rich Epoxy Coating*

Among metals and their compounds, zinc-rich organic coatings have been successfully applied since the 1930s because of their excellent resistance to chemical corrosion [10,11]. Open-circuit potential (OP) and electrochemical impedance spectroscopy (EIS) tests showed that a zinc-rich organic coating (80 wt% zinc particles) provided better cathodic protection for carbon steel [12]. Pandis et al. [10] found that commercial zinc-rich organic coatings (ZN691) could effectively prevent 304 L stainless steel from being corroded under highly acidic media (40% H_2SO_4, 40% HNO_3, and 37% HCl). The authors proved that the anti-corrosion performance of two commercial zinc-rich organic coatings (ZN691 and ANORZ-INC) was higher than that of an FeO-rich organic coating (TROPIS PRIMER CL). The anti-corrosion performance of zinc-rich organic coatings was remarkably improved when 0.5 wt% zinc fibers were added to the mixture system [13]. Due to the high calcite content, eggshell wastes were studied as additives to enhance the corrosion resistance properties of zinc-rich coatings [14]. Electrochemical analysis revealed that added eggshell wastes could enhance the anti-corrosion properties of zinc-rich epoxy coatings because the chloride ions were successfully prevented by eggshell nanoparticles (NPs) from contacting the substrate surface (Figure 1). Qi et al. [11] investigated the corrosion protection performances of zinc-rich coatings when spherical zinc particles were partially replaced by stainless steel flakes, and they considered the enhancement to be attributed to the synergistic effects of zinc and steel. Haddadi et al. [15] synthesized Zn-FALE (zinc cations/*Ferula Asafoetida* leaves extract) and revealed that the material significantly enhanced the barrier and active corrosion protection performance of epoxy coatings. Galvanized steel showed outstanding adhesion strength and corrosion resistance once its surface was treated by Zn-Al layered double hydroxide (LDH) grown in situ [16]. Rahmani et al. [17] used zinc cation and

3-nitrobenzoic acid as raw materials to treat carbon steel by a pulse-reverse electrodeposition method. The treated carbon steel was then laid over epoxy resin and showed stronger adhesion and anti-corrosion performance. Compared with the anti-corrosion performance and mechanical properties of epoxy coatings modified with nano hybrid ZnO particles, Kabaoglu et al. [18] found that a micro ZnO/Epoxy composite coating on steel plate had superior barrier protection. Superhydrophobic coating technology was also utilized to treat carbon steel, and the experimental results showed that a nano zinc oxide/epoxy coating significantly improved the corrosion protection of the substrate [5]. To enhance anti-corrosion performance, organotitanate-modified ZnO NPs were prepared and tested [19]. As the modified ZnO NPs could be dispersed in epoxy resin better than the unmodified ZnO NPs, the modified ZnO NPs improved the mechanical properties, chemical resistance, thermal stability, and anti-corrosion protection ability of epoxy resin coating [19]. Xavier [3] synthesized 3-amino-1,2,4-triazole-5-thiol-functionalized Al_2O_3-ZnO NPs, and the author found that epoxy resin with added Al_2O_3-ZnO NPs had excellent barrier strength and mechanical properties. Moreover, ZnO NPs were added to an organic–inorganic hybrid coating fabricated by bisphenol A diglycidyl ether epoxy resin with silica NPs [20]. The obtained dual-function coating showed higher anti-corrosion and photocatalysis activities. Nano ZnO and micro ZnO have different effects on epoxy coatings. Nano ZnO provides better dispersion and a higher surface area-to-volume ratio, enhancing corrosion protection through improved barrier properties and active inhibition. Micro ZnO, on the other hand, offers superior mechanical properties and easier processing. The choice between nano ZnO and micro ZnO depends on the specific application requirements, with nano ZnO being more suitable for high-corrosion environments and micro ZnO being more suitable for applications where mechanical strength is crucial.

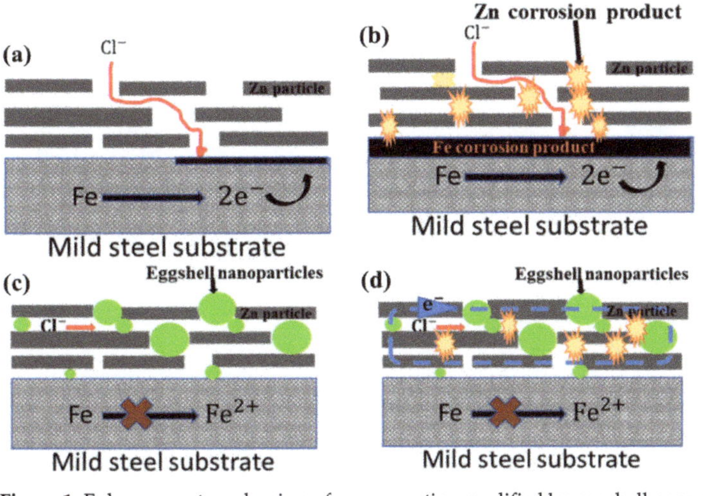

Figure 1. Enhancement mechanism of epoxy coating modified by eggshell nanoparticles ((**a**) ZREC with interstitial spaces for permeation of chloride species; (**b**) ZREC coated mild steel with zinc corrosion products; (**c**) ZENE-5-C coating showing treated eggshell nanoparticles within the interstitial spaces of zinc particles at the start of immersion; (**d**) ZENE-5-C coating showing electrical contact between zincparticles and treated eggshell nanoparticles with deposition of zinc corrosion products after 30 days immersion) [14].

2.2. Titanium-Modified Epoxy Coatings

Titanium compounds as additives in epoxy coatings have also been extensively studied for the anti-corrosion protection of steel [6,21–26]. Dagdag et al. [21] found that the anti-corrosion capability of carbon steel was enhanced in 3% NaCl solution once epoxy resin and

TiO_2 composite, serving as anti-corrosive materials, were coated. The authors explained that the fundamental anti-corrosion mechanism was TiO_2 blocking the surface micropores and hindering the penetration or diffusion of corrosive species. To future improve the corrosion protection, TiO_2 NPs were fabricated with poly-dimethylamino siloxane and applied to steel petroleum tanker trucks. The painted tanker displayed excellent anti-corrosion, mechanical, chemical, electrical, thermal, and UV resistance [22]. As a starting material, TiO_2 coupled with silane was added into epoxy coating to boost the corrosion resistance of electrical steel [26]. Compared to the conventional phosphate coating, the silane-modified TiO_2 had better dispersion stability [26]. In addition, a mixture of TiO_2 and graphite was added into epoxy resin to enhance the thermal resistance of the coating [27].

Recently, MXene polymer hybrid materials have attracted significant attention due to their outstanding resistance to permeation, diverse surface chemical properties, impressive mechanical properties, and metal-like electrical and thermal conductivity [28]. As a kind of MXene polymer hybrid material, epoxy resins with $Ti_3C_2T_x$ MXene were also investigated as an additive for high-performance organic coatings for steel. L-cysteine-modified $Ti_3C_2T_x$ MXene nanosheets were fabricated by Li's group [23] to improve the anti-corrosion performance of waterborne epoxy coatings. The results indicated that a good dispersion of $Ti_3C_2T_x$ MXene enhanced the corrosion resistance of epoxy resin coatings. Due to their excellent barrier performance and mechanical properties, $Ti_3C_2T_x$ (x = 0) MXene nanosheets were synthesized and thoroughly studied as additives in epoxy coatings (Figure 2) [25]. Experiments revealed that the synthesized Ti_3C_2 MXene@PANI (polyaniline) composites could increase the corrosion resistance by 1–2 orders of magnitude compared to pure waterborne epoxy coating in a 3.5 wt% NaCl solution. By a hierarchical design method, Zhang et al. [29] demonstrated that the thermal, tribological, and anti-corrosive performances of steel were significantly enhanced by a microcapsule-based epoxy coating modified with $Ti_3C_2T_x$. Moreover, $Ti_3C_2T_x$ MXenes, in combination with other treatment techniques, were applied to strengthen the anti-corrosion of other metal alloys [24].

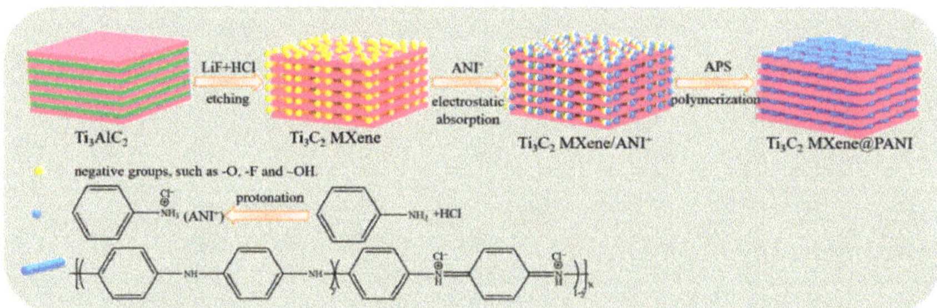

Figure 2. Synthesis of Ti_3C_2 MXene@PANI composites by etching Al layer of the precursor Ti_3AlC_2 using LiF and HCl treatment [25].

2.3. Silicon-Modified Epoxy Coatings

Given that silicon belongs to the metal-like elements, we classify silicon compounds as a type of metal–organic coating in this review. As effective organic coating modification reagents, organic silanes are widely used for doping epoxy coatings on steel. On the carbon steel surface, a multilayer silane-doped epoxy coating exhibited remarkable anti-corrosion and adhesion performance [30]. To enhance the corrosion resistance of steel fibers, core–shell structural PANI@SiO_2 was synthesized via an in situ polymerization method (Figure 3), and a 2 wt% PANI@SiO_2/epoxy coating showed the most excellent resistance to stress corrosion and electrochemical corrosion [31]. To reduce the delamination rate of the coating from the matrix, Fernández-Alvarez et al. [32] modified SiO_2 with calcium using an ion-exchange method for incorporation into organic coatings. The experimental data confirmed that this treatment method could strengthen the scratch resistance, universal hardness, and wear resistance. On the surface of steel, Fe-O-M (metal) and/or Fe-N-M bonds effectively resist steel corrosion. Atta et al. [33] investigated the performance of epoxy coatings separately hybridized with amino-functionalized calcined silica NPs and NIPAM-VTS-polysiloxane (N-isopropylacrylamide-co-vinyltrimethoxysilane). The amino-functionalized hybrid silica epoxy coatings were found to be superior in terms of adhesion, mechanical testing, and corrosion protection over those of hybrid materials without amino-functionalization. The authors also demonstrated that the amino-functionalized NIPAM-VTS-polysiloxane exhibited novel self-healing performance [34]. All data indicated that Fe-O-Si and Fe-N-Si bonds strengthened the adhesion strength of epoxy coating, resulting in enhanced anti-corrosion performance. Hybrid silica NPs were prepared with silica and 2-chlorothioxanthone under mild preparation conditions [35]. The synthesized hybrid NPs displayed higher photopolymerized efficiency and dispersion properties than pure 2-chlorothioxanthone, indicating that the coating had excellent UV-curing performance. Silica hybridized with chitosan was induced into epoxy resin to promote the active barrier effect on the metal surface. The data from FT-IR, UV–vis, and EIS analyses showed that the active barrier effect on the metal surface had significant anticorrosive activity. 3-(triethoxysilyl)propyl isocyanate was grafted onto silica surface modified by 1-(3′-aminopropyl) imidazole to prepare nanofillers [36]. The engineered nanofillers were then dispersed into epoxy resin to obtain epoxy nanocomposite coatings. The experimental results indicated that both the mechanical strength and corrosion resistance were improved. Through pre-electrodeposited silane treatment, the interaction of epoxy coating and steel was enhanced, indicating that the corrosion resistance was reinforced [37]. Zhu et al. [38] fabricated triazine-modified silica and significantly enhanced the anti-corrosion property of epoxy coatings on carbon steel. Zeng et al. [39] tested the anti-corrosion properties of durable superhydrophobic silica/epoxy resin coating applied to steel. The results clearly showed that the substrate was well protected in high-salt (3.5% NaCl solution) and high-H_2S environments (Figure 4). Superhydrophobic mesoporous silica (MCM-41) was added to epoxy coatings to alter the wettability and anti-corrosion behavior of the matrix surface [40]. The Al_2O_3-SiO_2 in MCM-41 was fabricated with poly(o-toluidine) and incorporated into epoxy resin. The modified resin NPs were painted onto carbon steel, and the corrosion tests showed that the Al_2O_3-SiO_2 improved the protective features of the coating. Sol–gel-based silicon coatings have gained attention for their excellent corrosion resistance and mechanical properties. So, sol–gel technology was employed to produce organic–inorganic coatings containing silica [41]. The investigations showed that the coating prepared using the sol–gel method exhibited good endurance capacity under high-temperature and salt spray conditions due to the formation of highly cross-linked networks, enhancing barrier properties and chemical resistance. These sol-gel coatings also exhibit good adhesion to steel surfaces and can be tailored to specific applications through the incorporation of different silicon compounds. Their advantages include improved thermal stability, UV resistance, and durability in harsh environments, making them a promising alternative to traditional epoxy coatings. To improve the solubility of zinc phosphate in epoxy resin, three mixtures of NaA zeolite and zinc monophosphate (1:1; 3:1 and 1:3) were studied [42]. EIS tests found that the epoxy coating on low-alloy 09G2S carbon steel had the highest corrosion resistance when the ratio of NaA zeolite to zinc monophosphate was 1:3.

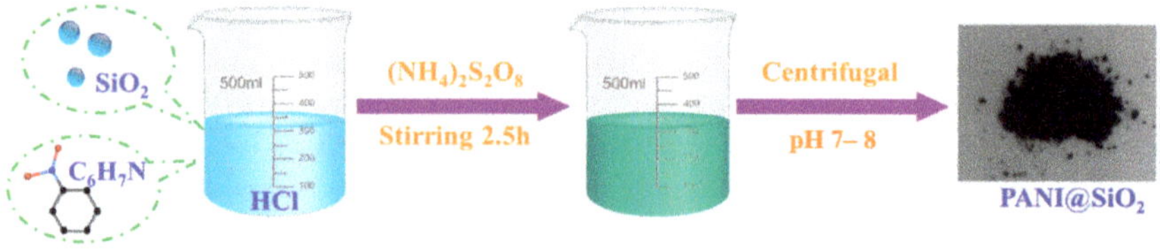

Figure 3. Synthesis of the core–shell structural PANI@SiO_2 [31].

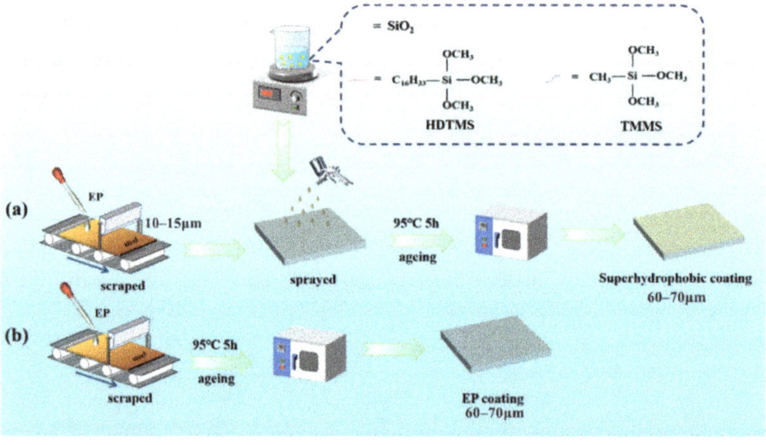

Figure 4. Preparation of superhydrophobic HDTMS@SiO_2 coating and EP coating on the Q235 carbon steel ((**a**) the superhydrophobic HDTMS@SiO2 coating and (**b**) the EP coating on the Q235 carbon steel substrate) [39].

As additives, natural minerals containing silicon were also investigated in epoxy coatings [43–45]. For example, kaolin mixed with graphite as an additive was utilized to enhance the thermal stability of epoxy coating on a steel surface [27]. The test results indicated that the substrate exhibited greater thermal resistance due to the formation of a homogeneous organic–ceramic coating. In another study, hydrophobic montmorillonite (HMT) clay was modified with silver and iron oxides to produce hybrid NPs [46]. When these modified HMT NPs were blended with epoxy resin, the resulting coating demonstrated extraordinary adhesion, mechanical strength, chemical resistance, and abrasion resistance under a high-salinity environment. Additionally, Beryl and Xavier [47] studied the electrochemical and mechanical properties of silanized clay-epoxy coatings on mild steel. All the test data revealed that the modified clay was effectively dispersed within the epoxy resin, leading to a dramatic improvement in the coating's anti-corrosion performance. The same research group also investigated the barrier, hydrophobic, and mechanical properties of epoxy coating mixed with silanized halloysite NPs on steel exposed to natural seawater [45]. The authors claimed that diethoxy(3-glycidyloxypropyl) methylsilane-modified clay epoxy coating exhibited stronger adhesive properties and a lower corrosion rate (0.0017 mm/year). Furthermore, the UV stability of epoxy coating was improved by incorporating halloysite nanotubes, encapsulated with organic UV stabilizers and lignin [44]. The epoxy system containing 2 wt% of encapsulated halloysite showed significantly higher UV stability compared to pure epoxy resin. In terms of other additives, phenyl-modified etched basalt was fabricated into an epoxy resin layer to improve the anti-corrosion performance of coatings [48]. The modified basalt constructed micro–nano capillary structures and generated π electrons, which contributed to the enhanced performance. Lastly, organically fabricated sepiolite nanofillers were added to epoxy resins to boost the anti-corrosion resistance of surface coatings for mild carbon steel [43].

2.4. Epoxy Coatings Modified by Other Metal Compounds

Due to their low costs, ease of synthesis, and chemical stability, cerium oxide (CeO_2) and polyaniline (PANI) have been incorporated into epoxy resin to enhance the anti-corrosive properties of organic coatings [7,49]. Fan et al. [49] prepared a CeO_2-PANI conductive polymer via in situ chemical oxidation polymerization. The prepared polymer was dispersed in F51 epoxy resin and sprayed onto a steel surface to boost its corrosion resistance. The authors highlighted that the synergistic interaction between PANI and Ce^{3+} ions was crucial for enhancing the steel's anti-corrosion properties. Similarly, Lei et al. [7] demonstrated that PANI/CeO_2 NPs exhibited excellent anti-corrosive performance for epoxy coating on carbon steel in 3.5 wt% NaCl solution. They attributed this to PANI's redox behavior, which improved the coating's self-healing ability. Dun et al. [50] developed CeO_2@BNNSs (BN, boron nitride) fillers (Figure 5) and dispersed them in epoxy resin to create high-quality coatings for steel surfaces. This significantly improved the corrosion resistance in 3.5 wt% NaCl solution. Wan et al. [51] synthesized novel PDA-BN@f-Al_2O_3 hybrids us γ-Aminopropyltriethoxysilane (KH550), Al_2O_3, polydopamine (PDA), and hexagonal boron-nitride (h-BN). Compared to individual Al_2O_3 and h-BN, PDA-BN@f-Al_2O_3 enhanced the barrier properties and corrosion resistance of epoxy coatings when mixed in. Micro and nano Al_2O_3 particles were also added to epoxy coatings to intensify the barrier and mechanical performance of steel [52]. The results showed that the tensile strength, strain, toughness, and flexural properties, respectively, were enhanced with adding 1 wt% micro Al_2O_3 before 1 wt% nano Al_2O_3 in the coating [52]. Magnetic Fe_3O_4 NPs modified with oleic acid were incorporated into epoxy resin to form in situ repairable coatings. Experiments confirmed that the Fe_3O_4 NPs prevented micro-channel formation and enhanced coating adhesion through target-driving and competitive wetting mechanisms [9]. Inspired by "Band Aid", this group modified epoxy binder with magnetic Fe_3O_4 NPs [53], showing high self-repair capability for organic coatings in seawater. Additionally, Ni-Fe LDHs partially replaced with MoO_4^{-2} efficiently improved the corrosion resistance of carbon steel (Q235) [54]. Nano MoO_3 treated with epoxy-silane was painted onto mild steel to elevate the anti-corrosion, adhesion, and mechanical properties [55]. After silane treatment, MoS_2 NPs were embedded into steel epoxy coatings [56]. The experiments indicated that MoS_2 with amino silane had better corrosion resistance than with neat epoxy silane. ZrO_2@rGO (reduced graphene oxide) was synthesized by a hydrothermal method and mixed into epoxy resin by a solvent-mixing method to create a hybrid coating [57], exhibiting excellent wear resistance and anti-corrosion performance. Multifunctional 2-amino-5-(methylthio)-1,3,4-thiadiazole (AMTTD) was utilized to modify ZrC [58]. The hybridized AMMTTD/ZrC was added into epoxy resin to increase the hydrophobicity, mechanical strength, and electrochemical properties of the coating in seawater. The anti-corrosion performance of epoxy coatings was enhanced by treating a steel surface with nanostructured samarium oxide-poly-dopamine film [59]. For 15CDV6 steel, electrocatalytically deposited cadmium improved the anti-corrosion of macromolecular epoxy coatings in 3 wt% NaCl solution by enhancing coating adhesion [60].

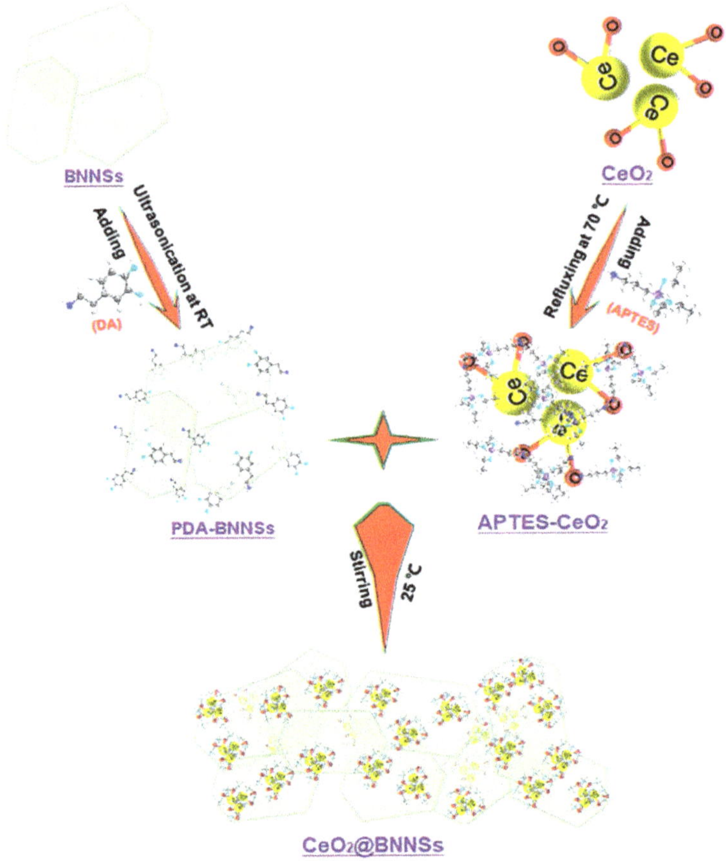

Figure 5. Synthesis route of CeO_2@BNNSs [50].

Table 1 provides a concise overview of the advantages and limitations of different corrosion protection methods discussed in this review. It is designed to help readers quickly compare and contrast the various approaches, facilitating informed decision-making for specific industrial applications.

Table 1. Summary of benefits and limitations of various corrosion protection methods involving metals and their compounds.

Corrosion Protection Method	Benefits	Limitations
Epoxy coatings modified by metals and their compounds	- Significantly enhance corrosion resistance - Improve barrier properties - Provide active corrosion protection	- Some metal compounds may have limited dispersibility - Potential for micro-crack formation - Cost considerations for certain metals
Zinc-rich epoxy coatings	- Excellent resistance to chemical corrosion - Good cathodic protection for carbon steel - Effective under highly acidic media	- Require specific application techniques - Cost of zinc particles - Environmental concerns with zinc disposal

Table 1. *Cont.*

Corrosion Protection Method	Benefits	Limitations
Titanium-modified epoxy coatings	- Enhanced corrosion protection in harsh environments - Improved thermal and UV resistance - Effective barrier against corrosive species	- Higher material costs - Complex processing requirements - Potential brittleness issues
Silicon-modified epoxy coatings	- Superior mechanical and thermal properties - Enhanced adhesion and corrosion resistance - Effective under high-temperature conditions	- Silane modification requires precise control - Limited availability of certain silicon compounds - Cost of specialized silicon additives
Epoxy coatings modified by other metal compounds	- Low cost and easy synthesis - Chemical stability - Synergistic effects with other additives	- Potential environmental and health concerns - Limited solubility in some coating systems - Performance depends on particle size and distribution

3. Epoxy Coatings Modified by Organic Compounds

3.1. Epoxy Coatings Modified by N Organic Compounds

As modifiers, amines are widely researched in organic coatings due to their exceptional reactivity with epoxy resin. Among these amines, PANI is more commonly used to construct the structure of the epoxy coating to improve corrosion resistance. Sun et al. [61] studied how PANI filler reacted with waterborne epoxy resin to fabricate anti-corrosion coatings on a 304 stainless steel surface. The corrosion protection performance of the prepared coatings was evaluated by EIS and potentiodynamic polarization in 3.5 wt% NaCl solution, with all data revealing that the substrate was effectively protected by the coating. Due to its high corrosion inhibitor loading rate, mesoporous polyaniline hollow spheres loaded with benzotriazole were synthesized and introduced into epoxy resin to obtain self-healing coating materials (Figure 6) [62]. After a systematic investigation, the experimental results indicated that not only did the synergistic effect of the passivation function of PANI contribute to enhancing the corrosion resistance of epoxy coating, but the corrosion inhibition effect of benzotriazole also played a very helpful role. Acid-doped PANI also enhanced the anti-corrosion properties of coatings [63,64]. Yao et al. [64] found that PANI doped with 2-hydroxyphosphonocarboxylic acid could reinforce the corrosion protection performance and self-healing function of epoxy coatings. Hu et al. [63] investigated the properties of PANI doped with hydrochloric acid, *p*-toluenesulfonic acid, and dodecylbenzene sulfonic acid, respectively. Epoxy resins mixed with the doped PANIs were coated on carbon steel surfaces and evaluated by electrochemical corrosion measurements in a 3.5% NaCl solution. The results indicated that the dodecylbenzene sulfonic acid-doped PANI epoxy coating showed the best barrier effect. Liu et al. [65] researched the effects of polyaniline-based (PANP) plates added into epoxy resin and found that only 1.0 wt% PANPs could dramatically improve the corrosion protective performance of the epoxy coating. Diraki and Omanovic [66] revealed that the corrosion protection of carbon steel surfaces was enhanced once the substrate was successively painted with PANI and epoxy resin. Using *p*-aminobenzoic acid as a linker, a PANI-BN powder was synthesized using PANI and BN as raw materials [67]. Electrochemical tests displayed that the PANI-BN epoxy coating on hot-dip galvanized steel had outstanding corrosion resistance in NaCl solution. Moreover, PANI has been used to modify metal oxides to improve the anti-corrosion performance of organic coatings [7,25,31,49]. This approach leverages the unique properties of PANI to enhance the protective capabilities of metal-oxide-based coatings, offering a promising avenue for developing advanced anti-corrosive materials.

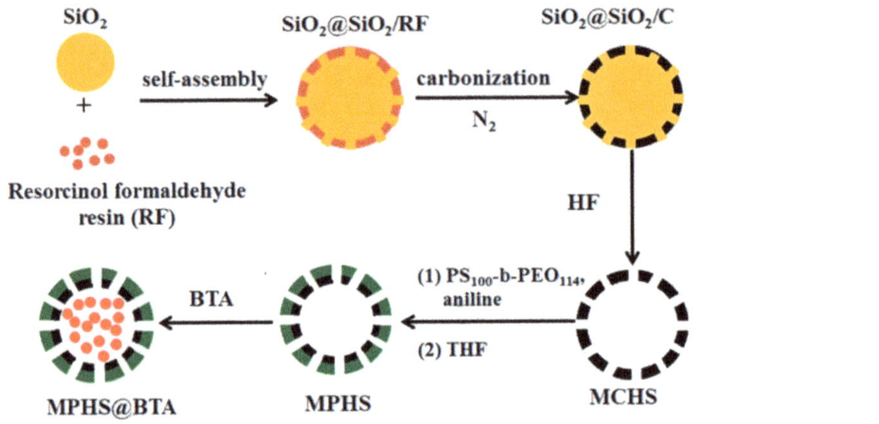

Figure 6. Synthesis of mesoporous polyaniline hollow spheres loaded with benzotriazole (MPHS@BTA) [62].

Doped PANI coatings show enhanced electrical conductivity and corrosion protection compared to undoped PANI coatings. Doping with acids or other agents improves the protonation state of PANI, leading to better charge transfer and inhibition of corrosive species. Doped PANI coatings also exhibit improved mechanical properties and adhesion to steel surfaces. The choice of dopant and doping level can be optimized to achieve the desired balance between conductivity, corrosion resistance, and processability.

Besides PANI, other nitrogen-containing organics were employed to enhance epoxy coating properties. Polyether amine D230 reacted with epoxy to form the poly(epoxy-amine) microcapsules (Figure 7), embedding liquid epoxy resin [68]. The resulting epoxy coating exhibited superior corrosion resistance due to its strong self-healing capability. Bi et al. [69] found that urea-formaldehyde could mitigate epoxy coating's negative impact on Q235 carbon steel, improving mechanical strength and chemical stability. Curing epoxy resin with methylene dianiline enhanced its adhesion to carbon steel [70]. EIS and potentiodynamic polarization (PDP) analysis showed that methylene dianiline addition boosted the anti-corrosion properties of the coating. Liu et al. [71] synthesized the COF material LZU-1 via a reaction between *p*-phenylenediamine and pyrogallol. LZU-1 was combined with benzotriazole (BTA) to create LZU-1@BTA, which was dispersed in epoxy resin to improve its anti-corrosion properties. Tetraaniline (ATA) doped with hydrophobic polyhedral oligomeric silsesquioxane (POSS) and protonic acid was designed and synthesized, respectively [72]. EIS and salt spray tests revealed that the coating containing with 0.5 wt% POSS–ATA had the highest resistance among the tested systems. Atta et al. [73] successfully applied imidazolium cationic ionic liquids to modify the thermomechanical, toughness, and anti-corrosion characteristics of epoxy coating, as the fabricated epoxy coating's curing networks formed more easily than commercial epoxy/amine systems. Xia et al. [74] studied the physicochemical properties of poly(arylene ether nitrile)-reinforced epoxy coating, finding that the coating's mechanical and corrosion protection capabilities were enhanced when poly(arylene ether nitrile) was incorporated. As a carbon steel anti-corrosive coating, triglycidyl ether triethoxy triazine epoxy resin cured with 1,6-diaminohexane and 4,4-diaminodiphenyl methane was studied in 3.5 wt% NaCl solution [75]. The authors concluded that the 1,6-diaminohexane-cured epoxy had higher protective efficiency than the 4,4-diaminodiphenyl-methane-cured one. Isophorone diisocyanate microcapsules, prepared by interfacial polymerization and incorporated into epoxy resin, yielded an anti-corrosive coating [76]. The coating cured on carbon steel displayed high damage healing owing to a protective polymeric barrier layer. Adding isophorone diisocyanate microcapsules and pH-sensitive cerium tri(bis(2-ethylhexyl)phosphate) particles to epoxy coating enhanced steel plate's corrosion resistance through the combined protective effects of each component [77].

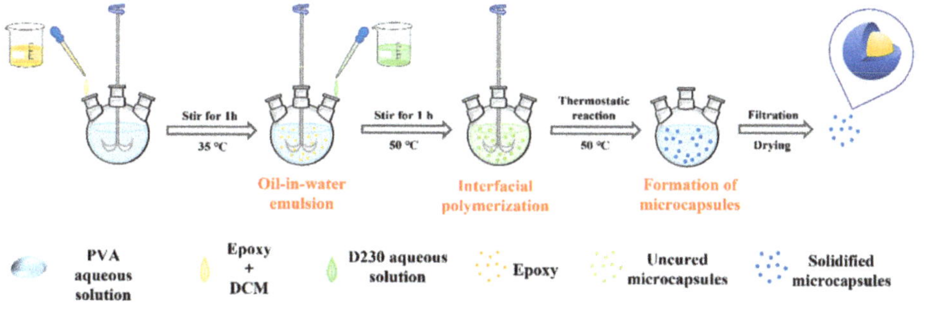

Figure 7. Synthesis of epoxy microcapsules [68].

3.2. Epoxy Coatings Modified by Natural Organic Compounds

To promote the sustainability of coating materials, bio-based alternatives are of great interest for both industry and academia [78,79]. Plant oil, a widely available natural resource, has been studied as an additive or modifier in organic coatings. To develop a fully organic coating system, Ammar et al. [80] used epoxidized soybean oil to modify epoxy resin. Their research revealed that the fabricated coating system had an optimal curing level, better wettability, and good physical–mechanical properties, indicating its potential to protect the matrix from corrosion. Li et al. [81] prepared tung oil embedded in urea-formaldehyde microcapsules via in situ polymerization, which reacted with epoxy resin to form a self-healing coating. EIS tests and a salt spray experiment demonstrated that the coating had good anti-corrosion performance. Veedu et al. [82] proved that *Ixora* leaf extract could intensify the corrosion resistance of epoxy coatings, as the interaction of Fe and phytochemicals at the metal–electrolyte interface provided promising inhibition efficiency and stability against metal dissolution. They also investigated the properties of *Gliricidia sepium* leaf extract and silica hybrid [83], suggesting the -OH interacting with the metal surface promoted the corrosion protection of the epoxy matrix. Qi et al. [84] found that tannin could improve the anti-corrosion performance of Zn^{2+} and Ce^{3+}. The authors identified that the synergistic effect of the co-deposited metal hydroxides/oxides and uniform iron–tannin complexes were key factors. Teijido et al. [85] synthesized epoxy coatings with various epoxide and hydroxyl groups using epoxidized soybean oil and tannic acid as natural substrates. These liquid mixtures, when painted onto a carbon steel surface, exhibited excellent anti-corrosion activity. Epoxy-Zn coatings loaded with 0.5 wt% oil palm frond cellulose nanocrystal (OPF-CNC) showed a 36% improvement in performance compared to unloaded cellulose coatings [86]. The epoxy–lignin coating was prepared using acetylated lignin for carbon steel corrosion protection [87]. Experimental data indicated that the acetylated lignin improved the modified coating's corrosion resistance. Nunes et al. [88] synthesized a novel organic coating from low-cost cashew nut shell liquid. Compared to an uncoated steel electrode, the coated ones had a lower corrosion current density and less negative corrosion potential. Similar results on cashew nut shell liquid were reported by the Silva group [89]. Abbout et al. [90] studied the corrosion protection of carbon steel surfaces with galactomannan-modified epoxy coatings. The modified coating effectively protected the matrix in an acidic medium. Natural urushiol (Ur) was reacted with mica to prepare functional mica nanosheets (MNs) via catechol group interactions (Figure 8) [91]. The modified epoxy coating's anti-corrosion performance improved due to the superhydrophobic properties and high dispersion of MNs@Ur. Additionally, biocide and silanized epoxy were used to synthesize a hybrid coating [92]. When applied to carbon steel, this coating demonstrated excellent anti-corrosion activity.

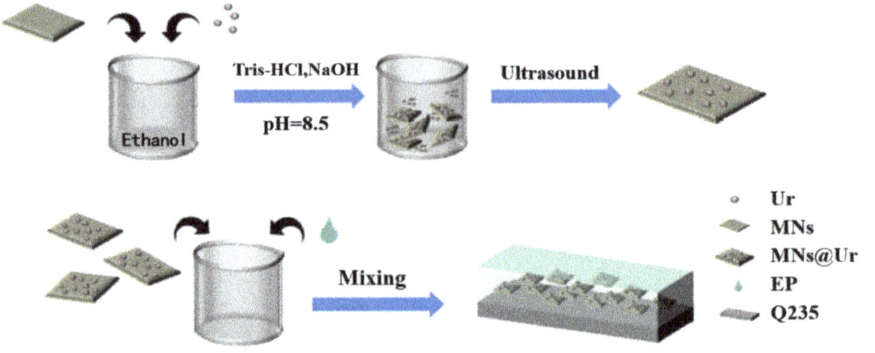

Figure 8. Preparation of MNs@Ur and MNs@Ur/EP composite coatings [91].

Natural organic compounds are renewable, biodegradable, and have a lower environmental impact compared to synthetic additives. Tannins [84], for example, are abundant in nature and can form stable complexes with metal ions, enhancing corrosion protection while reducing the need for harmful chemicals. Lignin [87] and cashew nut shell liquid [88,89] also provide sustainable alternatives for improving coating performance, aligning with the growing demand for eco-friendly materials in industrial applications.

3.3. *Organic Coatings Modified by Other Organic Compounds*

In addition to nitrogen-containing organics and natural products, other organics have also been utilized to modify epoxy resin for high-performance coatings. Zhou et al. [93] successfully synthesized acrylate copolymers via radical polymerization and incorporated these copolymer fillers into epoxy resin to prepare high-performance epoxy coatings. The results indicated that the anti-corrosion performance and self-healing characteristics of the epoxy coatings were significantly improved, attributed to the acrylate copolymer reinforcing the coating's tightness, adhesion, hydrophobicity, and weatherability. To boost corrosion resistance in a marine environment, Chen et al. [94] developed a textured epoxy resin coating using surface texture preparation technology and reduced air bubble formation via vacuum treatment, thereby enhancing the corrosion protection of marine structural steel.

Organic-compound-modified epoxy coatings offer numerous advantages, yet their preparation methods and required time can vary significantly. Coatings modified with polyether amine D230 or those utilizing in situ polymerization of PANI typically require longer preparation times due to their complex synthesis processes. Conversely, coatings incorporating natural organic compounds, such as epoxidized-soybean-oil- or tung-oil-based microcapsules, can be prepared more quickly. These natural compounds involve simpler modification steps and more straightforward coating processes, making them more efficient in terms of preparation time.

4. Organometallic-Compound-Modified Epoxy Coatings

4.1. Epoxy Coatings Modified by Metal–Organic Frameworkds

Metal–organic frameworks (MOFs), a class of organometallic compounds, have broad applications in catalysis, separation, and drug delivery systems [95–97]. MOFs can enhance the adhesion of epoxy coatings to steel through several mechanisms. Their porous structure and high surface area allow for better interaction with the steel surface, forming strong chemical bonds. MOFs can also act as fillers, improving the mechanical interlocking between the coating and the steel. Additionally, MOFs release corrosion inhibitors that further strengthen the adhesion by forming protective layers on the steel surface. This combination of physical and chemical interactions results in enhanced coating adhesion and durability. ZIF-8 (zeolitic imidazolate framework-8) was synthesized from 2-methylimidazole and zinc nitrate tetrahydrate and then combined with graphene oxide (GO) sheets to form GO@ZIF-8 particles [98]. The results showed that adding GO improved the epoxy coating's corrosion resistance by 70%. Xiong et al. [43] modified salicyldehyde@ZIF-8 fillers anchored on GO with 3-aminopropyltriethoxysilane. The filler improved the corrosion protection property of the epoxy coating because of its excellent dispersion stability. Zhang et al. [99] found that incorporating ZIF-8 with lubrication oil boosted the epoxy coating's anti-corrosion performance. He et al. [100] synthesized PPy@ZIF-8 (PPy, polypyrrole) NPs via in situ polymerization and added them to epoxy resin, producing a long-term anti-corrosion coating. The long-term efficiency of the corrosion protection of the epoxy coating on steels was attributed to the self-healing behavior of the PPy@ZIF-8 particles, which facilitated the formation of a passive iron oxide and $Zn(OH)_2$ deposition layer. Wan et al. [101] synthesized 5-aminotetrazole@ZIF-7@zinc gluconate/boron nitride NPs via a one-pot method using 5-aminotetrazole, zinc gluconate, benzimidazole, and boron nitride as starting materials. Owing to the synergetic effects of passive and active protection, the synthesized MOF NPs demonstrated a significant enhancement in the anti-corrosion performance of waterborne epoxy coating. Hasanzadeh and Saadatabadi [102] studied the corrosion protection properties of ZIF-67-decorated g-C_3N_4 nanosheets. Their findings revealed that the corrosion rate of mild steel was reduced by 50%, while its corrosion resistance was enhanced 5-fold compared to the blank sample. Zr-UIO-66 MOF, Zr-UIO-NH_2-MOF, and Zr-NH_2-UIO-MOF were developed and utilized as novel functional anti-corrosive fillers [103]. The test results indicated that these three Zr-MOFs could effectively improve the anti-corrosion performance of epoxy coating in 3.5% NaCl solution across various pH values (pH = 2, 7.5, and 12). Hsia et al. [104] fabricated 2D Zr-MOFs, which were incorporated into epoxy resin to boost the anti-corrosive activity of the coating. Functioning as a filler, MIL-88A (Fe) was used to modify epoxy resin so as to reinforce its anti-corrosion activity [105]. EIS and salt spray test results confirmed that the MIL-88A (Fe) particles possessed excellent barrier properties. In particular, salt spray tests showed that the adhesion loss of the epoxy coatings decreased from 31.4% to 9.4% with the addition of MIL-88A (Fe). Zhang et al. [106] developed BTA/PPy/MIL-88(Fe) through a successive oxidation and loading procedure (Figure 9). The synthesized MOFs were applied to strengthen the anti-corrosion performance of an epoxy coating. Keshmiri [107] designed and fabricated GO@Ce-MOF using GO and Ce-MOF as raw materials, after which an epoxy coating was modified by the fabricated GO@Ce-MOF filler. EIS and salt spray tests indicated that the GO@Ce-MOF/epoxy resin could enhance the anti-corrosion property of the coating. Compared to the blank sample, the anti-corrosion performance was boosted by 500%. Cu-MOFs and Zn-MOFs with unsaturated metal sites were subjected to hydrophobic modification using octadecylphosphonic acid (OPA) to prepare MOFs/OPA hydrophobic composites. The two MOFs were blended with epoxy resin to formulate hydrophobic acid-modified MOF epoxy coatings [108]. These coatings exhibited satisfactory corrosion resistance, which was attributed to the strong barrier and shielding effects. Recently, the corrosion protection properties of epoxy coating have been improved through the use of HKUST-1-BTA@e-Mica two-dimensional nanocomposites as fillers, which is credited to the enhanced mechanical properties of the coating [109].

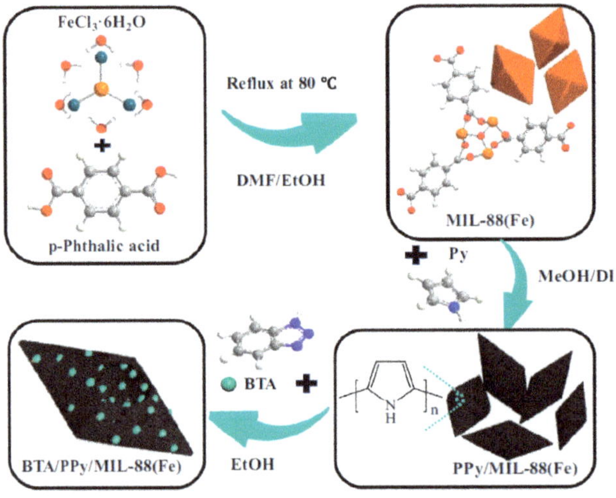

Figure 9. Fabrication of BTA/PPy/MIL-88(Fe) composite material [106].

4.2. Epoxy Coatings Modified by Other Organometallic Compounds

To enhance the corrosion resistance of coatings, other organometallic compounds have also been employed as fillers to modify epoxy coatings. Attaei et al. [77] modified epoxy resin with isophorone diisocyanate and pH-sensitive cerium tri(bis(2-ethylhexyl)phosphate, and the resulting coating exhibited the dual functions of self-healing and corrosion resistance. Cu(II) and Zr(IV) complexed with metformin and 2,2-bipyridine were incorporated into epoxy resin to formulate multi-functional coatings [110]. Salt spray and acid spot tests revealed that the obtained coatings, particularly the modified Cu(II) epoxy coating, offered excellent corrosion protection properties. Carbon steel was treated via a pulse-reverse electrodeposition method in a bath containing 3-nitrobenzoic acid and zinc nitrate, after which the treated carbon steel was coated with epoxy resin [17]. The experimental results indicated that the adhesion strength and anti-corrosion performance of the epoxy coating on the treated matrix were improved. Given the strong chelation ability of Ce^{3+} with organics, Butt et al. [111] designed and synthesized ODA/PDA/APT-Ce^{3+} (OPA–Ce^{3+}) coatings using octadecylamine (ODA) and cerium nitrate as raw materials (Figure 10). The synthesized coating exhibited superhydrophobic, self-healing, and self-cleaning properties, suggesting its potential for effectively protecting stainless steel from corrosion. Chopra et al. [112] investigated the corrosion inhibitive effect of ZA (acetoxime derivative of zinc chloride). They found that the epoxy coating modified with ZA possessed excellent barrier characteristics due to the formation of Zn–O–C and O–Zn–O linkages. Peng et al. [113] studied the impact of lanthanum 4-hydroxycinnamate and 3-(4-methylbenzoyl)propanoate on the corrosion resistance of epoxy coatings. EIS analysis evidenced that the rare earth carboxylate could offer anti-corrosion inhibition for steel. Similarly, Zhang et al. [114] demonstrated that rare earth carboxylate-containing coatings exhibited outstanding corrosion protection properties. To address the low anti-corrosion efficacy of inorganic cerium salts for carbon steel, Ren et al. [115] developed an organic cerium phosphate complex as a rapid repair agent. The corrosion resistance of the formed film on Q235 steel was enhanced by the repair capabilities of the organic cerium complex.

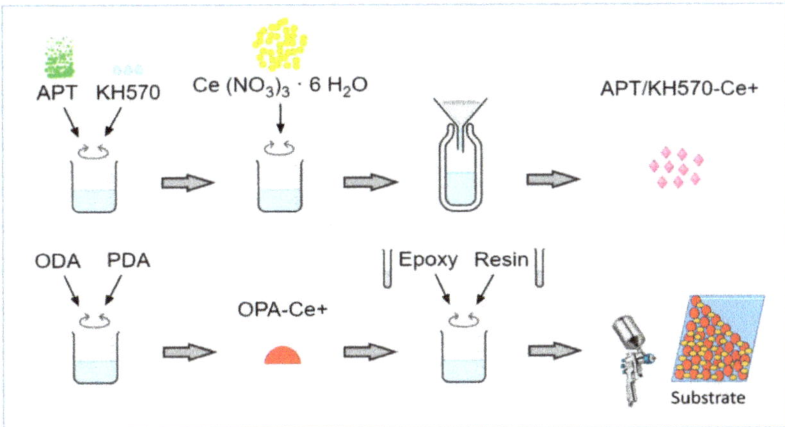

Figure 10. Preparation of OPA–Ce^{3+} epoxy coating [111].

5. Carbon-Material-Modified Epoxy Coatings

5.1. Graphene-Modified Epoxy Coatings

Recently, the application of graphene and graphene oxides (GOs) in the modification of epoxy coatings has garnered significant attention [116–119]. Yan et al. [119] fabricated an ideal Ti_3C_2/graphene hybrid using Ti_3AlC_2 and graphene as raw materials. The graphene could decrease the microstructural defects and reinforce the adhesion of the epoxy coating [120]. Compared to the pure epoxy coating, the graphene-modified epoxy coating had higher anti-corrosion capacity [120]. As an additive, the Ti_3C_2/graphene hybrid enhanced the anti-corrosive properties of the organic coating because of its synergistic effect and thermal conductivity. Ye et al. [121] verified that a coating modified with graphene had excellent self-healing capacity due to the spontaneous release from a graphene-based nanocontainer. Madhusudhana et al. [116] investigated the anti-corrosion ability of functionalized graphene oxide–epoxy coating on mild steel in 3.5 wt% NaCl solution. An epoxy phenolic novolac polymer was modified by GO, which was fabricated by refluxing a mixture of epoxy and phenol formaldehyde amine. The results indicated that a corrosion inhibition efficiency of 99.99% was achieved with the 0.2 wt% functionalized GO-grafted epoxy phenolic novolac coating. Ramezanzadeh et al. [98] developed ZIF-8-decorated GO to enhance the anti-corrosion performance of an epoxy coating. Compared to the blank sample, the corrosion resistance of the coating modified with GO@ZIF-8 was improved by 70%. Silanized Cr_3C_2 was encapsulated with GO and incorporated as a nanofiller into an epoxy coating [122]. EIS testing revealed that the modified epoxy coating exhibited significantly higher corrosion resistance compared to the unmodified epoxy coating. Organic-functionalized GOs have been extensively investigated due to their remarkable corrosion protection properties [123,124]. To further enhance the anti-corrosion performance, GO was functionalized with phytic acid and subsequently incorporated into a water-borne epoxy coating [125]. Electrochemical measurements and salt spray tests implied that the corrosion resistance was remarkable, which was primarily attributed to the excellent dispersity of the functionalized GO within the coating matrix. Zhu et al. [126] explored the anti-corrosion performance of polypyrrole-functionalized GO synthesized via an in situ polymerization method. EIS analysis indicated that the functionalized GO (at a concentration of 0.05%) coating exhibited significant corrosion protection performance. GO was functionalized using 3-amino-1, 2, 4-triazole-5 thiol and was used to enhance the corrosion protection performance of an epoxy coating on mild steel [124]. Compared to the neat epoxy coating, the modified matrix exhibited higher film resistance, charge transfer resistance, and low capacitance. GO functionalized with 1-butyl-3-methylimidazolium acetate was uniformly dispersed within an epoxy coating, thereby strengthening the anti-corrosive properties of the modified epoxy coating on carbon steel [123]. Electrochemical studies evidenced that the modified coating achieved a higher corrosion protection efficiency (about 99%) in 3.5 wt% NaCl solution compared to the blank sample. Xiang et al. [127] synthesized GO grafted with 4-amino-antipyrine (AAP) via a hydrothermal method. All data suggested that the synergistic effect of physical shielding and corrosion inhibition imparted by GO@AAP could significantly enhance the anti-corrosion performance of waterborne epoxy coating on Q235 mild steel in a 3.5 wt% NaCl solution. Li et al. [128] employed amino-modified graphene oxide, synthesized using urea and melamine, to bolster the corrosion protection of epoxy coatings on a Q235 steel surface. Diraki and Omanovic [129] fabricated a double-layer PANI-GO/epoxy coating to protect carbon steel in a saltwater environment. Specifically, conductive PANI loaded with GO was initially coated onto the carbon steel, followed by coverage with a commercial epoxy coating to construct the double-layer structure. The experimental results demonstrated that the fabricated coating possessed high anti-corrosion resistance, as the PANI-GO layer effectively hindered the transport of hydrated corrosive ions to the carbon steel surface. To address the issues of poor dispersion and high mechanical damage associated with GO, Zhang et al. [117] synthesized covalent organic framework (COF)-modified GO to further elevate the anti-corrosion performance of epoxy coatings (Figure 11). Subsequently, the synthesized GO/COF was loaded with benzotriazole (BTA) to produce GO/COF-BTA filler. EIS testing revealed that the epoxy coating reinforced with GO/COF@BTA-2% exhibited superior corrosion protection and self-healing ability, attributed to its good compatibility and higher crosslinking density and the controllable release of BTA. The application of reduced GO (rGO) was also explored to enhance the anti-corrosion performance of epoxy coatings [118,130]. Kooshksara and Mohammadi [130] treated multi-layered GO using an in situ solvothermal reduction method to obtain rGO. Long-term salt spray and EIS studies confirmed that rGO enhanced the corrosion protection of epoxy coatings owing to its high-quality dispersion. Additionally, rGO functionalized with PANI and was employed to improve the corrosion protection of epoxy coatings [118]. The authors revealed that the high surface area and remarkable structural barrier of the functionalized rGO provided excellent protective coverage in a 3.5 wt% NaCl solution. Moreover, graphene composited with metal oxide or MOFs has been applied to enhance the anti-corrosion performance of epoxy coatings [6,57,131].

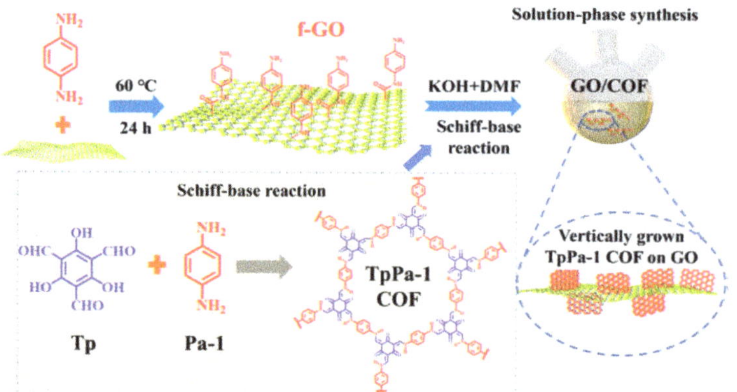

Figure 11. Synthesis of GO/COF [117].

5.2. Epoxy Coatings Modified by Other Carbon Materials

Except for graphene and graphene oxide, some other carbon martials have been explored to enhance the corrosion protection of epoxy coatings. Chen et al. [132], for the first time, utilized g-C_3N_4 to disperse graphene in an epoxy solution via an ultrasonic vibration method. EIS and salt spray tests revealed that the modified epoxy coating exhibited enhanced anti-corrosion performance, primarily attributed to the strengthened barrier properties of g-C_3N_4 and graphene. Recently, Hasanzadeh and SaadatAbadi [102] found that ZIF-67-decorated g-C_3N_4 nanosheets could significantly improve the self-healing capacity and corrosion resistance of epoxy coatings. Graphitic carbon nitride was employed to encapsulate MnO_2 modified by benzidine within pure epoxy resin, and the resulting decorated epoxy coating demonstrated high corrosion protection efficacy owing to the increased adhesive strength [133]. As an anti-corrosive additive, PNAI/CNT (CNT, carbon nanotube) hybrids were functionalized with 1-methyl-3-butyl imidazolium tetrafluorate [134]. The findings suggested that the epoxy coating loaded with 1 wt% of the synthesized PANI/CNT hybrid presented outstanding anti-corrosion properties. An epoxy coating was modified with functionalized multi-walled carbon nanotube/polyindole fillers and tested on mild steel in a 3.5 wt.% NaCl solution [135]. The investigations indicated that the coating containing 0.25 wt% fillers exhibited excellent anti-corrosion and barrier protection properties. Furthermore, boron nitride modified by amino-functionalized carbon dots was also incorporated into waterborne epoxy resin to enhance its anti-corrosive properties (Figure 12) [136]. The synergistic effects of these composites markedly bolstered the anti-corrosive ability of the epoxy coating.

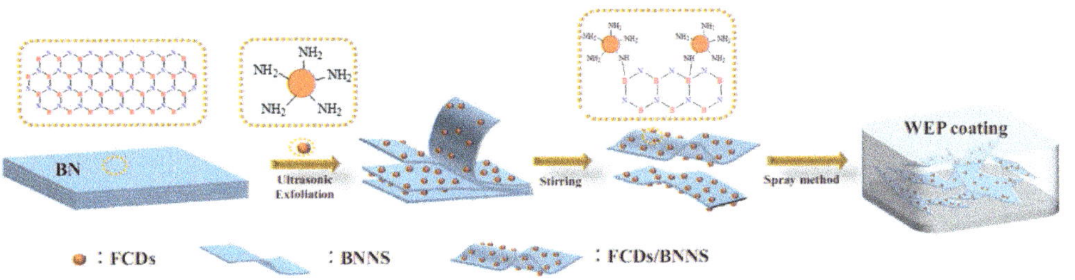

Figure 12. Preparation of waterborne epoxy coating modified by amino-functionalized carbon dots\boron nitride nanosheets [136].

When considering the large-scale introduction of new carbon-material-modified epoxy coatings, multiple indicators suggest a favorable trajectory. For instance, materials such as g-C_3N_4 and PANI/CNT hybrids have demonstrated significant potential in improving corrosion resistance. Moreover, the development of boron nitride modified with amino-functionalized carbon dots presents a promising avenue for enhancing barrier properties. However, it is crucial to acknowledge that challenges like cost-effective synthesis and consistent industrial-scale performance must be overcome. In summary, while certain materials are nearing commercial viability, others remain in research phases, signaling a gradual yet steady upward trend in their large-scale application.

6. Conclusions and Outlook

As promising corrosion-resistant materials, organic epoxy coatings modified by various additives will be researched in-depth and applied widely in industry in the future. Our review highlights the diversity and innovation in modifying epoxy coatings with different categories of additives, including metals and their compounds, organic compounds, organometallic compounds, and carbon materials. Each category offers unique advantages and addresses specific corrosion challenges, expanding the scope of protective coating solutions.

Zinc-rich organic epoxy coatings, when modified with organic compounds or metal oxides, continue to be industrially attractive due to their robust performance and established application basis. Among the various modifiers, nitrogen-containing organic compounds, such as PANI, stand out because they form M-N bonds with metal surfaces, enhancing coating adhesion and preventing steel corrosion. The hybrid use of organics and metal oxides not only blocks corrosive agents but also boosts coating corrosion resistance through M-O-Fe or M-N-Fe bond formation, a strategy that has garnered significant research interest.

Metal–organic compounds, particularly MOFs, have shown great potential in reinforcing the anti-corrosive activity of epoxy coatings. Their structural tunability and compatibility with epoxy resins make them promising candidates for developing high-performance modified coatings. Additionally, natural materials, with their wide range of sources and dual organic–inorganic properties, offer sustainable alternatives for enhancing the anti-corrosion properties of epoxy coatings, though further research is needed to identify the key active components.

Carbon materials, including graphene and graphene oxides, will continue to be thoroughly researched as fillers for their ability to hybridize with other substances and improve coating performance.

While composite epoxy resin coatings offer improved performance, they also face challenges such as dispersion difficulties and cost considerations. Some additives, especially nanomaterials, tend to aggregate, leading to uneven distribution and reduced effectiveness. Proper dispersion techniques and surface modifications are essential to address this issue. Cost-effectiveness is another concern, as some advanced additives and modification methods can increase production costs. Balancing performance improvements with economic viability is crucial for the widespread adoption of these coatings in industrial applications.

Future research in modified organic epoxy resin coatings for steel could explore several promising directions. The development of more efficient and cost-effective dispersion techniques would enhance the performance of these coatings. Investigating the long-term durability and environmental impact of composite coatings is also essential. Additionally, the integration of smart materials and self-healing technologies could further improve corrosion protection. Exploring new additive combinations and hybrid systems may lead to innovative solutions for specific industrial challenges. Finally, advancing the understanding of the fundamental mechanisms behind the improved properties of composite coatings will guide the design of next-generation materials.

Composite epoxy resin coatings modified with various additives show great potential for enhancing steel corrosion resistance in diverse industrial environments. This review highlights the diversity and innovation in modifying epoxy coatings with different categories of additives. Zinc-rich coatings modified with organic compounds or metal oxides continue to be industrially attractive. The hybrid use of organics and metal oxides not only blocks corrosive agents but also boosts coating corrosion resistance through bond formation, a strategy that has garnered significant research interest. Metal–organic compounds, particularly MOFs, have shown great potential in reinforcing the anti-corrosive activity of epoxy coatings. Natural materials offer sustainable alternatives for enhancing the anti-corrosion properties of epoxy coatings, though further research is needed to identify key active components. Carbon materials, including graphene and graphene oxides, will continue to be thoroughly researched as fillers for their ability to hybridize with other substances and improve coating performance. Future research should focus on developing more high-performance modified epoxy resins, leveraging the unique properties of these diverse additives to enhance steel corrosion resistance in various industrial environments.

Author Contributions: Conceptualization, J.H.; investigation, J.W.; resources, K.Y.; writing—original draft preparation, M.W.; writing—review and editing, Y.L.; supervision, Y.L.; funding acquisition, M.W. and Y.L. All authors have read and agreed to the published version of the manuscript.

Funding: This research was funded by The Science and Technology Research Program of Chongqing Municipal Education Commission grant number KJZD-M202303201 and The Science and Technology Research Program of Chongqing Municipal Education Commission KJZD-M202403201.

Institutional Review Board Statement: Not applicable.

Informed Consent Statement: Not applicable.

Data Availability Statement: No new data were created or analyzed in this study.

Acknowledgments: The personnel involved in this work also include Guibin Yu (Chongqing Chang'an Wangjiang Industrial Group Co., Ltd.), Kunfang Xu (Chongqing Chang'an Wangjiang Industrial Group Co., Ltd.), Yu Chang, Linyao Li, and Lijuan Dong.

Conflicts of Interest: Author Jianghua Hao was employed by the company Avic Xi'an Aircraft Industry Group Co., Ltd. Author Kun Yang was employed by the company Avic Shenyang Aircraft Corporation. The remaining authors declare that the research was conducted in the absence of any commercial or financial relationships that could be construed as a potential conflict of interest.

References

1. Cui, G.; Bi, Z.; Wang, S.; Liu, J.; Xing, X.; Li, Z.; Wang, B. A comprehensive review on smart anti-corrosive coatings. *Prog. Org. Coat.* **2020**, *148*, 105821. [CrossRef]
2. Kausar, A. Performance of corrosion protective epoxy blend-based nanocomposite coatings: A review. *Polym. Plast. Technol. Mater.* **2020**, *59*, 658–673. [CrossRef]
3. Xavier, J.R. Superior barrier, hydrophobic, and mechanical properties of the epoxy nanocomposite containing mixed metal oxides. *J. Adhes. Sci. Technol.* **2023**, *37*, 1394–1418. [CrossRef]
4. Kumari, S.; Saini, A.; Dhayal, V. Metal oxide based epoxy coatings for corrosion protection of steel. *Mater. Today Proc.* **2021**, *43*, 3105–3109. [CrossRef]
5. Yap, S.W.; Johari, N.; Mazlan, S.A.; Ahmad, S.N.A.S.; Arifin, R.; Hassan, N.A.; Johari, M.A.F. Superhydrophobic zinc oxide/epoxy coating prepared by a one-step approach for corrosion protection of carbon steel. *J. Mater. Res. Technol.* **2023**, *25*, 5751–5766. [CrossRef]
6. Guo, S.-Y.; Luo, H.-H.; Tan, Z.; Chen, J.-Z.; Zhang, L.; Ren, J. Impermeability and interfacial bonding strength of TiO_2-graphene modified epoxy resin coated OPC concrete. *Prog. Org. Coat.* **2021**, *151*, 106029. [CrossRef]
7. Lei, Y.; Qiu, Z.; Tan, N.; Du, H.; Li, D.; Liu, J.; Liu, T.; Zhang, W.; Chang, X. Polyaniline/CeO_2 nanocomposites as corrosion inhibitors for improving the corrosive performance of epoxy coating on carbon steel in 3.5% NaCl solution. *Prog. Org. Coat.* **2020**, *139*, 105430. [CrossRef]
8. Xavier, J.R. Electrochemical, mechanical and adhesive properties of surface modified NiO-epoxy nanocomposite coatings on mild steel. *Mater. Sci. Eng. B* **2020**, *260*, 114639.
9. Feng, Z.; Wan, R.; Chen, S.; Tang, X.; Ju, H.; Li, Y.; Song, G.-L. In-situ repair of marine coatings by a Fe_3O_4 nanoparticle-modified epoxy resin under seawater. *Chem. Eng. J.* **2022**, *430*, 132827. [CrossRef]
10. Pandis, P.K.; Papaioannou, S.; Siaperas, V.; Terzopoulos, A.; Stathopoulos, V.N. Evaluation of Zn- and Fe-rich organic coatings for corrosion protection and condensation performance on waste heat recovery surfaces. *Int. J. Thermofluids* **2020**, *3–4*, 100025. [CrossRef]
11. Qi, C.; Dam-Johansen, K.; Weinell, C.E.; Bi, H.; Wu, H. Enhanced anticorrosion performance of zinc rich epoxy coatings modified with stainless steel flakes. *Prog. Org. Coat.* **2022**, *163*, 106616. [CrossRef]
12. Xing, C.; Wang, W.; Qu, S.; Tang, Y.; Zhao, X.; Zuo, Y. Degradation of zinc-rich epoxy coating in 3.5% NaCl solution and evolution of its EIS parameters. *J. Coat. Technol. Res.* **2021**, *18*, 843–860. [CrossRef]
13. Qi, C.; Weinell, C.E.; Dam-Johansen, K.; Wu, H. Assessment of anticorrosion performance of zinc-rich epoxy coatings added with zinc fibers for corrosion protection of steel. *ACS Omega* **2023**, *8*, 1912–1922. [CrossRef]
14. Mamudu, U.; Omeiza, L.A.; Hussin, M.R.; Subramanian, Y.; Azad, A.K.; Alnarabiji, M.S.; Ebenso, E.E.; Lim, R.C. Recycled eggshell waste in zinc-rich epoxy coating for corrosion protection of mild steel in a controlled elevated temperature saline environment. *Prog. Org. Coat.* **2024**, *186*, 108025. [CrossRef]

15. Haddadi, S.A.; Ghaderi, S.; Sadeghi, M.; Gorji, B.; Ahmadijokani, F.; Ramazani, S.A.A.; Mahdavian, M.; Arjmand, M. Enhanced active/barrier corrosion protective properties of epoxy coatings containing eco-friendly green inorganic/organic hybrid pigments based on zinc cations/Ferula Asafoetida leaves. *J. Mol. Liq.* **2021**, *323*, 114584. [CrossRef]
16. Amanian, S.; Naderi, R.; Mahdavian, M. The role of an in-situ grown Zn-Al layered double hydroxide conversion coating in the protective properties of epoxy coating on galvanized steel. *J. Electrochem. Soc.* **2022**, *169*, 031511. [CrossRef]
17. Rahmani, M.H.; Naderi, R.; Mahdavian, M. Pulse-reverse electrodeposition of a conversion coating based on zinc cation and 3-nitrobenzoic acid on carbon steel to enhance adhesion and protective function of epoxy coating. *Prog. Org. Coat.* **2022**, *172*, 107124.
18. Kabaoglu, E.; Karabork, F.; Kayan, D.B.; Akdemir, A. Improvement of anti-corrosion performance (surface and near the cut edge) and mechanical properties of epoxy coatings modified with nano, micro and hybrid ZnO particles. *J. Compos. Mater.* **2023**, *57*, 451–463. [CrossRef]
19. Nguyen, C.T.; Ly, L.T.N.; Nguyen, T.X.; Dao, H.P.; Nguyen, S.A.; Tran, T.H.; Vu, T.Q.; Ngo, Q.T.C.; Nguyen, T.N.; Thai, H. The effect of organotitanate-modified zinc oxide nanoparticles on some characteristics and anticorrosion protection properties of epoxy coating. *J. Coat. Technol. Res.* **2024**, *22*, 255–268. [CrossRef]
20. Jia, Z.; Hong, R. Anticorrosive and photocatalytic properties research of epoxy-silica organic–inorganic coating. *Colloids Surf. A Physicochem. Eng. Asp.* **2021**, *622*, 126647.
21. Dagdag, O.; Guo, L.; Safi, Z.; Verma, C.; Ebenso, E.E.; Wazzan, N.; Masroor, S.; Haldhar, R.; Jodeh, S.; El Gouri, M. Epoxy resin and TiO_2 composite as anticorrosive material for carbon steel in 3% NaCl medium: Experimental and computational studies. *J. Mol. Liq.* **2020**, *317*, 114249. [CrossRef]
22. Fadl, A.M.; Abdou, M.I.; Hamza, M.A.; Sadeek, S.A. Corrosion-inhibiting, self-healing, mechanical-resistant, chemically and UV stable PDMAS/TiO_2 epoxy hybrid nanocomposite coating for steel petroleum tanker trucks. *Prog. Org. Coat.* **2020**, *146*, 105715. [CrossRef]
23. Li, S.; Huang, H.; Chen, F.; He, X.; Ma, Y.; Zhang, L.; Sheng, X.; Chen, Y.; Shchukina, E.; Shchukin, D. Reinforced anticorrosion performance of waterborne epoxy coating with eco-friendly L-cysteine modified $Ti_3C_2T_x$ MXene nanosheets. *Prog. Org. Coat.* **2021**, *161*, 106478. [CrossRef]
24. Yan, H.; Cai, M.; Wang, J.; Zhang, L.; Li, H.; Li, W.; Fan, X.; Zhu, M. Insight into anticorrosion/antiwear behavior of inorganic-organic multilayer protection system composed of nitriding layer and epoxy coating with $Ti_3C_2T_x$ MXene. *Appl. Surf. Sci.* **2021**, *536*, 147974. [CrossRef]
25. Li, C.; Xu, J.; Xu, Q.; Xue, G.; Yu, H.; Wang, X.; Lu, J.; Cui, G.; Gu, G. Synthesis of Ti_3C_2 MXene@PANI composites for excellent anticorrosion performance of waterborne epoxy coating. *Prog. Org. Coat.* **2022**, *165*, 106673. [CrossRef]
26. Kim, B.G.; Choi, H.-H.; Park, H.-Y.; Kwon, M.S.; Byeun, Y.-K.; Kang, S.; Jung, Y.-G.; Son, J.-H.; Yang, S. Preparation and characterization of organic–inorganic hybrid coatings for improving the insulation properties of electrical steel. *J. Coat. Technol. Res.* **2023**, *20*, 1383–1393. [CrossRef]
27. Mastalska-Popławska, J.; Kadac, K.; Izak, P.; Gierej, M.; Stempkowska, A.; Góral, Z. The influence of ceramic additives on intumescence and thermal activity of epoxy coatings for steel. *J. Appl. Polym. Sci.* **2021**, *138*, 49914. [CrossRef]
28. He, X.; Cui, C.; Chen, Y.; Zhang, L.; Sheng, X.; Xie, D. MXene and polymer collision: Sparking the future of high-performance multifunctional coatings. *Adv. Funct. Mater.* **2024**, *34*, 2409675. [CrossRef]
29. Zhang, Y.; Si, C.; Zhang, Z.; Li, L.; Fan, X.; Zhu, M. Hierarchical design of microcapsules-based epoxy resin coating enhanced with $Ti_3C_2T_x$ for improving thermal, tribological, anti-corrosive performance. *Carbon* **2024**, *228*, 119379. [CrossRef]
30. Pessoa, M.O.; Freitas, B.R.; de Oliveira Braga, J.; Santos, S.L.B.S.; da Silva, B.P.; Capelossi, V.R.; Cotting, F. Anticorrosion and adhesion performance of a monolayer and double layer silane-epoxy coating systems applied on carbon steel. *Surf. Coat. Technol.* **2024**, *485*, 130909. [CrossRef]
31. Liu, L.; Zhao, M.; Pei, X.; Liu, S.; Luo, S.; Yan, M.; Shao, R.; Sun, Y.; Xu, W.; Xu, Z. Improving corrosion resistance of epoxy coating by optimizing the stress distribution and dispersion of SiO_2 filler. *Prog. Org. Coat.* **2023**, *179*, 107522.
32. Fernández-Álvarez, M.; Velasco, F.; de la Fuente, D.; Bautista, A. Powder organic coatings functionalized with calcium ion-exchanged silica corrosion inhibitors. *Surf. Coat. Technol.* **2024**, *477*, 130377.
33. Atta, A.M.; Ezzat, A.O.; El-Saeed, A.M.; Tawfeek, A.M.; Sabeela, N.I. Self-healing of chemically bonded hybrid silica/epoxy for steel coating. *Prog. Org. Coat.* **2020**, *141*, 105549.
34. Li, X.; Liu, Z.; Hong, P.; Chen, L.; Liu, X. Synthesis of organic and inorganic hybrid nanoparticles as multifunctional photoinitiator and its application in UV-curable epoxy acrylate-based coating systems. *Prog. Org. Coat.* **2020**, *141*, 105565.
35. Ma, I.W.; Sh, A.; Bashir, S.; Kumar, S.S.; Ramesh, K.; Ramesh, S. Development of active barrier effect of hybrid chitosan/silica composite epoxy-based coating on mild steel surface. *Surf. Interfaces* **2021**, *25*, 101250.
36. Yang, W.; Feng, W.; Liao, Z.; Yang, Y.; Miao, G.; Yu, B.; Pei, X. Protection of mild steel with molecular engineered epoxy nanocomposite coatings containing corrosion inhibitor functionalized nanoparticles. *Surf. Coat. Technol.* **2021**, *406*, 126639.

37. Geng, Y.; Liu, Y.; Liu, A.; Li, S.; Zhang, H. Improved interfacial interactions and corrosion resistance of epoxy coated reinforcement by pre-electrodeposited silane layer. *Prog. Org. Coat.* **2022**, *173*, 107171.
38. Zhu, H.; Liu, J.; Lu, X.; Wang, D.; Geng, T.; Feng, L.; Liang, D.; Ma, X.; Hu, Z. Wettability and anticorrosion behavior of organic-inorganic hybrid superhydrophobic epoxy coatings containing triazine corrosion inhibitor loaded in mesoporous molecular sieve. *J. Taiwan Inst. Chem. Eng.* **2022**, *141*, 104604.
39. Zeng, Q.; Kang, L.; Fan, J.; Song, L.; Wan, S.; Liao, B.; Guo, X. Durable superhydrophobic silica/epoxy resin coating for the enhanced corrosion protection of steel substrates in high salt and H_2S environments. *Colloids Surf. A Physicochem. Eng. Asp.* **2022**, *654*, 130137.
40. El-Lateef, H.M.A.; Khalaf, M.M. Fabrication and characterization of alumina-silica/poly(o-toluidine) nanocomposites as novel anticorrosive epoxy coatings films on carbon steel. *Microchem. J.* **2020**, *158*, 105129.
41. Rêgo, G.C.; Bandeira, R.M.; van Drunen, J.; Tremiliosi-Filho, G.; Casteletti, L.C. Multi-layer organic-inorganic hybrid anticorrosive coatings for protection of medium carbon steel. *Mater. Chem. Phys.* **2023**, *303*, 127841. [CrossRef]
42. Zin, I.M.; DatskoKhlopyk, B.M.O.P.; Sobodosh, N.Y.; Korniy, S.A. Effect of zeolite-phosphate anti-corrosion pigment on protective properties of epoxy coating on carbon steel. *Mater. Sci.* **2024**, *59*, 658–665. [CrossRef]
43. Xiong, H.; Qi, F.; Zhao, N.; Yuan, H.; Wan, P.; Liao, B.; Ouyang, X. Effect of organically modified sepiolite as inorganic nanofiller on the anti-corrosion resistance of epoxy coating. *Mater. Lett.* **2020**, *260*, 126941. [CrossRef]
44. Nikafshar, S.; McCracken, J.; Dunne, K.; Nejad, M. Improving UV-Stability of epoxy coating using encapsulated halloysite nanotubes with organic UV-Stabilizers and lignin. *Prog. Org. Coat.* **2021**, *151*, 105843. [CrossRef]
45. Xavier, J.R.; Raja Beryl, J.; Ravisankar, N. A study on the influence of silanized clay on the barrier, hydrophobic and mechanical properties of epoxy coated steel in natural seawater. *J. Adhes.* **2023**, *99*, 1889–1915. [CrossRef]
46. Atta, A.M.; Ezzat, A.O.; El-Saeed, A.M.; Wahby, M.H.; Abdallah, M.M.S. Superhydrophobic organic and inorganic clay nanocomposites for epoxy steel coatings. *Prog. Org. Coat.* **2020**, *140*, 105502. [CrossRef]
47. Beryl, J.R.; Xavier, J.R. Electrochemical and mechanical studies of epoxy coatings containing eco-Friendly nanocomposite consisting of silane functionalized clay-epoxy on mild steel. *J. Bio- Tribo-Corros.* **2020**, *6*, 126. [CrossRef]
48. Jing, Y.; Meng, F.; Wang, F.; Liu, L. Design of an anticorrosion/bactericidal dual functional organic coating based on the slippery liquid-infused porous surface. *Appl. Surf. Sci.* **2023**, *639*, 158214.
49. Fan, W.; Wang, H.; Wang, C.; Liu, Z.; Zhu, Y.; Li, K. Epoxy coating capable of providing multi-component passive film for long-term anti-corrosion of steel. *Appl. Surf. Sci.* **2020**, *521*, 146417. [CrossRef]
50. Duan, H.; Ji, J.; Cao, C.; Bai, W.; Song, Z.; Xue, Y.; Tang, C. Enhanced anti-corrosion performance of carbon steels via CeO_2@BNNSs/epoxy resin composite coatings. *Macromol. Chem. Phys.* **2023**, *224*, 2300006. [CrossRef]
51. Wan, P.; Zhao, N.; Qi, F.; Zhang, B.; Xiong, H.; Yuan, H.; Liao, B.; Ouyang, X. Synthesis of PDA-BN@f-Al_2O_3 hybrid for nanocomposite epoxy coating with superior corrosion protective properties. *Prog. Org. Coat.* **2020**, *146*, 105713.
52. Saber, D.; Alghtani, A.H.; Ahmed, E.M.; Felemban, B.F.; Ali, H.T.; Megahed, M.; El-Aziz, K.A. Enhancement of barrier and mechanical performance of steel coated with epoxy filled with micron and nano alumina fillers. *Mater. Res.* **2022**, *25*, e20210413.
53. Feng, Z.; Huang, J.; Guo, H.; Zhang, X.; Li, Y.; Fang, B.; Li, Y.; Song, G.L.; Liu, J. A magnetic "Band-Aid" incorporated with Fe_3O_4 NPs modified epoxy binder for in-situ repair of organic coating under seawater. *Colloids Surf. A Physicochem. Eng. Asp.* **2023**, *676*, 132317.
54. Peng, G.; Qiao, Q.; Huang, K.; Wu, J.; Wang, Y.; Fu, X.; Zhang, Z.; Fang, T.; Zhang, B.; Huang, Y.; et al. Ni-Fe-MoO_4^{2-} LDHs/epoxy resin varnish: A composite coating on carbon steel for long-time and active corrosion protection. *Prog. Org. Coat.* **2020**, *140*, 105514.
55. Xavier, J.R. Investigation on the anticorrosion, adhesion and mechanical performance of epoxy nanocomposite coatings containing epoxy-silane treated nano-MoO_3 on mild steel. *J. Adhes. Sci. Technol.* **2020**, *34*, 115–134.
56. Asemabadi, Z.; Behzadnasab, M.; Mohammadloo, H.E.; Ramezanpour, J.; Zargarian, S. Investigating the role of silane treatment in enhancing the anti-corrosion properties of MoS_2 particle-embedded epoxy coatings on steel. *Surf. Coat. Technol.* **2024**, *477*, 130348.
57. Zhou, S.; Yan, J.; Yan, H.; Zhang, Y.; Huang, J.; Zhao, G.; Liu, Y. ZrO_2-anchored rGO nanohybrid for simultaneously enhancing the wear resistance and anticorrosion performance of multifunctional epoxy coatings. *Prog. Org. Coat.* **2022**, *166*, 106795.
58. Xavier, J.R. Improvement in Corrosion Resistance and Mechanical properties of epoxy coatings on steel with the addition of thiadiazole treated ZrC. *J. Mater. Eng. Perform.* **2023**, *32*, 3980–3994.
59. Bahremand, F.; Shahrabi, T.; Ramezanzadeh, B. Epoxy coating anti-corrosion properties enhancement via the steel surface treatment by nanostructured samarium oxide-poly-dopamine film. *J. Hazard. Mater.* **2021**, *403*, 123722.
60. Dagdag, O.; El Harfi, A.; Safi, Z.; Guo, L.; Verma, C.; Ebenso, E.E.; Wazzan, N.; El Gouri, M. Fabrication on designing of a macromolecular epoxy resin as anti-corrosive coating material for electrocatalytically deposited cadmium on 15CDV6 steel in 3% NaCl solution. *J. Mater. Res. Technol.* **2020**, *9*, 5549–5563.
61. Sun, M.; Ma, Z.; Li, A.; Zhu, G.; Zhang, Y. Anticorrosive performance of polyaniline/waterborne epoxy/poly(methylhydrosiloxane) composite coatings. *Prog. Org. Coat.* **2020**, *139*, 105462.

62. Hao, Y.; Sun, W.; Jiang, L.; Cui, J.; Zhang, Y.; Song, L.; Zhang, Y. Self-healing effect of epoxy coating containing mesoporous polyaniline hollow spheres loaded with benzotriazole. *Prog. Org. Coat.* **2021**, *159*, 106445.
63. Hu, C.; Li, T.; Yin, H.; Hu, L.; Tang, J.; Ren, K. Preparation and corrosion protection of three different acids doped polyaniline/epoxy resin composite coatings on carbon steel. *Colloids Surf. A Physicochem. Eng. Asp.* **2021**, *612*, 126069.
64. Yao, Y.; Sun, H.; Zhang, Y.; Yin, Z. Corrosion protection of epoxy coatings containing 2-hydroxyphosphonocarboxylic acid doped polyaniline nanofibers. *Prog. Org. Coat.* **2020**, *139*, 105470.
65. Liu, T.; Wei, J.; Ma, L.; Liu, S.; Zhang, D.; Zhao, H. Effect of polyaniline-based plate on the anticorrosion performance of epoxy coating. *Prog. Org. Coat.* **2021**, *151*, 106109.
66. Diraki, A.; Omanovic, S. Smart PANI/epoxy anti-corrosive coating for protection of carbon steel in sea water. *Prog. Org. Coat.* **2022**, *168*, 106835.
67. Sang, Y.; Liu, Q.; Wang, S.; Dong, S.; Fan, Y.; Zhao, X.; Li, S. Synthetic polyaniline-boron nitride-aqueous epoxy resin composite coating for improving the corrosion resistance of hot-dip galvanized steel plates. *Appl. Surf. Sci.* **2022**, *592*, 153229.
68. Liu, T.; Zhao, Y.; Deng, Y.; Ge, H. Preparation of fully epoxy resin microcapsules and their application in self-healing epoxy anti-corrosion coatings. *Prog. Org. Coat.* **2024**, *188*, 108247.
69. Bi, Y.; Li, J.; Zhuang, J.; Zhang, P.; Yang, J.; Liao, B.; Wang, Q.; Li, W. Unraveling the detrimental effect of polymeric microcapsules on the corrosion performance of epoxy resin coatings. *Prog. Org. Coat.* **2024**, *194*, 108620.
70. Dagdag, O.; Hsissou, R.; El Harfi, A.; Berisha, A.; Safi, Z.; Verma, C.; Ebenso, E.E.; Touhami, M.E.; El Gouri, M. Fabrication of polymer based epoxy resin as effective anti-corrosive coating for steel: Computational modeling reinforced experimental studies. *Surf. Interfaces* **2020**, *18*, 100454.
71. Liu, T.; Liu, Y.; Qu, D.; Chen, S.; Chen, W.; Han, L.; Qiu, Z.; Zhu, L.; Chen, M. Smart self-healing coating based on the covalent organic frameworks (COF LZU-1) for corrosion protection of steel. *Colloids Surf. A Physicochem. Eng. Asp.* **2024**, *685*, 133246. [CrossRef]
72. Liu, T.; Zhao, H.; Li, J.; Zhang, D.; Zheng, W.; Wang, L. POSS-tetraaniline based giant molecule: Synthesis, self-assembly, and active corrosion protection of epoxy-based organic coatings. *Corros. Sci.* **2020**, *168*, 108555. [CrossRef]
73. Atta, A.M.; Azzam, E.M.S.; Alenezi, K.M.; El Moll, H.; Mechi, L.; El-Sofany, W.I. New epoxy and hardener system based on an imidazolium ionic liquid as an anticorrosive coating for steel in the marine environment. *ACS Omega* **2023**, *8*, 16315–16326. [CrossRef]
74. Xia, Y.; Tong, L.; Feng, X.; Zhang, S.; Xiang, H.; He, Y.; Liu, X. An investigation on the mechanical and corrosion protection properties of poly(arylene ether nitrile) reinforced epoxy coating. *Prog. Org. Coat.* **2024**, *192*, 108463. [CrossRef]
75. Hsissou, R.; Benhiba, F.; Echihi, S.; Benzidia, B.; Cherrouf, S.; Haldhar, R.; Alvi, P.A.; Kaya, S.; Serdaroğlu, G.; Zarrouk, A. Performance of curing epoxy resin as potential anticorrosive coating for carbon steel in 3.5% NaCl medium: Combining experimental and computational approaches. *Chem. Phys. Lett.* **2021**, *783*, 139081. [CrossRef]
76. Attaei, M.; Calado, L.M.; Taryba, M.G.; Morozov, Y.; Shakoor, R.A.; Kahraman, R.; Marques, A.C.; Montemor, M.F. Autonomous self-healing in epoxy coatings provided by high efficiency isophorone diisocyanate (IPDI) microcapsules for protection of carbon steel. *Prog. Org. Coat.* **2020**, *139*, 105445. [CrossRef]
77. Attaei, M.; Calado, L.M.; Morozov, Y.; Taryba, M.G.; Shakoor, R.A.; Kahraman, R.; Marques, A.C.; Montemor, M.F. Smart epoxy coating modified with isophorone diisocyanate microcapsules and cerium organophosphate for multilevel corrosion protection of carbon steel. *Prog. Org. Coat.* **2020**, *147*, 105864. [CrossRef]
78. Robert, T.; Eschig, S.; Sangermano, M.; Ocepek, M. Biobased aromatic building blocks for coating applications. *Curr. Opin. Green Sustain. Chem.* **2024**, *49*, 100962. [CrossRef]
79. Sreehari, H.; Sethulekshmi, A.S.; Saritha, A. Bio epoxy coatings: An emergent green anticorrosive coating for the future. *Macromol. Mater. Eng.* **2022**, *307*, 2200004. [CrossRef]
80. Ammar, S.; Iling, A.W.M.; Ramesh, K.; Ramesh, S. Development of fully organic coating system modified with epoxidized soybean oil with superior corrosion protection performance. *Prog. Org. Coat.* **2020**, *140*, 105523. [CrossRef]
81. Li, J.; Shi, H.; Liu, F.; Han, E.-H. Self-healing epoxy coating based on tung oil-containing microcapsules for corrosion protection. *Prog. Org. Coat.* **2021**, *156*, 106236.
82. Veedu, K.K.; Mohan, S.; Somappa, S.B.; Gopalan, N.K. Eco-friendly anticorrosive epoxy coating from Ixora leaf extract: A promising solution for steel protection in marine environment. *J. Clean. Prod.* **2022**, *340*, 130750. [CrossRef]
83. Veedu, K.K.; Banyangala, M.; Kalarikkal, T.P.; Somappa, S.B.; Gopalan, N.K. Green approach in anticorrosive coating for steel protection by *Gliricidia sepium* leaf extract and silica hybrid. *J. Mol. Liq.* **2023**, *369*, 120967.
84. Qi, C.; Li, Z.; Bi, H.; Dam-Johansen, K. Towards sustainable steel corrosion protection: Expanding the applicability of natural hydrolyzable tannin in epoxy coatings via metal complexation. *Electrochim. Acta* **2024**, *497*, 144546. [CrossRef]
85. Teijido, R.; Monteserin, C.; Blanco, M.; Odriozola, J.L.L.; Olabarria, M.I.M.; Vilas-Vilela, J.L.; Lanceros-Méndez, S.; Zhang, Q.; Ruiz-Rubio, L. Exploring anti-corrosion properties and rheological behaviour of tannic acid and epoxidized soybean oil-based fully bio-based epoxy thermoset resins. *Prog. Org. Coat.* **2024**, *196*, 108719.

86. Azani, N.F.S.M.; Hussin, M.H. Comparison of cellulose nanocrystal (CNC) filler on chemical, mechanical, and corrosion properties of epoxy-Zn protective coatings for mild steel in 3.5% NaCl solution. *Cellulose* **2021**, *28*, 6523–6543.
87. Diógenes, O.B.F.; de Oliveira, D.R.; da Silva, L.R.R.; Linhares, B.G.; Mazzetto, S.E.; Lomonaco, D.; Araujo, W.S. Acetylated lignin as a biocomponent for epoxy coating—Anticorrosive performance analysis by accelerated corrosion tests. *Surf. Coat. Technol.* **2023**, *474*, 130116.
88. Nunes, M.S.; Bandeira, R.M.; Figueiredo, F.C.; Junior, J.R.D.S.; de Matos, J.M.E. Corrosion protection of stainless steel by a new and low-cost organic coating obtained from cashew nutshell liquid. *J. Appl. Polym. Sci.* **2022**, *140*, e53420.
89. da Silva, L.R.R.; Avelino, F.; Diogenes, O.B.F.; Sales, V.D.O.F.; da Silva, K.T.; Araujo, W.S.; Mazzetto, S.E.; Lomonaco, D. Development of BPA-free anticorrosive epoxy coatings from agroindustrial waste. *Prog. Org. Coat.* **2020**, *139*, 105449. [CrossRef]
90. Abbout, S.; Hsissou, R.; Chebabe, D.; Erramli, H.; Hajjaji, N. Investigation of the anti-corrosion properties of Galactomannan as additive in epoxy coatings for carbon steel: Rheological and electrochemical study. *Inorg. Chem. Commun.* **2021**, *134*, 108971.
91. Xiao, X.; Ye, Z.; Meng, G.; Gu, L. Mussel-inspired preparation of superhydrophobic mica nanosheets for long-term anticorrosion and self-healing performance of epoxy coatings. *Prog. Org. Coat.* **2023**, *178*, 107456.
92. Balakrishnan, A.; Govindaraj, S.; Dhaipule, N.G.K.; Thirumalaisamy, N.; Anne, R.S.; Sublime, N.; Philip, J. Enhancing microbiologically influenced corrosion protection of carbon steels with silanized epoxy-biocide hybrid coatings. *Environ. Sci. Pollut. Res.* **2024**, *31*, 13302–13326.
93. Zhou, Y.; Chen, G.; Yan, S.; Ni, C.; Yu, L.; Li, X. Epoxy composite coating with excellent anticorrosion and self-healing properties based on acrylate copolymers. *Prog. Org. Coat.* **2022**, *172*, 107098. [CrossRef]
94. Chen, P.; Wang, G.; Li, J.; Zhang, M.; Qiao, X. Preparation of textured epoxy resin coatings for excellent hydrophobicity and corrosion resistance. *Prog. Org. Coat.* **2023**, *175*, 107312.
95. Mallakpour, S.; Nikkhoo, E.; Hussain, C.M. Application of MOF materials as drug delivery systems for cancer therapy and dermal treatment. *Coord. Chem. Rev.* **2022**, *451*, 214262.
96. Ma, S.; Han, W.; Han, W.; Dong, F.; Tang, Z. Recent advances and future perspectives in MOF-derived single-atom catalysts and their application: A review. *J. Mater. Chem. A* **2023**, *11*, 3315–3363.
97. Liu, K.G.; Bigdeli, F.; Panjehpour, A.; Jhung, S.H.; Al Lawati, H.A.J.; Morsali, A. Potential applications of MOF composites as selective membranes for separation of gases. *Coord. Chem. Rev.* **2023**, *496*, 215413.
98. Ramezanzadeh, M.; Ramezanzadeh, B.; Mahdavian, M.; Bahlakeh, G. Development of metal-organic framework (MOF) decorated graphene oxide nanoplatforms for anti-corrosion epoxy coatings. *Carbon* **2020**, *161*, 231–251.
99. Zhang, Y.; Tian, H.; Sui, X.; Wang, X.; Zhou, F.; Zhang, X. The Improved antiwear and anticorrosion properties of epoxy resin with metal–organic framework ZIF-8 containing lubrication oil. *Langmuir* **2022**, *38*, 10649–10661.
100. He, Z.; Lin, H.; Zhang, X.; Chen, Y.; Bai, W.; Lin, Y.; Jian, R.; Xu, Y. Self-healing epoxy composite coating based on polypyrrole@MOF nanoparticles for the long-efficiency corrosion protection on steels. *Colloids Surf. A Physicochem. Eng. Asp.* **2023**, *657*, 130601.
101. Wan, S.; Chen, H.; Ma, X.; Chen, L.; Lei, K.; Liao, B.; Dong, Z.; Guo, X. Anticorrosive reinforcement of waterborne epoxy coating on Q235 steel using NZ/BNNS nanocomposites. *Prog. Org. Coat.* **2021**, *159*, 106410. [CrossRef]
102. Hasanzadeh, A.; SaadatAbadi, A.R. ZIF-67 metal-organic framework decorated g-C3N4 nanosheets with inbuilt organic-inorganic corrosion inhibitor for self-healing epoxy anti-corrosion coating. *Surf. Coat. Technol.* **2024**, *492*, 131218. [CrossRef]
103. Hosseinpour, A.; Abadchi, M.R.; Mirzaee, M.; Tabar, F.A.; Ramezanzadeh, B. Recent advances and future perspectives for carbon nanostructures reinforced organic coating for anti-corrosion application. *Surf. Interfaces* **2021**, *23*, 100994.
104. Hsia, H.-H.; Chen, Y.-L.; Tai, Y.-T.; Tian, H.-K.; Kung, C.-W.; Liu, W.R. Two-dimensional metal–organic frameworks/epoxy composite coatings with superior O_2/H_2O resistance for anticorrosion applications. *ACS Appl. Mater. Interfaces* **2024**, *16*, 41421–41434. [CrossRef] [PubMed]
105. Alipanah, N.; Yari, H.; Mahdavian, M.; Ramezanzadeh, B.; Bahlakeh, G. MIL-88A (Fe) filler with duplicate corrosion inhibitive/barrier effect for epoxy coatings: Electrochemical, molecular simulation, and cathodic delamination studies. *J. Ind. Eng. Chem.* **2021**, *97*, 200–215. [CrossRef]
106. Zhang, M.; Zhang, Y.; Chen, Y.; Tian, X.; Liu, L.; Wang, Y.; Guo, R.; Yan, H. Dual-inhibitor composite BTA/PPy/MIL-88(Fe) for active anticorrosion of epoxy resin coatings. *J. Ind. Eng. Chem.* **2023**, *119*, 660–673.
107. Keshmiri, N.; Najmi, P.; Ramezanzadeh, M.; Ramezanzadeh, B. Designing an eco-friendly lanthanide-based metal organic framework (MOF) assembled graphene-oxide with superior active anti-corrosion performance in epoxy composite. *J. Clean. Prod.* **2021**, *319*, 128732.
108. Wei, R.; Liu, Z.; Wei, W.; Wang, S.; Lv, Y.-J.; Han, G.-C. Anticorrosion performance of hydrophobic acid-modified-MOFs/epoxy coatings. *Colloid Interface Sci. Commun.* **2022**, *46*, 100580.
109. Li, K.; Yu, Z.; Tang, J.; Guo, S.; Peng, B.; Yu, P.; Lin, B. +Improving the corrosion protection properties of epoxy resin coating by using HKUST-1-BTA@e-Mica two-dimensional nanocomposites. *Colloids Surf. A Physicochem. Eng. Asp.* **2024**, *701*, 134869. [CrossRef]

110. Fadl, A.M.; Sadeek, S.A.; Magdy, L.; Abdou, M.I.; El-Shiwiniy, W.H. Multi-functional epoxy composite coating incorporating mixed Cu(II) and Zr(IV) complexes of metformin and 2,2\-bipyridine as intensive network cross-linkers exhibiting anti-corrosion, self-healing and chemical-resistance performances for steel petroleum platforms. *Arab. J. Chem.* **2021**, *14*, 103367.
111. Butt, A.M.; Wang, Y.; Ma, H.; Li, H. The preparation of cerium nitrate and attapulgite based superhydrophobic epoxy coatings for the corrosion protection of Q355 mild steel surface. *Surf. Coat. Technol.* **2023**, *473*, 129977.
112. Chopra, I.; Ola, S.K.; Gopalakrishnan, S.; Dhayal, V. Tailoring epoxy coating with acetoxime derivative of zinc for advanced anticorrosive performance on mild steel: Experimental and computational insights. *J. Mol. Model.* **2023**, *29*, 300.
113. Peng, Y.; Hughes, A.E.; Mardel, J.I.; Deacon, G.B.; Junk, P.C.; Catubig, R.A.; Forsyth, M.; Hinton, B.R.W.; Somers, A.E. Dual function of rare earth carboxylate compounds on the barrier properties and active corrosion inhibition of epoxy coatings on mild steel. *Prog. Org. Coat.* **2023**, *185*, 107870.
114. Zhang, Q.; Shen, M.; Liu, X.; Fu, H.; Yang, W.; Zhao, J.; Wang, J.; Du, Y.; Ma, C. Construction and utilization of rare earth complexes as efficient corrosion inhibitor in epoxy coating. *Prog. Org. Coat.* **2024**, *186*, 108024.
115. Ren, P.; Li, J.; Wang, L.; Guo, H.; Lei, B.; Feng, Z.; Meng, G. Organo-cerium as a quick repair agent for coating damage on carbon steel. *J. Mater. Eng. Perform.* **2023**, *32*, 9755–9764.
116. Madhusudhana, A.M.; Mohana, K.N.S.; Hegde, M.B.; Nayak, S.R.; Rajitha, K.; Swamy, N.K. Functionalized graphene oxide-epoxy phenolic novolac nanocomposite: An efficient anticorrosion coating on mild steel in saline medium. *Adv. Compos. Hybrid Mater.* **2020**, *3*, 141–155.
117. Zhang, C.; Li, W.; Liu, C.; Zhang, C.; Cao, L.; Kong, D.; Wang, W.; Chen, S. Effect of covalent organic framework modified graphene oxide on anticorrosion and self-healing properties of epoxy resin coatings. *J. Colloid Interface Sci.* **2022**, *608*, 1025–1039.
118. Mourya, P.; Goswami, R.N.; Saini, R.; Ray, A.; Khatri, O.P. Epoxy coating reinforced with graphene-PANI nanocomposites for enhancement of corrosion-resistance performance of mild steel in saline water. *Colloids Surf. A Physicochem. Eng. Asp.* **2024**, *687*, 133500.
119. Yan, H.; Zhang, L.; Li, H.; Fan, X.; Zhu, M. Towards high-performance additive of Ti_3C_2/graphene hybrid with a novel wrapping structure in epoxy coating. *Carbon* **2020**, *157*, 217–233.
120. Zhang, J.; Kong, G.; Li, S.; Le, Y.; Che, C.; Zhang, S.; Lai, D.; Liao, X. Graphene-reinforced epoxy powder coating to achieve high performance wear and corrosion resistance. *J. Mater. Res. Technol.* **2022**, *20*, 4148–4160.
121. Ye, Y.; Chen, H.; Zou, Y.; Ye, Y.; Zhao, H. Corrosion protective mechanism of smart graphene-based self-healing coating on carbon steel. *Corros. Sci.* **2020**, *174*, 108825.
122. Xavier, J.R. Novel multilayer structural epoxy composite coating containing graphene oxide and silanized chromium carbide for the protection of steel structures. *J. Coat. Technol. Res.* **2022**, *19*, 1713–1730.
123. Dhongde, N.R.; Baranwal, P.K.; Rajaraman, P.V. Functionalization of graphene oxide with an ionic liquid (1-butyl-3-methylimidazolium acetate): Preparation of epoxy-based coating on carbon steel for anticorrosive applications. *J. Appl. Polym. Sci.* **2023**, *140*, e54026.
124. Vinodhini, S.P.; Xavier, J.R. Evaluation of corrosion protection performance and mechanical properties of epoxy-triazole/graphene oxide nanocomposite coatings on mild steel. *J. Mater. Sci.* **2021**, *56*, 7094–7110.
125. Zhou, X.; Huang, H.; Zhu, R.; Chen, R.; Sheng, X.; Xie, D.; Mei, Y. Green modification of graphene oxide with phytic acid and its application in anticorrosive water-borne epoxy coatings. *Prog. Org. Coat.* **2020**, *143*, 105601.
126. Zhu, Q.; Li, E.; Liu, X.; Song, W.; Li, Y.; Wang, X.; Liu, C. Epoxy coating with in-situ synthesis of polypyrrole functionalized graphene oxide for enhanced anticorrosive performance. *Prog. Org. Coat.* **2020**, *140*, 105488.
127. Xiang, Q.; Qiang, Y.; Guo, L. Designing a novel GO@AAP reinforced epoxy coating for achieving the long-term corrosion protection of steel substrate. *Prog. Org. Coat.* **2023**, *174*, 107293.
128. Li, X.; Chen, X.; Chen, J.; Huo, D.; Gao, X.; Dong, J.; Yin, Y.; Liu, J.; Nan, D. Development of an epoxy resin-based anticorrosive coating enhanced by monolayer amino-modified graphene oxide. *Prog. Org. Coat.* **2024**, *194*, 108624.
129. Diraki, A.; Omanovic, S. Anticorrosive properties of the double-layer PANI-(graphene oxide)/epoxy coating in protecting carbon steel in saltwater. *J. Coat. Technol. Res.* **2023**, *20*, 995–1006.
130. Kooshksara, M.M.; Mohammadi, S. Investigation of the in-situ solvothermal reduction of multi-layered Graphene oxide in epoxy coating by acetonitrile on improving the hydrophobicity and corrosion resistance. *Prog. Org. Coat.* **2021**, *159*, 106432.
131. Xiong, L.; Yu, M.; Li, Y.; Kong, X.; Li, S.; Liu, J. Modified salicylaldehyde@ZIF-8/graphene oxide for enhancing epoxy coating corrosion protection property on AA2024-T3. *Prog. Org. Coat.* **2020**, *142*, 105562.
132. Chen, C.; He, Y.; Xiao, G.; Zhong, F.; Xia, Y.; Wu, Y. Graphic C_3N_4-assisted dispersion of graphene to improve the corrosion resistance of waterborne epoxy coating. *Prog. Org. Coat.* **2020**, *139*, 105448.
133. Xavier, J.R.; Bhaskar, R.; Subramanian, S. Multifunctional graphitic carbon nitride/manganese dioxide/epoxy nanocomposite coating on steel for enhanced anticorrosion, flame retardant, mechanical, and hydrophobic properties. *J. Ind. Eng. Chem.* **2024**, *134*, 514–536.

134. Souto, L.F.C.; Soares, B.G. Polyaniline/carbon nanotube hybrids modified with ionic liquids as anticorrosive additive in epoxy coatings. *Prog. Org. Coat.* **2020**, *143*, 105598.
135. Nayak, S.R.; Mohana, K.N.S.; Hegde, M.B.; Rajitha, K.; Madhusudhana, A.M.; Naik, S.R. Functionalized multi-walled carbon nanotube/polyindole incorporated epoxy: An effective anti-corrosion coating material for mild steel. *J. Alloys Compd.* **2021**, *856*, 158057.
136. Wan, S.; Chen, H.; Cai, G.; Liao, B.; Guo, X. Functionalization of h-BN by the exfoliation and modification of carbon dots for enhancing corrosion resistance of waterborne epoxy coating. *Prog. Org. Coat.* **2022**, *165*, 106757.

materials

MDPI

Article

Correlation between the Chemical Structure of (Meth)Acrylic Monomers and the Properties of Powder Clear Coatings Based on the Polyacrylate Resins

Katarzyna Pojnar [1] and Barbara Pilch-Pitera [2,*]

[1] Doctoral School of Engineering and Technical Sciences, Rzeszow University of Technology, ul. Powstańców Warszawy 12, 35-959 Rzeszów, Poland; d521@stud.prz.edu.pl

[2] Department of Polymers and Biopolymers, Faculty of Chemistry, Rzeszow University of Technology, ul. Powstańców Warszawy 6, 35-959 Rzeszów, Poland

* Correspondence: barbpi@prz.edu.pl; Tel.: +48-17-86-51-700; Fax: +48-17-85-43-655

Abstract: This paper presents studies on the influence of the chemical structure of (meth)acrylic monomers on the properties of powder coatings based on polyacrylate resins. For this purpose, a wide range of monomers were selected—2-hydroxyethyl methacrylate (HEMA), methyl methacrylate (MMA), *n*-butyl acrylate (*n*BA), *tert*-butyl acrylate (*t*BA), dodecyl acrylate (DA), ethyl acrylate (EA) and benzyl acrylate (BAZ)—for the synthesis of the polyacrylate resin. The average molecular mass and molecular mass distribution of the synthesized resins were measured by gel permeation chromatography (GPC). The glass transition temperature (T_g) and viscosity of polyacrylate resins were determined by using differential scanning calorimetry (DSC) and a Brookfield viscometer. These parameters were necessary to obtain information about storage stability and behavior during the application of powder clear coatings. Additionally, DSC was also used to checked the course of the low-temperature curing reaction between the hydroxyl group contained in the polyacrylate resin and the blocked polyisocyanate group derived from a commercial agent such as Vestanat B 1358/100. The properties of the cured powder clear coatings were tested, such as: roughness, gloss, adhesion to the steel surface, hardness, cupping, scratch resistance, impact resistance and water contact angle. The best powder clear coating based on the polyacrylate resin L_HEMA/6MMA/0.5*n*BA/0.5DA was characterized as having good scratch resistance (550 g) and adhesion to the steel surface, a high water contact angle (93.53 deg.) and excellent cupping (13.38 mm). Moreover, its crosslinking density (CD) and its thermal stability was checked by using dynamic mechanical analysis (DMA) and thermogravimetric analysis (TGA).

Keywords: polyacrylate resin; powder coating properties; low-temperature coatings

1. Introduction

Currently, specific customer requirements and EU directives are increasing, necessitating the development of new products that meet the new requirements set for the paints and coatings market [1–3]. The rising concerns over energy consumption, high emissions, harmful and toxic gases, liquid coatings containing large amounts of volatile organic compounds (VOCs), the utilization of petroleum-based raw materials in large quantities and other significant issues, highlight the need for new approaches. Powder coatings, in comparison to liquid paints, do not contain VOCs, and their application meets requirements such as high economy, efficiency, sustainability and quality (for example: the new powder coatings line developed by Ecoline company, Poland) [4,5]. Depending on the raw materials used for the powder coating composition, it is possible to obtain a finished product cured at low temperature or UV cured, thereby reducing production costs and energy consumption [6]. Moreover, many researchers have conducted studies on the use of biobased or renewable products. Li et al. described biobased UV-curable (powder) coating resins based on limone-derived polycarbonates obtained from orange oils and carbon dioxide [7]. The UV-cured powder coatings used limone-derived polycarbonates as binders characterized high transparency, good acetone resistance, high pencil hardness (H–2 H) and high König hardness (174–199 s).

The most common resins used for powder coatings are based on epoxy and polyester [8–10]. Due to compatibility issues with other resins such as acrylic, polyacrylic resins are less frequently used. Separate production lines are required to produce powder coatings based on polyacrylate resins. Despite this disadvantage, acrylic powder coatings are characterized by high physical and mechanical properties, very good gloss and color durability, making them ideal for outdoor use compared to polyesters or epoxy resin. These properties result from the chemical structure of (meth)acrylic monomers. Polyacrylic resin can contain reactive hydroxyl-, carboxyl- or glycidyl groups [11]. Polyacrylates containing hydroxyl groups exhibit excellent weather durability, but their drawback is limited impact resistance. However, the final properties of the coating are also influenced by the appropriate crosslinking agent. A commonly use curing agent, hydroxyl functional acrylic resin, is a blocked polyisocyanate (PIC). Powder coating based on acrylic-resin-containing hydroxyl groups can cure below 160 °C creating low-temperature systems [12]. Acrylic resins containing glycidyl groups offer several benefits, including effective corrosion protection, resilience against weathering and high adhesion [13]. Moreover, epoxy functional acrylic resin can be crosslinked by UV radiation in the presence of cationic photoinitiators [14]. The possibility of UV curing in a shorter time, like low-temperature systems, allows the application of powder coatings to heat-sensitive surfaces such as wood, MDF board or even polymer composite [15]. Okada et al. described an acrylic/polyester hybrid powder coating. These coatings are characterized by good appearance and mechanical properties such as toughness and flexibility. However, a notable drawback of these coatings is their insufficient resistance to weathering [16]. Generally, acrylic monomers, thanks to the reactive double bonds, offer a wide range of opportunities. It is also possible to use acrylic monomers for Atom Transfer Radical Polymerization (ATRP) or even for the production of self-healing coatings, but everything depends on the appropriative selection of the monomers used [17].

The primary issue addressed in this article is the weak flexibility of acrylic resin. However, the selection of raw materials with an appropriate structure may alter the interactions between the polymer chains and allow a compromise between these properties to be achieved. There are studies on this topic in the scientific literature. Shi et al. synthesized hyperbranched polyurethane acrylate, which was blended with epoxy acrylate and tripropylene glycol diacrylate (TPGDA) to prepare UV-cured films [18]. They found that the incorporation of hyperbranched polyurethane acrylate could enhance the mechanical strength to 19 MPa, elongation at break to 310% and impact resistance to 26.9 MJ m^{-3} of UV-cured films simultaneously. However, the synthesis of hyperbranched polyurethane acrylate often requires complex and sophisticated synthesis routes [19]. Coatings with low cupping tend to crack faster. The brittleness of the coating may cause damage, leading to corrosion and lack of tightness. Li et al. described a flexible and self-healing epoxy acrylic coating by introducing dynamic disulfide bonds and dangling chains into the resin structure. Furthermore, nano-SiO_2 was sprayed to construct the microsurface structure to endow the coating super-hydrophobicity [20].

Obtaining innovative, specific acrylic resins can significantly expand their applications and meet the requirements set by customers and legal regulations. Nevertheless, formulating a material that can exhibit all desired properties and functions, especially if they are opposing (e.g., flexibility and hardness), remains challenging. Therefore, further research in this area needed, especially since few studies on acrylic resins for powder coating have been published in the scientific literature.

Therefore, the aim of this work is to investigate the correlation between the chemical structure of appropriate (meth)acrylic monomers and the properties of powder clear coatings. Gel permeation chromatography (GPC) and differential scanning calorimetry (DSC) were used to determine the molar mass distribution and the glass transition temperature T_g. Moreover, the viscosity was measured to evaluate the storage stability and behavior during the application of powder coatings. Nuclear magnetic resonance (NMR) was used to confirm the chemical structure of the resins.

The visual and mechanical properties of cured powder coatings were investigated using tests such as roughness, gloss, adhesion to the steel surface, hardness, cupping, scratch resistance, impact resistance and water contact angle. The crosslinking density and thermal stability were checked using dynamic mechanical analysis (DMA) and thermogravimetric analysis (TGA). Based on the conducted research, coatings with optimal properties were selected and correlated with the structure of monomers used for the synthesis of the resin.

2. Experimental Section

2.1. Materials

Materials and reagents used for the synthesis of polyacrylate resins:

- 2-hydroxyethyl methacrylate (HEMA) (Sigma Aldrich, Darmstadt, Germany);
- Methyl methacrylate (MMA) (Sigma Aldrich, Darmstadt, Germany);
- *n*-butyl acrylate (*n*BA) (Sigma Aldrich, Darmstadt, Germany);
- *tert*-butyl acrylate (*t*BA) (Sigma Aldrich, Darmstadt, Germany);
- Dodecyl acrylate (DA) (Sigma Aldrich, Darmstadt, Germany);
- Ethyl acrylate (EA) (Sigma Aldrich, Darmstadt, Germany);
- Ethyl methacrylate (EMA) (Sigma Aldrich, Darmstadt, Germany);
- Benzyl acrylate (BAZ) (Sigma Aldrich, Darmstadt, Germany);
- Free radical initiator of polymerization: azobisisobutyronitrile (AIBN) (Sigma Aldrich, Darmstadt, Germany).

Raw materials used for the powder clear coating preparation:

- Vestanat®B 1358/100 (Evonic Degussa, Marl, Germany);
- Degassing agent: benzoin (Sigma Aldrich, Darmstadt, Germany);
- Flow control agent: Byk 368P (Byk-Chemie, Wesel, Germany).

2.2. Synthesis of Polyacrylate Resins

For each synthesis, methyl methacrylate and 2-hydroxyethyl methacrylate were used. Depending on the type of resins, an appropriate amount of (meth)acrylate monomers were added (Table 1).

Table 1. Qualitative and quantitative composition of the synthesized acrylic resins.

Resin Symbol	2-hydroxyethyl methacrylate (HEMA)	methyl methacrylate (MMA)	n-butyl acrylate (nBA)	tert-butyl acrylate (tBA)	ethyl acrylate (EA)	ethyl methacrylate (EMA)	benzyl acrylate (BAZ)	dodecyl acrylate (DA)
HEMA/6MMA/*n*BA	1	6	1	-	-	-	-	-
HEMA/6MMA/*t*BA	1	6	-	1	-	-	-	-
HEMA/6MMA/EA	1	6	-	-	1	-	-	-
HEMA/6MMA/EMA	1	6	-	-	-	1	-	-
HEMA/6MMA/BAZ	1	6	-	-	-	-	1	-
HEMA/6MMA/DA	1	6	-	-	-	-	-	1
HEMA/6MMA/0.5*n*BA/0.5DA	1	6	0.5	-	-	-	-	0.5

As an initiator of free radical polymerization in bulk of acrylate monomers, 1.7% azabisisobutyronitrile (AIBN) was used. All compounds were placed in a three-necked flask equipped with a reflux condenser, a thermometer, a magnetic stirrer and a nitrogen inlet (the reaction was sensitive to oxygen inhibition). The reaction mixture was heated at a temperature of 80 °C. Polymerization reaction commenced when the viscosity experienced a rapid increase. Subsequently, the non-solidified resin mixture was poured onto a PTFE mold for the purpose of solidification. To complete the polymerization process, the PTFE mold containing the resin was sealed and positioned in an oven at 80 °C for one hour. Following this, the mold was allowed to cool and the solidified resin was subjected to grinding. The synthesized resins were designated based on the molar ratio and names of the monomers utilized; for instance, HEMA/6MMA/*n*BA denotes a resin synthesized from the HEMA, MMA and *n*BA monomers in a molar ratio of 1:6:1.

2.3. Preparation of Powder Clear Coatings Based on Synthesized of Polyacrylate Resins

For the preparation of powder coatings, the following materials were used: Vestanat B 1358/100 (oxime blocked polyisocyanate, based on isophorone diisocyanate), 1% of benzoin and 2% of Byk 368P. The amount of Vestanat B 1358/100 was calculated based on the hydroxyl number of the acrylic resin (L_{OH} = 40 mgKOH/g), so that the molar ratio of the -NCO groups to the -OH groups was 1:1. The prepared composition was mixed, ground and extruded in a co-rotating twin-screw mini-extruder EHP 2 × 12 Sline from Zamak (Cracow, Poland). The temperature in the extruder was maintained as follows: zone I-80 °C, zone II-90 °C, zone III-100 °C and adapter-110 °C. The rotational speed of screw was set to 150 rpm. In the next step, the extrudate was cooled, pulverized and sieved in a 100 µm sieve. The final powder coatings were applied to the metal plates. Prior to application, the metal plates were cleaned in acetone, degreased, and passivated using zirconium phosphate conversion ESKAPHOR Z 2000C (Haug Chemie, Sinsheim, Germany). The powder coatings were applied by electrostatic gun PEM X-1 controlled by EPG Sprint X (CORONA) from Wagner (Wagner, Altstätten, Switzerland). The coatings were curing at 160 °C for 15 min. The obtained powder coatings were named according to the names of the resin used, e.g., L_HEMA/6MMA/*n*BA means a coating made from the resin HEMA/6MMA/*n*BA.

3. Measurements

3.1. Gel Permeation Chromatography (GPC)

Gel permeation chromatography (GPC) (ViscoTec, Töging a. Inn, Germany) was conducted utilizing RI detector Shodex RI-71, a Shimadzu LC-20AD isocratic pump, a ViscoTec degassing system, a PSS SDV guard column, and PSS SDV 100, 1000 and 10,000 Å columns with a grain diameter of 5 µm, packed with styrene divinylbenzene-type gel were used. All samples were dissolved in tetrahydrofuran (HPLC grade) containing 5 mmol/L of the resin at a temperature of 22 °C. The analysis was carried out by interpreting results based on the conventional calibration of columns with polystyrene standards. Prior to analysis, solutions were shaken for 24 h at ambient temperature and subsequently filtered through syringe filters with a diaphragm (PTFE) of 0.25 µm.

3.2. Viscosity

The viscosity was measured at 140 °C using a cone-plate CAP 2000+ viscometer (AMETEK Brookfield, Middleboro, MA, USA) equipped with a cone no. 6, according to PN-EN ISO 2884-1.

3.3. Differential Scanning Calorimetry (DSC)

Thermal properties of resins and powder compositions were examined using a Mettler Toledo 822e calorimeter (Mettler Toledo, Columbus, OH, USA) equipped with Stare System software. The calibration of the DSC apparatus (Mettler Toledo r, Columbus, OH, USA) was conducted using indium and zinc standards supplied by Mettler Toledo (Columbus, OH, USA). Accurate weighing of sample (10 mg) was performed with a precision of 0.00001 g. Subsequently, the samples were hermetically sealed in standard 40 μL aluminum crucibles and positioned in the measuring chamber. The measurements were conducted in the temperature range from 0 to 200 °C, with a heating rate of 20 K min^{-1} and maintaining an atmosphere of nitrogen with a flow rate of 60 mL/min. This experimental setup facilitated the comprehensive investigation of the thermal behavior of the resin and powder compositions.

3.4. ^{1}H-NMR Spectroscopy

The nuclear magnetic resonance (^{1}H-NMR) spectra were acquired using a Bruker Avance II 500 MHz (Bruker BioSpin, Rheinstetten, Germany). This device featured a 5 mm nitrogen-cooled dual (BB-1H) cryoprobe. Tetramethylsilane served as the standard for internal referencing and chemical shift values were reported in parts per million (ppm). Deuterated chloroform ($CDCl_3$) was used as a solvent. NMR Topspin 2.1 pl 8 software (Topspin 2.1 pl 8, Bruker BioSpin, Rheinstetten, Germany) was used.

3.5. Polymerization Test

Polymerization test was conducted in accordance with the technical requirements of the QUALICOAT [21]. The assessment of the sample took place after subjecting the coatings to rubbing with a swab soaked in methyl ethyl ketone (MEK). The swab was moved lightly back and forth 30 times in each direction. The classification of the sample was based on following criteria:

1. The coating is matt and soft.
2. The coating is matt and can be scratched with a nail.
3. Slight loss of gloss.
4. No noticeable changes.
5. The polymerization test was performed twice for each coating.

3.6. Flowability

In accordance with PN-EN ISO 8130-11, the evaluation of paint flow from an inclined surface was carried out [22]. Cavities with a depth of 6.5 mm and a diameter of 25 mm were stamped into the steel test plates. Then, a paint sample of 0.5 g was placed in the cavity of the plate and positioned at 60° from horizontal in an oven heated to 160 °C for 15 min. After this time, the distance between the bottom edge of the cavity and the furthest point where the melted powder paint flowed was determined.

3.7. Thickness and Gloss

The gloss and thickness test were conducted using a gloss meter, specifically the micro-Tri-gloss-μ with a thickness measurement function, manufactured by BYK-Gardner (Geretsried, Germany). The assessment followed to the standards outlined in PN-EN ISO 2813 for gloss and PN-EN ISO 2808 for thickness [23,24]. Gloss was evaluated by measuring the intensity of light reflected from the coating at an incident angle of 60°. The thickness analysis was carried out using the same device, equipped with a built-in Fe/NFe sensor. Results for gloss and thickness were obtained by averaging measurements from ten trials for each sample.

3.8. Roughness

Measurements were carried out at a LT = 5600 mm and LC = 0.800 × 5N by the use of a Mar SurfPS1 apparatus from Mahr GmbH (Göttingen, Germany.), according to the PN-EN ISO 12085 standard [25]. The test focused on two roughness parameters, R_a and R_z. R_a represents the arithmetic mean deviation from the baseline expressed in micrometers, while R_z signifies the arithmetic mean of the five highest profile peaks decreased by the arithmetic mean of the five lowest profile depths. Roughness assessments were performed at ten different locations on the surface of the same coating and the average of ten measurements was taken as the final result.

3.9. Adhesion to the Steel Surface

The adhesion-to-the-steel-substrate test was conducted following the PN-EN ISO 2409 standard, utilizing the cross-cut method [26]. A specialized multi-cut tool equipped with six cutters, manufactured by Byk Gardner (Geretsried, Germany), was employed for this purpose. The cuts were made perpendicular to each other, forming a grid of squares with a 2 mm spacing. After the cutting process, any dust generated was brushed off the coating surface using a brush. Subsequently, a 50 mm wide adhesive tape with standardized peel force was applied to the surface. Upon breaking the tape, the appearance of the grid was assessed. The surface of the coatings was visually examined and classified on a 0–5 six-point scale, where 0 indicated no traces other than knife marks, and 5 represented almost complete or complete detachment of the coating. The test was repeated twice for the same coatings to ensure consistency in the results.

3.10. Hardness

Measurements of the coating hardness were carried out according to the standard PN-EN ISO 1522 using König Pendulum manufactured by BYK-Gardner (Geretsried, Germany) [27]. The relative hardness was calculated by dividing the arithmetic mean of the number of pendulum oscillations for the tested sample by glass constant, which is 171 pendulum oscillations. Three measurements were made for the same coating.

3.11. Scratch Resistance

The scratch resistance test was executed following the PN-EN ISO 1518 standard [28]. The cured coating was positioned on a test table that moved with increasing loads on the needle until the coating experienced a scratch. After the scratch occurred, the value of the needle load was recorded. The measurements were repeated four times using the same composition and the final result was determined based on the repeated measurements.

3.12. Water Contact Angle (WCA)

The water contact angle (WCA) was determined following the standard PN-EN ISO 19403-6:2020-08, employing an optical goniometer OCA15 EC from DataPhysics (Filderstadt, Germany) [29]. Drop contours were measured using the SCA20U computer software (Ravelin, London, United Kingdom) and the contact angle was subsequently calculated. Measurements were conducted at ten different locations on the surface of the same coating to capture variability. The final result was determined by averaging of these measurements to ensure a comprehensive representation of the coating's water contact angle.

3.13. Impact Resistance

To determine the resistance of coatings to impacts, the SP1890 tester by TQC (Capelle aan den IJssel, Netherlands) was used in accordance with the PN-EN ISO 6272-1 standard [30]. The device consists of a tube with a marked height scale, a weight dropping from a specified height according to the standard onto a matrix where the test sample was placed. The test involved deformation of a thin sheet metal, on which the coating was applied, by the falling weight. Impact resistance was measured by assessing cracking or detachment of the deformed coating.

3.14. Cupping

The results were acquired through a manual SP4300 tester by TQC, following the PN-EN ISO 1520 standard [31]. The test involved pressing a spherical drawing punch into a clamped sheet until a crack appeared. The result was read at the point of coating cracking. To assess the repeatability of the results, three measurements were conducted on the same cured coating.

3.15. Dynamic Mechanical Analysis (DMA)

Dynamic mechanical analysis (DMA) measurements were conducted using DMA/SDTA861e unit from Mettler Toledo (Mettler-Toledo, Columbus, OH, USA). The analysis was performed in compressing mode with a constant frequency of 1 Hz, spanning a temperature range of 0–200 °C (heating rate 3 °C/min). The force amplitude was maintained at a maximum of 0.4 N and displacement amplitude at maximum 2.5 μm. The samples, with a diameter of approximately 15 mm, thickness 2.5 mm, geometry factor 14,15 1/m and weight 0.5 g were formed using a SPECA hydraulic press.

3.16. Thermogravimetric Analysis (TGA)

Thermogravimetric analysis was performed using Mettler Toledo TGA/DSC1 instrument (Mettler-Toledo, Columbus, OH, USA). The analyses were conducted in nitrogen in the temperature range of 25 to 700 °C, with a heating rate of 10 °C/min. The measurement conditions were as follows: sample weight ~5 mg, gas flow rate 50 cm^3/min and 150 μL open alumina pan.

4. Results and Discussion

4.1. Choice of (Meth)Acrylic Monomers

The chemical structure of (meth)acrylate monomers plays a crucial role in determining the properties of resins and their suitability for various coating applications. The structure of these monomers, along with their molar ratio, affects factors such as thermal stability, rheology and other characteristics of the final product. With a wide variety of monomers available, there exists an extensive range of possible polyacrylate resin compositions. Hence, it is essential to comprehend how the chemical structure of (meth)acrylic monomers influences the resulting properties. Table 2 illustrates the impact of selected monomers on the resin performance.

The most commonly used monomer is methyl methacrylate (MMA) [32]. The presence of the -CH_3 groups in the α-position of vinyl groups increases stiffness and improves the physical and mechanical properties of the resin [33]. For this reason, the second monomer often used for coating applications is n-butyl acrylate (*n*BA). The four-carbon aliphatic chain increases flexibility [34]. This increased flexibility would allow the application of this polymer on surfaces of various shapes, such as roofs or elements for agricultural and automotive equipment.

The use of the butyl acrylate isomer in the form of tert-butyl acrylate (*t*BA) also alters resin properties. The packed, tetrahedral structure of tert-butyl substituents promotes greater chain mobility, which increases hydrophobicity, reduces the glass transition temperature (T_g) and maintains brilliant color [35,36]. Ethyl methacrylate (EMA) is often employed in the production of plastics with better resistance to weather conditions, while ethyl acrylate (EA) is more frequently used in various sectors of the chemical industry [37,38].

In addition to monomers containing aliphatic substituents in the ester group, a monomer containing an aromatic ring was also selected. Therefore, benzyl acrylate (BAZ) was chosen. The presence of the aromatic ring enhances mechanical strength and hardness. However, a disadvantage of using this monomer is the yellowing of coatings, which can lead to a decrease in adhesive and mechanical properties [39].

To optimize mechanical properties, dodecyl acrylate is an often-used monomer in acrylic resins. Due to the presence of long alkyl chain in ester groups, DA improves the tensile strength of polymers [40]. Additionally, the nonpolar long chain increases the hydrophobic properties [41].

Furthermore, monomers containing functional groups such as hydroxyl, carboxyl or epoxy groups are needed to ensure crosslinking ability. The 2-hydroxyethyl methacrylate was used in this work. The presence of polar hydroxyl groups increases hydrophilic properties, allowing for the creation of amphiphilic polymers [42]. Depending on the chemical structure of these monomers, properties related to elasticity or hardness can be controlled, and the appropriate curing agent can be selected to enable the crosslinking process [43,44].

Table 2. The summarizing of different types of (meth)acrylate monomers used and their influence on resin properties.

Name of Monomer	Symbol	Chemical Structure	Resin Properties	Reference
methyl methacrylate	MMA	$H_2C=C(CH_3)-C(=O)-O-CH_3$	- good thermal stability - high Young's Modulus - low elongation at break - high scratch resistance - good chemical resistance	[32,33]
n-butyl acrylate	*n*BA	$H_2C=CH-C(=O)-O-CH_2-CH_2-CH_2-CH_3$	- high flexibility - low thermal stability	[34]
tert-butyl acrylate	*t*BA	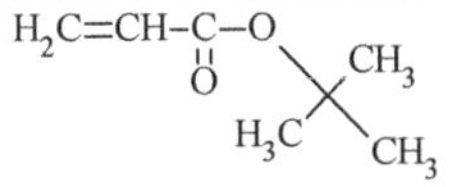	- improves impact resistance - more thermally stable than *n*PBA - hydrophobic properties	[35,36]
ethyl acrylate	EA	$H_2C=CH-C(=O)-O-CH_2-CH_3$	- hydrophobic properties	[37]
ethyl methacrylate	EMA	$H_2C=C(CH_3)-C(=O)-O-CH_2-CH_3$	- better resistance to weather conditions than PEA	[38]
benzyl acrylate	BAZ	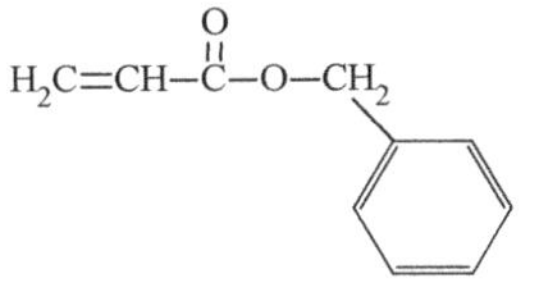	- increasing strength and hardness - decreasing adhesive properties	[39]
dodecyl acrylate	DA	$H_2C=CH-C(=O)-O-(CH_2)_{11}-CH_3$	- high hydrophobicity - high flexibility	[40,41]
2-hydroxyethyl methacrylate	HEMA	$H_2C=C(CH_3)-C(=O)-O-CH_2-CH_2-OH$	- hydrophilic properties - improves mechanical properties (e.g., flexibility)	[42]

4.2. Characterization of poly(meth)acrylate Resins

To characterize the obtained acrylic resins, gel permeation chromatography (GPC) and nuclear magnetic resonance (1H-NMR) spectra as well as viscosity and T_g measurements were performed. Gel permeation chromatography (GPC) was used to determine the average molecular masses of acrylic resins (Table 3). The average molecular mass number of obtained resins ranged from 6 670 to 9 820 Da. The resin HEMA/6MMA/DA was characterized by the lowest number average molecular mass, while the resin containing *t*BA exhibited the highest number average molecular mass. These values of molecular mass depend on the reactivity of individual monomers, with *t*BA having higher reactivity than DA.

Table 3. Average molecular masses of synthesized acrylic resins.

Resin Symbol \ Mass Unit	Average Molecular Mass Number (M_n) [Da]	Average Molecular Mass Weight (M_w) [Da]	Molecular Mass Z-Average (M_z) [Da]
HEMA/6MMA/*n*BA	8860	53,610	22,760
HEMA/6MMA/*t*BA	9820	55,500	18,520
HEMA/6MMA/EA	8240	34,930	15,030
HEMA/6MMA/EMA	7420	17,360	36,270
HEMA/6MMA/BAZ	7540	13,910	21,180
HEMA/6MMA/DA	6670	39,410	23,550
HEMA/6MMA/0.5*n*BA/0.5DA	8020	53,320	21,700

The high molecular weight of the resin used for powder coatings is desired to obtained good properties of the finished product [45]. However, excessively high molecular weight can lead to problems with the preparation of powder coatings due to excessively high viscosity. Conversely, a resin with too low a molecular mass may exhibit too low a viscosity and T_g, resulting in the powder grains sticking together during storage. Therefore, the glass transition temperature and viscosity are crucial parameters. These parameters allow for predicting the correlation between the resin structure and the properties of the powder coatings.

If the resin has low viscosity and T_g, the powder may stick together, making it impossible to apply the powder coating using an electrostatic gun. Conversely, if the viscosity and T_g of the acrylic resin are too high, problems may arise with the melting and flow of powder coatings, resulting in poor final product properties. Table 4 presents the T_g values and viscosity of acrylic resins. The recommended T_g of powder coatings is approximately 50 °C [46]. The resin HEMA/6MMA/0.5*n*BA/0.5DA exhibited the most optimal T_g value (T_g = 50.31 °C), which positively influenced the final properties of the powder coatings. Additionally, the viscosity of this resin had a beneficial effect on the flow and leveling properties of the powder coatings, as surfaces based on this resin did not exhibit defects such as orange peel or craters. However, monomers such as *t*BA or BAZ had an adverse effect on the formation of powder coatings.

Table 4. Glass transition temperature (T_g) and viscosity of obtained acrylic resins.

Resin Symbol	Glass Transition Temperature (T_g) [°C]	Viscosity [Pa*s]
HEMA/6MMA/*n*BA	54.03	23.85
HEMA/6MMA/*t*BA	78.29	22.20
HEMA/6MMA/EA	54.19	19.00
HEMA/6MMA/EMA	52.93	24.30
HEMA/6MMA/BAZ	42.62	36.15
HEMA/6MMA/DA	39.40	17.25
HEMA/6MMA/0.5*n*BA/0.5DA	50.31	20.73

The chemical structure of the synthesized acrylic resins was confirmed using ^{1}H-NMR spectra (Figure 1).

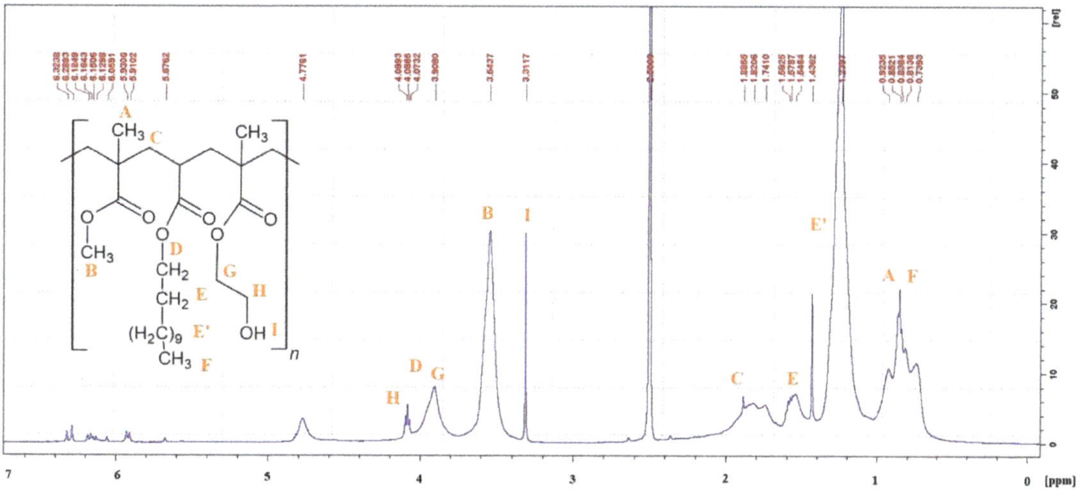

Figure 1. ^{1}H-NMR spectrum of HEMA/6MMA/DA resin.

The signal in the range of 0.60–1.00 ppm (assigned as 'A') is characteristic for the hydrogen atoms of the methyl groups derived from the main polymer chain [47]. In this region it is also visible that the signal comes from the protons of the methyl group at the end of the dodecyl and butyl substituent marked as 'F'. In the range of 3.60 ppm, the protons of the methyl groups derived from MMA ($-OC\underline{H_3}$) are observed (assigned as 'B'). The methylene groups adjacent to oxygen ($-OCH_2-$) assigned at 'D', 'G' and 'H' derived from HEMA, BA and DA are visible in the range of 3.80–4.10 ppm [48]. The hydrogen atom ('I') derived from the hydroxyl group is seen at 3.15 ppm. Signals in the range of 1.10–1.70 ppm come from protons of aliphatic methylene groups assigned as 'E' and 'E″' of BA and DA. Signals originating from protons of methylene groups forming the main polymer chain (assigned as 'C') are visible at 1.70–1.90 ppm [47,49]. A common problem with acrylic resins synthesized in bulk polymerization is the presence of small amounts of free monomers. In the spectrum ^{1}H NMR (Figure 1), a small number of unreacted monomers can be seen in the range of 5.60–6.40 ppm. In order to remove free monomers, the acrylate resin was heating at 80 °C by 1 h in oven.

4.3. The Crosslinking Process and Properties of Powder Coatings

After obtaining acrylic resin, the next step was the selection of an appropriate crosslinking agent. Due to the presence of hydroxyl groups in the resin, a blocked polyisocyanate named Vestanat® B 1358/100 by Evonic was suitable. To assess the course of the crosslinking reaction, a DSC analysis was conducted for the uncured powder coatings. To assess the course of the crosslinking reaction, a DSC analysis was conducted for the uncured powder coating compositions (Figure 2).

In order to verify the complete crosslinking, two heating cycles were conducted. In the range of 45–65 °C, the glass transition of the powder coating was observed. During the first heating cycle, in the range of 110–160 °C, a noticeable broad exothermic peak suggests the progress of the powder coating curing process. To confirm the complete curing process of the coatings, a second heating cycle was performed. The absence of a signal in the range of 110–160 °C during the second heating cycle indicates the complete occurrence of this process in the first cycle. This temperature range confirms the attainment of a low-temperature curing powder coating (160 °C for 15 min). The crosslinking process of other samples was very similar, which indicates that the influence of the chemical composition of the resin had no significant impact on this process. Additionally, the crosslinking of the powder coatings was confirmed by using the polymerization test performed according to Qualicoat [21]. After taking the test, no noticeable changes were observed. Moreover, the samples, after the curing stage, exhibited no defects such as orange peel or craters. Subsequently, the powder coatings underwent further tests, the results of which are provided in Table 5.

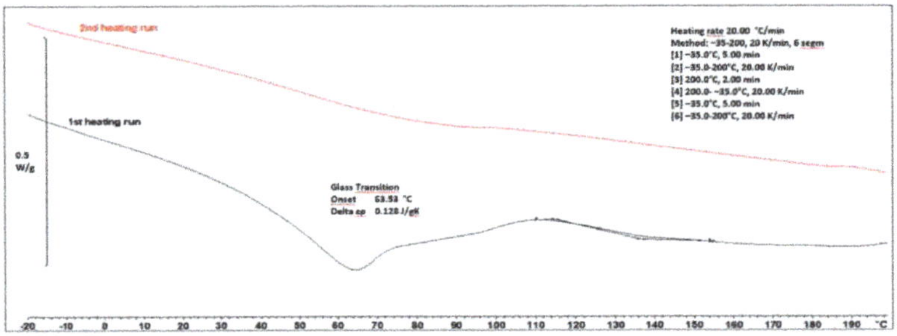

Figure 2. The DSC curves of L_HEMA/6MMA/0.5*n*BA/0.5DA.

Table 5. Summary of physical and mechanical parameters of the powder coatings.

Physical or Mechanical Parameter	Powder Coating Symbol						
	L_HEMA/6MMA/*n*BA	L_HEMA/6MMA/*t*BA	L_HEMA/6MMA/EA	L_HEMA/6MMA/EMA	L_HEMA/6MMA/BAZ	L_HEMA/6MMA/DA	L_HEMA/6MMA/0.5*n*BA.0.5DA
Flowability [cm]	1.30	0.90	0.95	0.75	2.20	5.00	4.40
Roughness: *Ra* *Rz*	0.69/2.74	1.79/8.73	2.88/14.12	6.33/11.73	1.42/6.79	1.09/5.11	0.42/2.05
Gloss 60 °C [GU]	83.45	34.73	34.24	11.73	63.12	64.62	79.63
Adhesion to the steel substrate [0—good 5—bad]	0	0	1	1	1	0	0
Relative hardness [-]	0.54	0.56	0.55	0.51	0.67	0.33	0.50
Scratch resistance [g]	500	450	250	250	550	300	550
Contact angle [deg]	85.40	83.93	87.91	83.91	87.02	93.14	93.53
Impact resistance [J/cm^2]	15	10	15	10	20	25	30
Cupping [mm]	5.43	4.37	5.12	5.07	9.96	11.18	13.38

The thickness of the obtained powder coatings ranged from 60 to 80 μm, meeting the Qualicoat requirements [21].

The study on the influence of chemical structure on the properties of powder coatings focused on the monomers such as *n*BA, *t*BA, EA, EMA, BAZ and DA. HEMA and MMA monomers were present in all samples in the same amount. HEMA was responsible for the crosslinking process and the MMA contributed to the stability of composition. The flow properties of uncured powder compositions were examined. Samples L_HEMA/6MMA/*t*BA, L_HEMA/6MMA/EA and L_HEMA/6MMA/EMA showed very poor flowability. This reduced flowability led to an increased roughness and decreased gloss of the cured coatings. Additionally, these coatings exhibited weak properties such as adhesion and scratch resistance compared to other powder coatings. The presence of tert-butyl substituents into *t*BA increase roughness and hardness but decreases flexibility causing a problem with brittleness of coatings compared with *n*BA, which contains a linear substituent with the same number of carbon atoms. A similar stiffening effect was observed when comparing the effects of EA and *n*BA (samples L_HEMA/6MMA/EA and L_HEMA/6MMA/*n*BA). The shorter aliphatic chain ethyl substituent on the ester group in EA increased roughness while reducing gloss, scratch resistance and adhesion to the steel. Comparing the effects of EA and EMA, the stiffening effect of the presence of a methyl group in the α-position of the vinyl bond in EMA (sample L_HEMA/6MMA/EA and L_HEMA/6MMA/EMA) was evident, causing worse flowability and gloss of the coating. Therefore, the presence of a methyl group in the α-position of the vinyl bond is important in designing powder coatings based on acrylic resins. To compare the effect of a substituent in the ester group with aliphatic and aromatic structure, BAZ was used. Coatings based on BAZ exhibited good flowability, the highest relative hardness, cupping and were classified as "semi-gloss" coatings. While styrene is a common monomer used in commercial coatings due to its cost-effectiveness and enhancement of mechanical properties, it suffers from brittleness and lack of flexibility [50].

The introduction of DA into the resin was aimed at addressing the flexibility issue associated with typical acrylic resins. However, the long chain attached to the ester group significantly improved flowability and cupping but decreased to low the glass transition temperature (T_g), which led to a problem with the storage and application of powder coatings. Therefore, a resin containing *n*BA and DA (HEMA/6MMA/0.5*n*BA/0.5DA) was developed. The coating based on this resin exhibited favorable properties including good flowability, gloss, adhesion and relative hardness. Additionally, it demonstrated the highest contact angle (93.53 degrees) and scratch resistance (550 g), excellent cupping (13.38 mm) and improved impact resistance compared to other samples. To further understand this behavior, crosslinking density and thermal analysis were performed.

The crosslinking density (v_e) was determined using dynamic mechanical analysis (DMA). This technique measures the mechanical forces applied in a function of temperature in a polymer. The dependence of storage modulus E′ on temperature for sample L_HEMA/6MMA/0.5*n*BA/0.5DA is presented in Figure 3.

The modulus in the rubbery plateau (E'_{min}) was determined based on this dependence. Crosslinking density v_e as the number of moles of effective network chains per volume of 1 m^3 (mol/m^3) was determined using equation (1). In this equation, R is the gas constant (8.314 J/mol K) and T is the temperature (K) at which E'_{min} was determined.

$$v_e = \frac{E'}{3RT} \tag{1}$$

The crosslinking density of the L_HEMA/6MMA/0.5*n*BA/0.5DA coating was determined to be v_e =111.2 mol/m^3, with values reported in the literature for powder coatings [40]. Furthermore, the L_HEMA/6MMA/0.5*n*BA/0.5DA coating exhibited a high storage modulus, confirming its excellent mechanical properties. It is known that higher storage modulus and higher T_g are indicative of better properties.

Additionally, the thermal stability of these powder coatings was assessed via thermogravimetric analysis (TGA). The coating L_HEMA/6MMA/0.5*n*BA/0.5DA displayed decomposition in three stages in the temperature range of 160–470 °C, as illustrated in Figure 4.

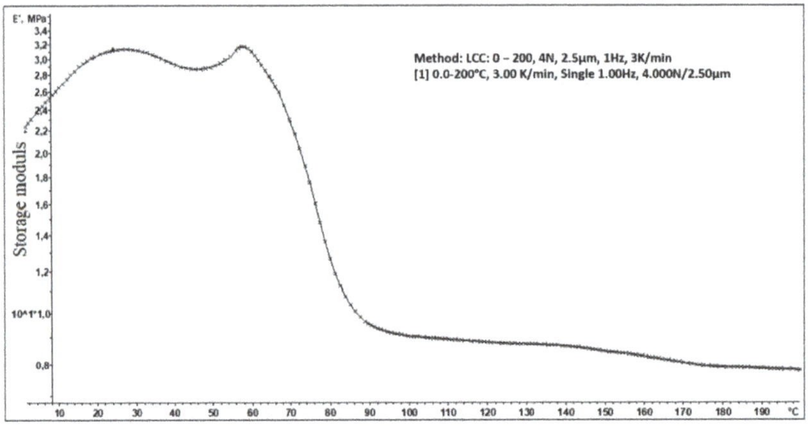

Figure 3. DMA curves of L_HEMA/6MMA/0.5*n*BA/0.5DA.

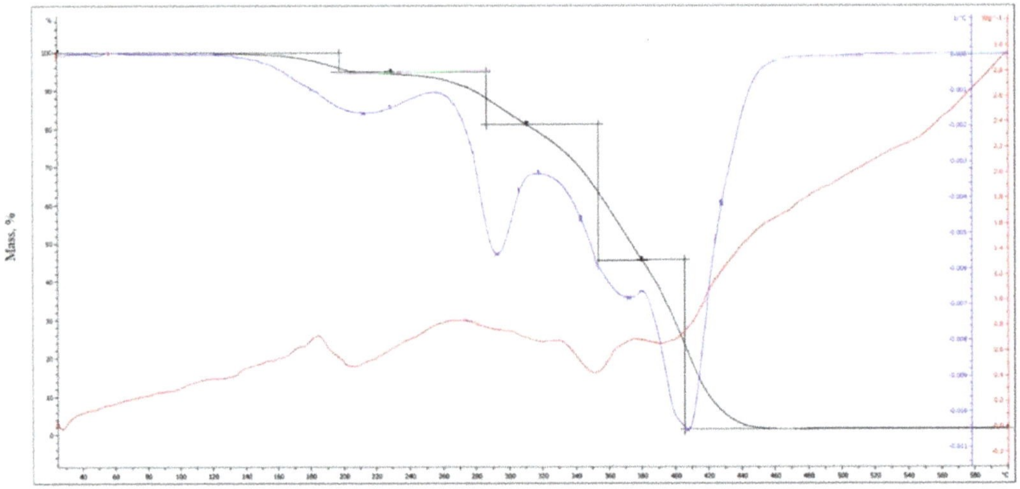

Figure 4. TG, DTG and DTA thermograms of L_HEMA/6MMA/0.5*n*BA/0.5DA sample.

The first step occurs at the temperature of the maximum mass loss rate of T_{max1} = 213 °C and is related to the deblocking of polyisocyanate. The second step occurs at the temperature of the maximum mass loss rate of T_{max2} = 293 °C and is related to the scission of the urethane bond, while the third step (T_{max3} = 404 °C) is associated with the degradation of the acrylic resin segments. The mass losses at the first, second and third stages of degradation amount to 8.9%, 19.1% and 70.7%, respectively. The temperature of 5% mass loss is 213 °C. On Figure 5 is shown thermograms of the L_HEMA/6MMA/*n*BA sample.

In the comparison of thermograms of the L_HEMA/6MMA/0.5*n*BA/0.5DA sample, no significant difference is visible. The degradation process of the powder coatings initiates with the breakdown of the weakest bonds, namely the urethane bonds formed between the polyisocyanate and the resin. This underscores the importance of understanding the bond strengths within the polymer matrix, as it dictates the onset and progression of degradation during thermal treatment.

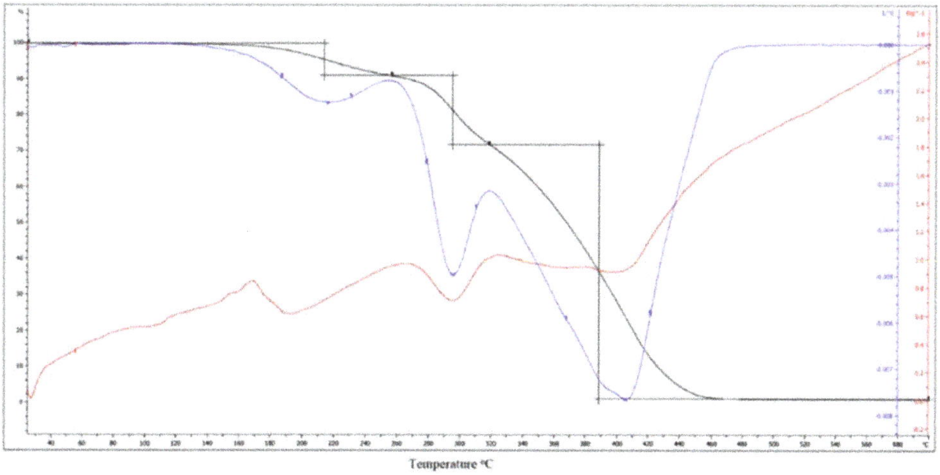

Figure 5. TG, DTG and DTA thermograms of L_HEMA/6MMA/*n*BA sample.

5. Conclusions

The study focused on investigating the impact of acrylic monomer chemical structure on the properties of powder coatings formulated with polyacrylate resin. Gel permeation chromatography (GPC) revealed an average molecular weight of approximately 7000 Da for the resins, a typical value for acrylic resins used in powder coatings. Parameters such as glass transition temperature (T_g) and viscosity were determined for the polyacrylate resins, both of which were influenced by the resin's chemical structure. These parameters play crucial roles in the formulation, storage, application and final properties of powder coatings. Understanding the correlation between the chemical structure, viscosity and T_g of the resin provided valuable insights for optimizing powder coating formulations.

Dynamic scanning calorimetry (DSC) was employed not only to determine the T_g of the resins but also to examine the crosslinking process. The coatings underwent crosslinking at 160 °C, enabling low-temperature curing and allowing for application on heat-sensitive materials. This low-temperature curing capability broadens the range of substrates onto which powder coatings can be applied, enhancing their versatility and utility in various industries.

The powder coatings based on the resins containing *t*BA, EA and EMA were characterized by high roughness and poor physical–mechanical properties. The best properties were noted for coating L_HEMA/6MMA/0.5*n*BA/0.5DA, which was characterized by good impact resistance, scratch resistance (550 g) and adhesion to the steel surface, a high water contact angle (93.53 deg.) and excellent cupping (13.38 mm). These results highlight the effectiveness of the resin formulation in achieving desirable properties, making it a promising candidate for various coating applications.

This powder coating is also characterized by high crosslinking density (111.2 mol/m^3) and good thermal stability.

Indeed, enhancing the elasticity of acrylic resins while maintaining other essential properties expands their potential applications, particularly in coating structures with irregular shapes or sharp edges. In such cases, excessively rigid coatings are prone to cracking, compromising both the aesthetic appearance and protective function of the coating. By increasing elasticity, acrylic resins can better accommodate substrate movements and deformations, thereby minimizing the risk of cracking and ensuring the long-term durability of the coating. This advancement allows for broader utilization of acrylic-based powder coatings in diverse industries, including those requiring coatings for challenging surface geometries and environmental conditions.

Author Contributions: Conceptualization, K.P. and B.P.-P.; methodology, K.P. and B.P.-P.; validation, K.P. and B.P.-P.; formal analysis, K.P. and B.P.-P.; investigation, K.P.; resources, K.P.; data curation, K.P. and B.P.-P.; writing—original draft preparation, K.P.; writing—review and editing B.P.-P.; visualization, K.P.; supervision, B.P.-P.; project administration, B.P.-P.; funding acquisition, B.P.-P. All authors have read and agreed to the published version of the manuscript.

Funding: This research was funded by Minister of Science and Higher Education Republic of Poland within the program "Regional Excellence Initiative". grant number RCD.RB.24.002.

Institutional Review Board Statement: Not applicable.

Informed Consent Statement: Not applicable.

Data Availability Statement: The original data presented in the study are included in the article. Additional raw data supporting the conclusions of this article will be made available by the authors on request.

Acknowledgments: The authors would like to thank Evonic Degussa G.m.b.H and BYK-Chemie G.m.b.H for sending free samples of raw materials. This research was financed by the Minister of Science and Higher Education Republic of Poland within the program "Regional Excellence Initiative".

Conflicts of Interest: The authors declare that they have no known competing financial interests or personal relationships that could have appeared to influence the work reported in this paper. Data availability: Data will be made available on request.

References

1. Jones, F.N.; Nichols, M.E.; Pappas, S.P. *Organic Coatings, Science and Technology*; Wiley: Hoboken, NJ, USA, 2017.
2. Directive 2004/42/EC of the European Parliament and of the Council of April 21, 2004 on the Limitation of Emissions of Volatile Organic Compounds Due to the Use of Organic Solvents in Certain Paints and Varnishes and Vehicle Refinishing Products, and Amending Directive 1999/13/EC, Document 32004L0042. Available online: https://eur-lex.europa.eu/eli/dir/2004/42/oj (accessed on 4 February 2024).
3. Directive 2000/60/EC of the European Parliament and of the Council of 23 October 2000 Establishing a Framework for Community Action in the Field of Water Policy, Document 32000L0060. Available online: https://eur-lex.europa.eu/eli/dir/2000/60/oj (accessed on 4 February 2024).
4. Emmanouil, S. *Powder Coatings Chemistry and Technology*; Vincentz Network GmbH: Hanover, Germany, 2014.
5. Available online: https://ecoline.com.pl/ (accessed on 18 January 2024).
6. Pilch-Pitera, B. *Farby i Lakiery Proszkowe: Otrzymywanie, Formowanie, Nanoszenie i Ocena Właściwości*; Oficyna Wydawnicza Politechniki Rzeszowskiej: Rzeszów, Poland, 2015.
7. Li, C.; Johansson, M.; Buisjen, B. Limonene-derived polycarbonates as biobased UV-curable (powder) coating resins. *Prog. Org. Coat.* **2021**, *151*, 106073. [CrossRef]
8. Yang, M.S.; Huang, J.; Noël, J.J. A Mechanistic Study on the Anti-Corrosive Performance of Zinc-Rich Polyester/TGIC Powder Coatings. *Processes* **2022**, *10*, 1853. [CrossRef]
9. Ayrilmis, N. A review on electrostatic powder coatings for the furniture industry. *Int. J. Adhes. Adhes.* **2022**, *113*, 103062. [CrossRef]
10. Malshe, V.C.; Waghoo, G. Weathering study of epoxy paint. *Prog. Org. Coat.* **2004**, *51*, 267. [CrossRef]
11. Zhou, Z.; Xu, W.; Fan, J.; Ren, F.; Xu, C. Synthesis and characterization of carboxyl group-containing acrylic resin for powder coatings. *Prog. Org. Coat.* **2008**, *62*, 179–182. [CrossRef]
12. Czachor-Jadacka, D.; Pilch-Pitera, B.; Byczynski, Ł.; Kisiel, M.; Zioło, A. Hydrophobic polyurethane powder clear coatings with lower curing temperature: Study on the synthesis of new blocked polyisocyanates. *Prog. Org. Coat.* **2021**, *159*, 106402. [CrossRef]
13. Fernández-Álvarez, M.; Velasco, F.; Bautista, A. Epoxy powder coatings hot mixed with nanoparticles to improve their abrasive wear. *Wear* **2020**, *448*, 203211. [CrossRef]
14. Sangermano, M.; Roppolo, I.; Chiappone, A. New Horizons in Cationic Photopolymerization. *Polymers* **2018**, *10*, 136. [CrossRef] [PubMed]
15. Sangermano, M. Advances in cationic photopolymerization. *Pure Appl. Chem.* **2012**, *84*, 2089. [CrossRef]
16. Okada, K.; Yamaguchi, K.; Takeda, H. Acrylic/polyester hybrid powder coating system having excellent weather durability. *Prog. Org. Coat.* **1998**, *34*, 169. [CrossRef]
17. Hughes, A.E.; Johnston, P.; Simons, T.J. Chapter 9—Self-healing coatings. In *Recent Advances in Smart Self-Healing Polymers and Composites*, 2nd ed.; Woodhead Publishing: Oxford, UK, 2022; pp. 217–270. [CrossRef]
18. Tong, J.; Xie, S.; Miao, J.T.; Luo, J.; Liu, R. Preparation of UV-cured polyurethane-urea acrylate coatings with high hardness and toughness. *Prog. Org. Coat.* **2024**, *186*, 107969. [CrossRef]

19. Wang, K.S.; Cheng, Y.W.; Lin, H.Y.; Chen, M.H.; Yeh, S.C.; Huang, Y.C.; Wu, C.H.; Jeng, R.J.; Liu, T.Y. Synthesis of dendritic urethane acrylates for fabricating a robust honeycomb-like structure acting for SERS detection. *Prog. Org. Coat.* **2023**, *184*, 107840. [CrossRef]
20. Li, L.; Zhang, Z.; Wang, M.; Zhang, Y.; Lu, X. UV-curable epoxy acrylate composite coatings with high flexibility, super-hydrophobicity, wear-resistance and self-healing property. *Prog. Org. Coat.* **2023**, *182*, 107649. [CrossRef]
21. Qualicoat. *Specifications for a Quality Label for Liquid and Powder Organic Coatings on Aluminium for Architectural Applications*, 16th ed.; Qualicoat: Zurich, Switzerland, 2019.
22. *PN-EN ISO 8130-11*; Coating Powders—Part 11: Inclined-Plane Flow Test. ISO: Geneva, Switzerland, 2019.
23. *PN-EN ISO 2813*; Paints and Varnishes—Determination of Specular Gloss of Non-Metallic Paint Films at 20 Degrees, 60 Degrees and 85 Degrees. ISO: Geneva, Switzerland, 2014.
24. *PN-EN ISO 2808*; Paints and Varnishes—Determination of Film Thickness. ISO: Geneva, Switzerland, 2019.
25. *PN-EN ISO 12085*; Geometrical Product Specifications (GPS)—Surface Texture: Profile Method—Part 2: Terms, Definitions and Surface Texture Parameters. ISO: Geneva, Switzerland, 2018.
26. *PN-EN ISO 2409*; Paints and Varnishes—Cross-Cut Test. ISO: Geneva, Switzerland, 2013.
27. *PN-EN ISO 1522*; Paints and Varnishes—Pendulum Damping Test. ISO: Geneva, Switzerland, 2011.
28. *PN-EN ISO 1518*; Paints and Varnishes—Scratch Hardness Tests—Part 1: Constant-Loading Method. ISO: Geneva, Switzerland, 2016.
29. *PN-EN ISO 19403-6:2020-08*; Paints and Varnishes—Wetting—Part 3: Determination of Surface Tension of Liquids by the Hanging Drop Method. ISO: Geneva, Switzerland, 2020.
30. *PN-EN ISO 6272-1*; Paints and Varnishes—Rapid-Deformation (Impact Resistance) Tests—Part 1: Falling-Weight Test, Large-Area Indenter. ISO: Geneva, Switzerland, 2011.
31. *PN-EN ISO 1520*; Paints and Varnishes—Cupping Test. ISO: Geneva, Switzerland, 2019.
32. Ali, U.; Juhanni, K.; Karim, A.; Buang, N.A. A Review of the Properties and Applications of Poly (Methyl Methacrylate) (PMMA). *Polym. Rev.* **2015**, *55*, 678–705. [CrossRef]
33. Takács, E.; Wojnárovits, L. Comparison of the reactivity of acrylate and methacrylate monomers. *Radiat. Phys. Chem.* **1995**, *46*, 1007–1010. [CrossRef]
34. Ibrahim, S.; Lotfy, S. Properties of butyl acrylate polymers synthesized by radiation and miniemulsion polymerization techniques as flexible coating for packaging materials. *J. Vinyl Addit. Technol.* **2020**, *27*, 1. [CrossRef]
35. Xingjiang, W.; Hong, R.; Meng, J.; Cheng, R.; Zhu, Z.; Wu, G.; Li, Q.; Wang, C.F.; Chen, S. Hydrophobic Poly(tert-butylacrylate)PhotonicCrystalstowardsRobust Energy-Saving Performance. *Angew. Int. Ed.* **2019**, *58*, 13556–13564. [CrossRef]
36. Martinez, G.; Sanchez-Chaves, M.; Madruga, E.L.; Fernandez-Monreal, C. Monomer reactivity ratios and microstructural analysis of 2-hydroxyethyl methacrylate–t-butyl acrylate copolymers. *Polymer* **2000**, *41*, 6021–6026. [CrossRef]
37. Saindane, P.; Jagtap, R.N. RAFT copolymerization of amphiphilic poly (ethyl acrylate-b-acrylic acid) as wetting and dispersing agents for water borne coating. *Prog. Org. Coat.* **2015**, *79*, 106–114. [CrossRef]
38. Hamlic, D.; Rodic, P.; Poberznik, M.; Jerb, M.; Kovec, J.; Milosev, I. The Effect of the Methyl and Ethyl Group of the Acrylate Precursor in Hybrid Silane Coatings Used for Corrosion Protection of Aluminium Alloy 7075-T6. *Coatings* **2020**, *10*, 172. [CrossRef]
39. Fang, C.; Zhou, F.; Zhu, X. The application research of benzyl methacrylate (BzMA) in acrylate latex pressure sensitive adhesives. *Int. J. Adhes. Adhes.* **2021**, *107*, 102861. [CrossRef]
40. Qiang, R.; Fanghong, G.; Dongliang, Z.; Xiaoxiao, S.; Jianbo, F.; Bibiao, J. Blends of high density polyethylene with branched poly(styrene-co-dodecyl acrylate): Rheological, mechanical, and thermal properties. *Appl. Polym. Sci.* **2008**, *109*, 1618–1624. [CrossRef]
41. Riemer, S.; Prevost, S.; Dzionara, M.; Appavou, M.S.; Schweins, R.; Gradzielski, M. Aggregation behaviour of hydrophobically modified polyacrylate—Variation of alkyl chain length. *Polymer* **2015**, *70*, 194–206. [CrossRef]
42. Dang, W.; Zhang, D.; Ha, Z.; Wang, J.; Xia, Y.; Chen, X.; Lei, L.; Shi, S. Emulsion polymerization of super-hydrophobic acrylate enabled by a novel "ferrying" strategy for developing waterborne coatings with reduced surface tension and glossiness. *Prog. Org. Coat.* **2024**, *186*, 108073. [CrossRef]
43. Popescu, D.; Hoogenboom, R.; Keul, H.; Moeller, M. Hydroxy functional acrylate and methacrylate monomers prepared via lipase—Catalyzed transacylation reactions. *J. Mol. Catal. B Enzym.* **2010**, *62*, 80–89. [CrossRef]
44. Pilch-Pitera, B.; Czachor, D.; Kowalczyk, K.; Pavlova, E.; Wojturski, J.; Florczak, Ł.; Byczyński, Ł. Conductive polyurethane-based powder clear coatings modified with carbon nanotubes. *Prog. Org. Coat.* **2019**, *137*, 105367. [CrossRef]
45. Soroush, M.; Grady, M.C. Chapter 1—Polymers, Polymerization Reactions, and Computational Quantum Chemistry. *Comput. Quantum Chem.* **2019**, *1*, 16. [CrossRef]
46. Massoumi, B.; Jaymand, M. Chemical and electrochemical grafting of polythiophene onto poly(methyl methacrylate), and its electrospun nanofibers with gelatin Ramis. *J. Mater. Sci. Mater. Electron* **2016**, *12*, 27. [CrossRef]
47. Pham, Q.T.; Petiaud, R.; Llauro-Darricades, M.F.; Waton, H. *Proton & Carbon NMR Spectra of Polymers*; CRC Press: Boca Raton, FL, USA, 1991.

48. Babac, C.; Utkan, G.; David, G.; Simionescu, B.; Piskin, E. Production of nanoparticles of methyl methacrylate and butyl methacrylate copolymers by microemulsion polymerization in the presence of maleic acid terminated poly(N-acetylethylenimine) macromonomers as cosurfactant. *Eur. Polym. J.* **2004**, *40*, 1947–1952. [CrossRef]
49. Chiantore, O.; Lazzari, M. Characterization of Acrylic Resins. *Int. J. Polym. Anal. Charact.* **1996**, *2*, 395–408. [CrossRef]
50. Kardar, P.; Ebrahimi, M.; Bastani, S. Curing behaviour and mechanical properties of pigmented UV-curable epoxy acrylate coatings. *Pigment Resin Technol.* **2014**, *43*, 177–184. [CrossRef]

materials

MDPI

Article

Thermally Sprayed Coatings for the Protection of Industrial Fan Blades

Maria Richert

Management Faculty, AGH University in Krakow, 30-067 Kraków, Poland; mrichert@agh.edu.pl

Abstract: This paper presents a study on thermally sprayed coatings. Coatings produced by high-velocity oxygen–fuel spraying HVOF and plasma spraying deposited on the A03590 aluminum casting alloy are tested. The subject of this research concerns coatings based on tungsten carbide WC, chromium carbide Cr_3C_2, composite coatings NiCrSiB + 2.5%Fe + 2.5%Cr, mixtures of tungsten and chromium powders WC-CrC-Ni, mixtures of carbide powders with the Cr_3C_2-NiCr + the composite 5% NiCrBSi and WC-Co + 5% NiCrBSi. The aim of this research is to find a coating most resistant to the erosive impact of particles contained in the medium centrifuged by industrial rotors. The suitability of the coating is determined by its high level of microhardness. The hardest coatings are selected from the coatings tested and subjected to abrasion tests against a sand particle impact jet and the centrifugation of a medium with corundum particles. It is found that the most favorable anti-erosion properties are demonstrated by a coating composed of a mixture of tungsten carbide and chromium carbide WC-CrC-Ni powders. It is concluded that the greatest resistance of this coating to the erosive impact of the particle jet results from the synergistic enhancement of the most favorable features of both cermets.

Keywords: thermal spraying; coatings; tungstem carbides; chromium carbides; wear erosion

1. Introduction

Anti-erosion and anti-corrosion coatings are commonly used to protect metallic surfaces [1]. They provide protection against the destructive effects of phenomena such as environmental influence, abrasion, particle impact, friction and corrosion. These phenomena often occur simultaneously. Various types of coatings are used to protect surfaces exposed to damage. Numerous friction pairs in machines, tools, devices or machine parts are covered with special anti-corrosion coatings or layers. The hardening of the surface itself increases the resistance to wear and friction, an example of which is steel nitriding [2]. Surface protection methods are individually selected for particular tools, devices and other friction pairs.

Anti-erosion coatings can be produced using various methods. One of them is pulsed magnetron sputtering [3]. Other methods include the vacuum cathodic arc deposition system (FCVA) [4], physical vacuum deposition (PVD) [5] and chemical deposition in vacuum (CVD) [6]. Various thermal spray methods are also used [7–10]. The latest development in surface engineering is high-entropic materials used as anti-erosion coatings [11]. Thermal spray methods are often used for anti-erosion and anti-corrosion surface protection due to their low costs and multiple variations in spray techniques. Thermal spray methods and their applications have been presented in many publications on surface protection [12–14]. There are several methods of applying thermal spray coatings, such as oxy fuel wire (OFW), electric arc wire (EAW), oxy fuel powder (OFP), plasma arc powder (PAP) and high-velocity oxy fuel (HVOF) [15–18].

Thermal spray coatings are a widespread surface engineering technique used all over the world. The coating can be deposited in a molten or plastic state through propulsion of particles or through the impact onto a workpiece. The particles of powders sprayed in thermal spray methods, e.g., CoCrNiWC and CoCr-based intermetallics, have a melting point of approximately 1230–1600 °C. In turn, particles of NiCrSiB have a melting point of approximately 1050 °C. The CrC–NiCr and CrC–Ni-based particles melt at a temperature of approximately 1930 °C, and WC–Co, WC–Ni and WC-based particles melt at approximately 1250–1480 °C. Ceramic particles melt at temperatures above 2000 °C. Thermal spray technologies must provide a sufficiently high temperature to melt the particles.

Plasma spraying uses a plasma-forming gas, which is usually argon with possible additions of other elements. The plasma-forming gas is ionized under the electric arc and turns into plasma. The plasma escapes through a hole in the torch front along with the heated metallic or ceramic powder. The plasma arc has a temperature in the range of 1173–3730 °C (4000 ÷ 12,000 K), but it can reach up to 14,730 °C (15,000 K), depending on the distance from the burner nozzle [19,20].

In the gas jet, individual particles achieve high speeds: approx. 160 m/s in flame spraying and arc spraying, approx. 200–250 m/s in plasma spraying, and over 330 m/s in supersonic spraying. The high temperature causes melting of the particles, but as they reach the surface, they often cool below the melting point, which is the reason for the presence of unmelted particles in the coating structure. The low heat capacity of the particles falling on the substrate causes them to cool very quickly. It was found that the sprayed object did not heat up to a temperature higher than 100 °C. As a result of this characteristic process, a specific structure of the sprayed coating is created. Layers with a strongly differentiated structure are formed, with grains flattened along the sprayed surface. The layer structure may contain voids, porosity, delamination, spalling, interface contamination, transverse cracks, inter-lamellar pores, intra-lamellar cracks, unmelted particles, pull-outs, oxides, oxide clusters, oxide stringers or metallic inclusions. A high density of artifacts in the layer structure weakens its properties, such as hardness or abrasion resistance.

A major challenge in surface engineering is protecting the surfaces of industrial fan blades. The blades are exposed to intense impacts from particles, dust and atmospheric corrosion. Carbide coatings, such as WC, WC-Co(Cr) and Cr-Cr_3C_2, are used for anti-erosion protection of the blade surfaces. These types of coatings are also used in friction pairs and as anti-wear coatings in elements exposed to particle impact [21]. In turn, TiN and TiCrN [22–27] coatings are most often used to protect turbine blades. Gas turbine blades can be coated with thermal spray technology [28].

The hardness of thermally sprayed coatings depends on the chemical composition of the coating and the spraying method. The hardness of the WC-WB-Co coating was found to be between 1200 and 1300 $HV_{0.1}$, and the hardness of the WC-FeCrAl coating was found to be between 1000 and 1100 $HV_{0.1}$ [29]. Microhardness values from HV = 650 to 1021 were obtained for WC-12Co (88wt%WC–12wt%Co) coatings containing phases WC, W_2C and Co sprayed using the HVOF method [30]. In the works of Myalska et al. [31] and Gaceb et al. [32], the hardness of the thermally sprayed NiCrBSiFeC coating was $HV_{0.1} = 753 \pm 115$. In turn, research by Rachidi et al. [33] showed that the hardness of the NiCrBSi coating by the thermal spraying technique using SuperJetEutalloy reached an average of 800$HV_{0.3}$, and the NiCrBSi—60% WC coating reached the hardness of 1200$HV_{0.3}$. In the work [34], Cr_3C_2-25NiCr and WC-10Co-4C coatings were tested. The hardness of the first coating fluctuating between 900–1050 $HV_{2.94N}$ and the second between 1000–1200 $HV_{2.94N}$. The hardness of the WC-CoCr coating produced using the high-velocity oxygen–fuel spraying (HVOF) method was above 1000 $HV_{0.3}$. The variation in microhardness was only 215.7 $HV_{0.3}$ [35]. The hardness of the WC-4Co-10Cr coating sprayed using the HVOF method reached values from 1177 to 1256 $HV_{0.3}$ [36]. The WC-12Co coating thermally sprayed onto the TC18 substrate achieved the hardness of 1018–1213 $HV_{0.3}$ [37]. Tungsten carbide has a very high hardness of approximately 2200–2800 HV, while the hardness of HVOF coatings made of WC-Co-based powders is reported in research works as significantly lower compared to the hardness of tungsten carbide, from 866 to 1226 HV [38]. The reduction in hardness is influenced by the mechanism of the thermal spray process, in which very high temperatures lead to decarbonization of WC carbide. The presence of the bonding phase, i.e., Co, Ni, Cr, which has a lower hardness than tungsten carbide, also reduces the hardness level of the sprayed coating. Hardness is also reduced by various artifacts occurring in the structure of the coatings. The presented data indicate that the microhardness range obtained for coatings based on tungsten carbide and chromium carbide ranges from 600 to 1300 HV01. The level of microhardness depends not only on the chemical composition of the coating but also on the method of coating deposition.

Research shows that hardness is related to erosion resistance. The erosion rates were found to increase with the increase in the alloy bulk hardness [39]. The wear resistance of the hard metal coatings is highly dependent on its composition, carbide grain size, porosity, toughness and hardness value [40]. In turn, the results of the work of Cesanek and Schubert show that the abrasion resistance is not directly proportional to hardness [41]. The effect of hardness on the abrasion resistance of particles was studied in the work of Sundararajan [42]. It was found that abrasion resistance has a different relationship with hardness. The impact of hardness on abrasion resistance and the impact of such parameters as impact velocity, impact angle and the size and shape of the particles were studied in the work of Oka and Hutchings [43]. It was found that hardness is an important parameter determining abrasive wear resistance. Sheldon also points out that Vickers hardness is probably one of the most important parameters in determining erosion resistance [44]. Studies on the influence of hardness, thickness of the protective coating and the coating structure on the erosion resistance of turbine blades are presented in the work by Wang et al. [45]. It was found that there is a critical coating thickness of at least 0.02 mm that is necessary to protect the surface. Cracks and discontinuities in the coating reduce erosion resistance. The key research issue is the possibility of assessing the relationship between the influence of hardness and the erosion resistance of the coating. The presented data indicate that there is no clear relationship between erosion resistance and the level of microhardness. Most studies indicated an increase in erosion resistance with an increase in coating microhardness. However, it was found that, as the size of the eroding particle increases and the speed of the abrasive stream increases, the nature of the wear of the cermet changes to a more plastic one. This means that there is a research gap regarding the impact of various environmental parameters on the coating's resistance to erosion in dynamic operating conditions.

In the work by Levy and Wang, it was determined that a small grain size, low porosity and the absence of cracks were the microstructural features that enhanced erosion resistance [46]. Hardness levels and the composition and distribution of the second phase, i.e., hard particles, had less effect on coating performance. In turn, the appropriate level of hardness of the HVOF coating ensures erosion resistance [47]. In the work by Wen, the resistance of steel to erosion by sand blasting [48] was examined. The results show that the increased hardness resulted in lower erosion rates because the increase in hardness provides resistance to penetration and results in a lower erosion rate. Thermally sprayed coatings consist of hard ceramic particles with a tough metallic binder. The ceramic particles provide excellent wear and corrosion resistance due to a high hardness and chemical stability. The metal matrix is usually ductile and provides good wetting and adhesion characteristics that, in return, give fracture toughness to the coating. This structure of coatings, especially those with almost nanometric grains, provides a very good resistance to abrasion and corrosion [49].

The great usefulness of thermal spray methods for anti-erosion and anti-corrosion surface protection was confirmed in the review work by Mehta et al. [50]. A study of the effect of particle impact velocity and the hardness of the material on the wear rate was presented in the work of Kosa and Göksenli [51]. Erosion resistance increased with increasing hardness. In turn, the study presented in the work [52] shows that an increase in plasticity leads to a greater resistance to the erosive impact of particles. This is the result of absorbing the kinetic energy of impacting particles, causing plastic deformation at the surface while maintaining the material within the fracture strain limits. It was observed that, under certain conditions, the erosion rate of WC/Co cermets decreases monotonically with the increase in the volume fraction of reinforcement [53]. However, it was not clear-cut because, with the increasing volume fraction of WC for a WC-Co cermet following an impact by a slurry of 100 μm silicon carbide particles at 40 ms^{-1} and at impact angles of 90°, the erosion increased to a certain WC content and then decreased.

The presented work focuses on issues related to the protection of the surfaces of technical fan blades exposed to the erosive impact of the industrial environment. There is a variable relationship between erosion resistance and hardness. Despite this, it should be noted that, in general, coatings are resistant to erosion precisely due to their high hardness. Therefore, the hardness parameter should be considered the most important in assessing the suitability of a coating for erosion protection. However, within the range of hardness obtained for cermetal coatings reported in the literature, research is needed to select individual types of coatings for specific operational purposes.

The aim of this research was to produce a thermally sprayed coating that would effectively protect the material of the rotor blades against damage caused by particles contained in the air exhausted by industrial fans. The rationale for undertaking this research was the possibility of reducing the weight of fan blades and, consequently, reducing energy consumption. Thermally sprayed coatings, due to their small thickness compared to, e.g., hard facing coatings, do not excessively load the blades and, therefore, consume less energy. At the same time, an appropriately selected chemical composition of the coating, ensuring a high hardness, should effectively protect the rotor blade material against damage. One of the most commonly used coatings for an anti-erosion surface protection are carbide coatings [54–57].

This work examined WC-Co and Cr_3C_2 coatings in a Co, Ni, Cr matrix and their various compositions, determining their hardness and structure. The high hardness was the basis for selecting the coatings that best met the desired feature of the resistance to particle impact in the medium centrifuged by fans. The use of diverse, new chemical compositions of coatings based on tungsten carbide and chromium carbide and a mixture of these carbides enabled the verification of the hypothesis, determining the possibility of increasing resistance to erosive wear through the synergistic effect of the carbide composition.

2. Materials and Methods

This study examined thermally sprayed coatings on a substrate made of A03590 casting aluminum alloy with the chemical composition shown in Table 1.

Table 1. Chemical composition of A03590 alloy.

Si	Cu	Mg	Mn	Ti	Zn	Fe	Al
8.5–10.5	0.3	0.25–0.4	0.25–0.5	0.15	0.2	0.6	rest

Carbide coatings based on the Cr_3C_2-NiCr powder and WC-Co powder were selected for testing using various application parameters. Coatings made of NiCrSiB composite materials with the addition of E = 2.5%Fe + 3.1%Si + 0.4%C + 2.1%Bi + Ni residue were also deposited. Coatings were also made from the following compositions: NiCrSiB + 5%Fe—nanoparticles, NiCrSiB + 5%Cr—nanoparticles, NiCrSiB + 2.5%Fe + 2.5%Cr—nanoparticles. In addition, a coating was produced from a mixture of Cr_3C_2 and WC-Co and an Inconel 625 (Tables 2 and 3). The differences in the chemical composition of the coatings were reflected in their different levels of microhardness. The microhardness value was adopted as a parameter for selecting the optimal coating, resistant to the erosive effects of the environment. The basic criterion for the suitability of the coatings was a high microhardness value. Microhardness was tested using the Haneman method on the cross-section of samples after grinding and polishing. Microhardness was measured on polished samples under a load of 200 G using a Microhardness measuring microscope (PMT-3)–Vickers microhardness tester (from Russian manufacturer). In addition to microhardness tests, the structure of the coatings was observed. Samples for structure observation were cut perpendicularly to the thermally sprayed surface using a Leco ball with a diamond disc cooled with oil diluted with water. The cut sections were sanded on 220–4000 grit sandpaper. Then, the samples were polished using an Al_2O_3 OPS suspension and diamond pastes with a gradation of 6 µm, 3 µm and 1 µm, manufactured by Struers (Kraków, Poland). Samples for observation using a scanning electron microscope (SEM) were embedded in a special conductive material before preparation for testing. The structure of the samples was examined using a STEREOSCAN 420 scanning electron microscope(Manufacturer CAMBRIDGE, UK). The chemical composition of the samples was examined using the EDX method with a scanning electron microscope. In the first stage of the research, the microhardness of the coatings was measured. Then, 7 coatings with the highest level of microhardness were selected and tested for resistance to the erosive impact of the environment with corundum particles under the operating conditions of an industrial fan rotor. The coatings were sprayed onto individual rotor blades. Each coating was deposited on one of the rotor blades. The test rotor operated at a speed of 3000 rotations per minute, and it was exposed to the veneer paper of panels that contained corundum. The operating time of the rotor in operational conditions was 2.5 months. After this period of operation, the rotor was dismantled, and the coatings were examined for wear and tear and assessed for pitting, cracks and damage. After assessing the tested rotor, the coating that showed the least damage was selected.

Table 2. Chemical composition of powders and parameters of thermal spray coating.

Powder Composition	Plasma Spray Parameters				Coating Thickness [µm]
	Ar [L/h]	Voltage [V]	Current [A]	Distance [mm]	
WC-Co 88/12	3000	52	500	90	300 ± 20
WC-Co 88/12 (+propane–butane cover)	3000	52	500	90	
NiCrSiB	2500	48	480	90	
NiCrSiB + 5% Fe	2500	48	480	90	
NiCrSiB + 5% Cr	2500	48	480	90	
NiCrSiB + 2.5% Fe + 2.5% Cr	2500	48	480	90	
Cr_3C_2—NiCr 75–25	3000	52	500	90	
Cr_3C_2—NiCr 75–25 -45+ (+propane–butane cover)	3000	52	500	90	
WC-Co + 5%(NiCrFeBSi) (+propane–butane cover)	3000	52	500	90	

Table 3. Chemical composition and parameters of HVOF deposition.

Powder Composition	HVOF Spray Parameters					Coating Thickness [µm]
	O_2 [L/min]	Kerosene [L/h]	N_2 [L/min]	Distance [mm]	Powder [g/min]	
WC-Co 88-12	944	25.5	9.5	370	92	300^{+80}
NiCrSiB	920	21.5	11	355	95	
NiCrSiB+5%Fe—nanopowder	920	21.5	11	355	95	
NiCrSiB+5%Cr—nanopowder	920	21.5	11	355	95	
NiCrSiB + 2.5%Fe + 2.5%Cr—nanopowder	920	21.5	11	355	95	

Table 3. *Cont.*

Powder Composition	HVOF Spray Parameters					Coating Thickness [μm]
	O_2 [L/min]	Kerosene [L/h]	N_2 [L/min]	Distance [mm]	Powder [g/min]	
Cr_3C_2-NiCr 75–25	850	24	9.5	370	65	
WC-Co + 5%(NiCrFeBSi)	944	25.5	9.5	370	92	
WC-CrC-Ni 73/20/7	944	24	10	370	75	
WC-Co-Cr 86/10/4 nanopowder	944	25.5	9.5	370	92	300^{+50}
Ni-Cr-B-Si	944	24	10	370	75	300^{+50}
Inconel 625	944	24	10	370	75	300^{+50}

3. Results

Figures 1 and 2 show the results of microhardness measurements of coatings applied to the AK9 substrate. The microhardness of coatings sprayed using the HVOF method is shown in Figure 1, and the microhardness of plasma sprayed coatings is shown in Figure 2. The microhardness of coatings sprayed using the HVOF method showed higher values than the microhardness of plasma spray coatings. The highest microhardness, among the coatings produced by the HVOF method, was achieved by the WC-Co-Ni coating sprayed from nanometric powders. In the case of plasma sprayed coatings, the highest hardness was demonstrated by the Cr_3C_2-NiCr +5%E composite coating produced in a propane–butane cover.

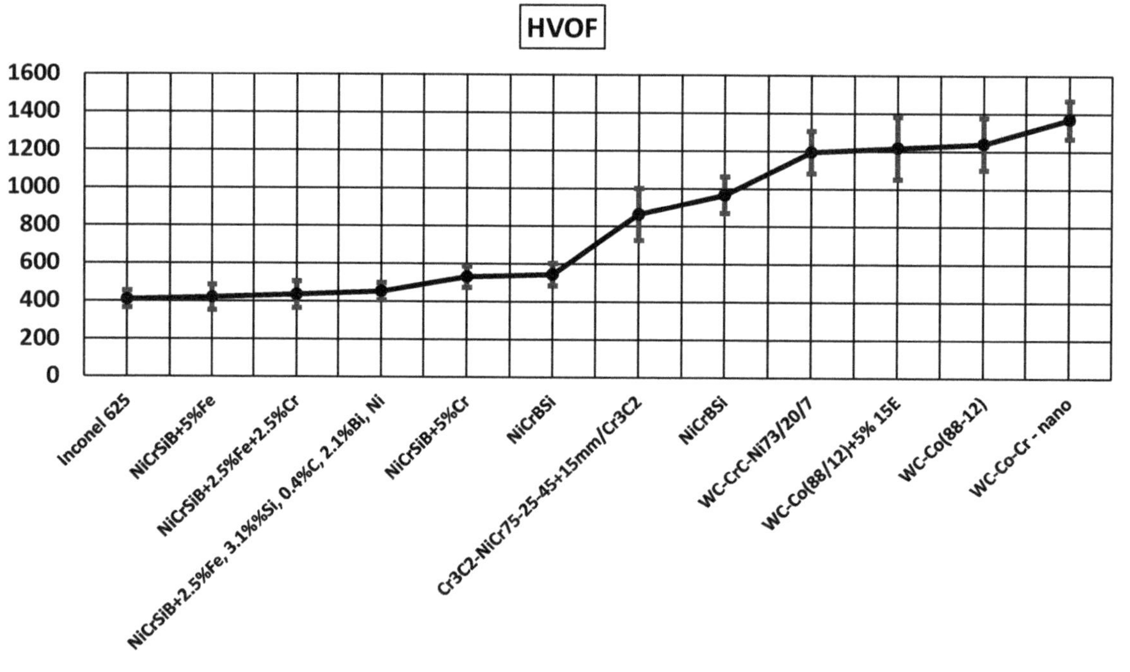

Figure 1. Microhardness of coatings produced by the HVOF method.

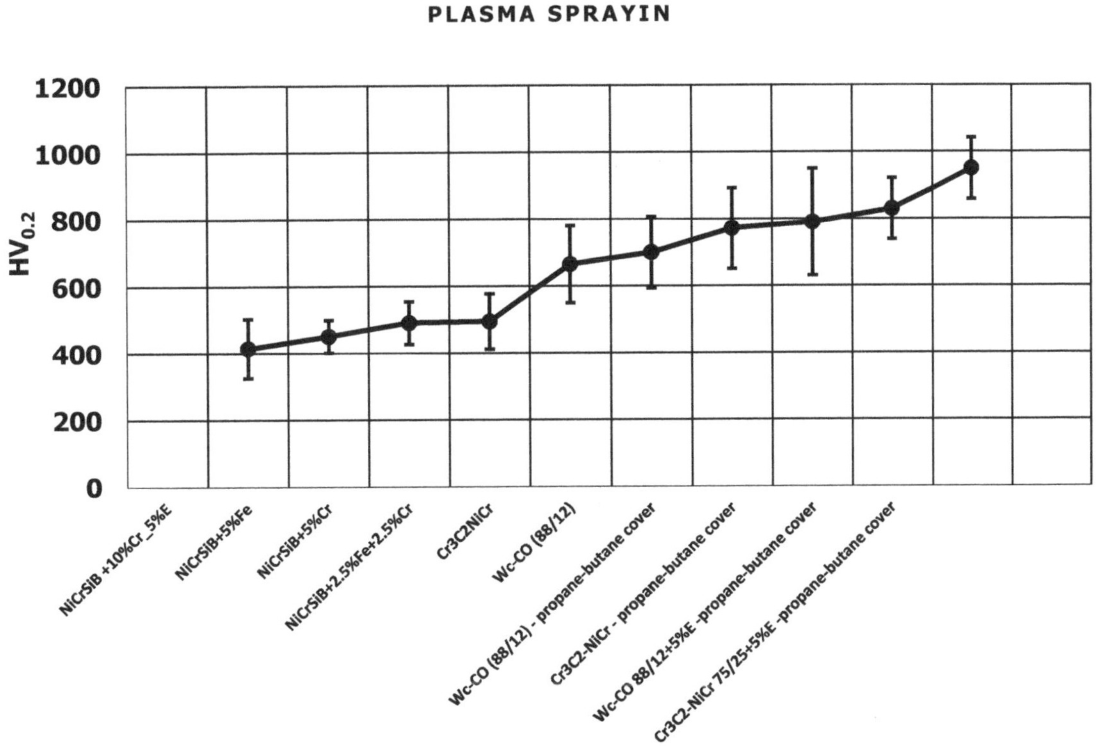

Figure 2. Microhardness of plasma spray coatings, E = 2.5%Fe + 3.1%Si + 0.4%C + 2.1%Bi + Ni rest.

Seven coatings with the highest microhardness were selected for further testing. The chemical composition of the selected coatings and the type of deposition process are presented in Table 4.

Table 4. Chemical composition of coatings deposited on rotor blades intended for industrial tests.

Rotor Blade No	Chemical Composition	Thermal Spray Method	$HV_{0.2}$
1	WC-Co-Cr	HVOF	1369 (±100)
2	WC-Co	HVOF	1240 (±138)
3	WC-Co + 5% NiCrBSi	plasma spray	1218 (±166)
4	WC-Co-CrC-Ni	HVOF	1195 (±112)
5	Cr_3C_2-NiCr + 5% NiCrBSi	plasma spray	967 (±98)
6	NiCrBSi	HVOF	950 (±92)
7	Cr_3C_2-NiCr	HVOF	863 ± (138)

The structures of the coatings with the highest microhardness are shown in Figures 3–9.

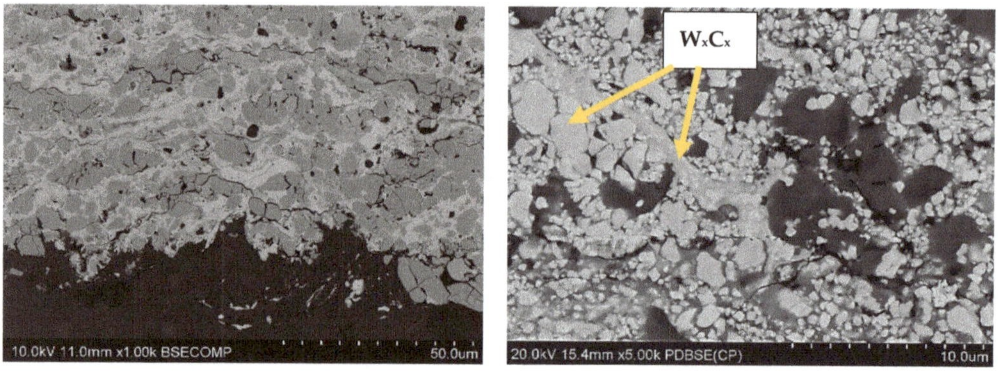

Figure 3. Coating No. 1: WC-Co-Cr, HVOF ($HV_{0.2}$ = 1369) (W_xC_x—possible W_2C, WC).

Figure 4. Coating No. 2: WC-Co, HVOF ($HV_{0.2}$ = 1240) (W_xC_x—possible W_2C, WC).

Figure 5. Coating No. 3: WC-Co + 5% NiCrSiB, plasma spray in a propane–butane shield ($HV_{0.2}$ = 1218), equiaxed W_xC_x carbides (possible W_2C, WC), NiCrSiB matrix.

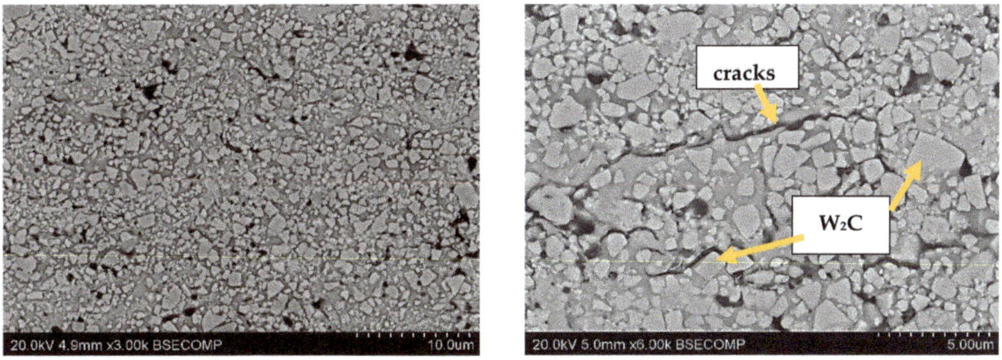

Figure 6. Coating No. 4: WC-Co-CrC-Ni, HVOF ($HV_{0.2}$ = 1195).

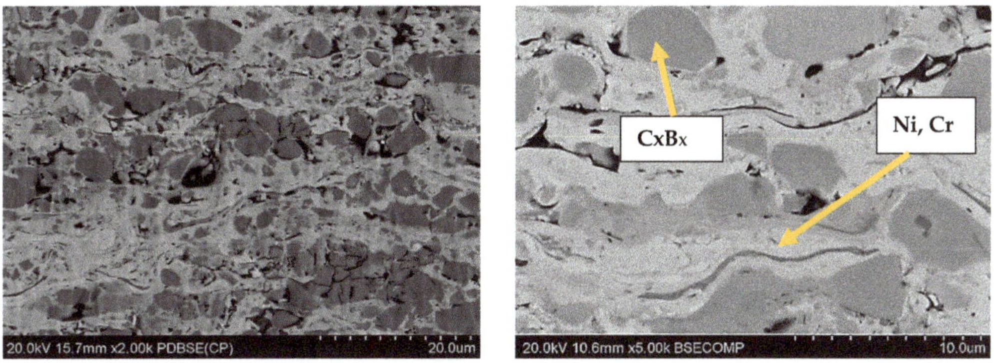

Figure 7. Coating No. 5: NiCrBSi, HVOF ($HV_{0.2}$ = 967) (Cr_xB_x—possible CrB, Cr_3B_2, Cr_3B).

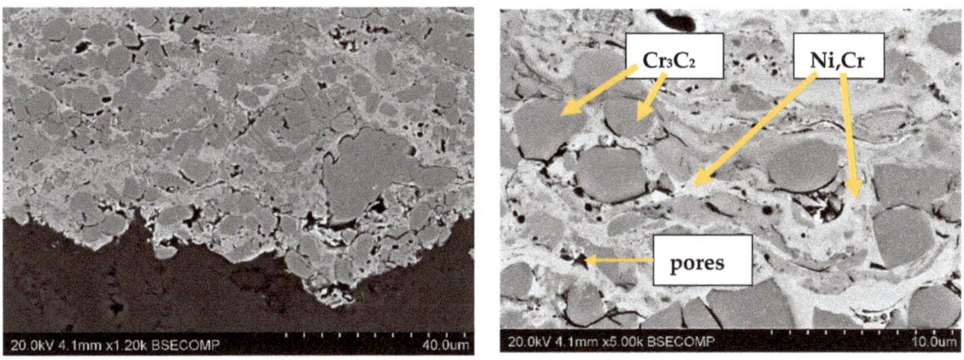

Figure 8. Coating No. 6: Cr_3C_2-NiCr + 5%NiCrBSi, plasma-spray from nanopowders in a propane–butane shield ($HV_{0.2}$ = 950).

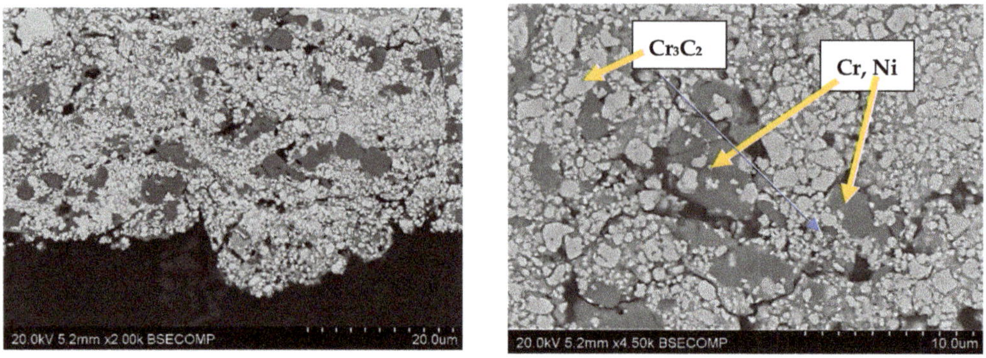

Figure 9. Coating No. 7: Cr_3C_2—NiCr, HVOF ($HV_{0.2}$ = 863).

Erosive wear resistance tests were carried out using a sanding device operating at an air pressure in the range of 1 ÷ 6 bar. The volume of material removed during the test was taken as a measure of the erosion resistance of the tested coatings (Figure 10).

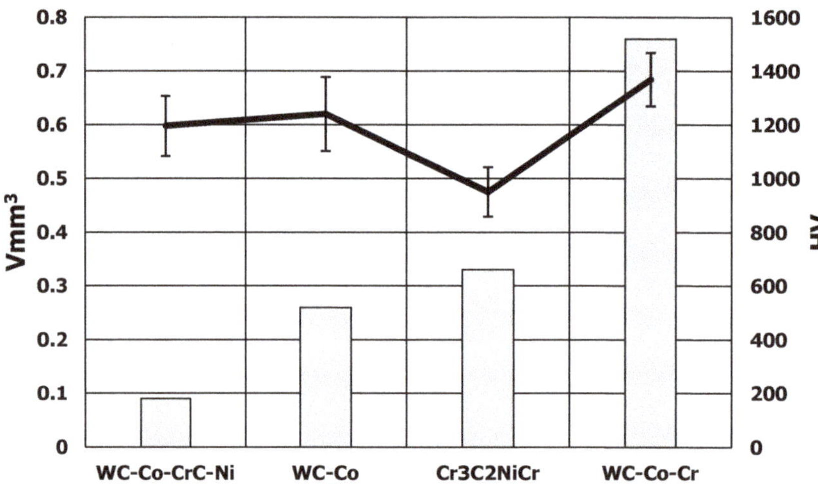

Figure 10. Comparison of the volume of erosion craters of thermally sprayed coatings against the microhardness results of the coatings.

The abrasion results of the Cr_3C_2NiCr coating did not correlate with the results of the tungsten carbide-based coatings. A correlation was obtained between the hardness of coatings based on WC and abrasion. The higher the hardness of the coating, the larger the craters and the volume of material removed. It seems that a higher plasticity of the coating ensures a lower degree of carbide knockout by the particles of the centrifuged medium. The WC-Co-CrC-Ni coating, sprayed using the HVOF method ($HV_{0.2}$ = 1195), showed the greatest resistance to the erosive impact of particles. The resistance of the coatings to the particle impact was on average 18 times greater than the resistance of the AK9 alloy substrate.

The coatings with the highest hardness were applied to the blades of the test rotor, which operated in industrial conditions and were exposed to corundum particles. In Figure 11, the surface profiles of the coatings applied to the rotor blades tested in industrial conditions are presented. All profiles show pitting craters caused by the impact of the hard particles of the centrifuged medium.

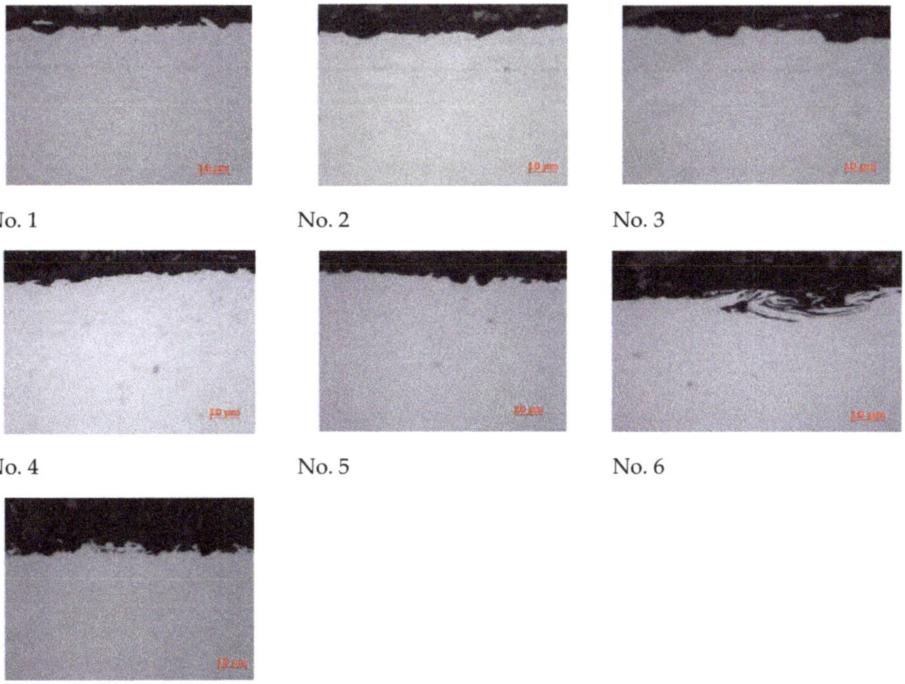

No. 1 No. 2 No. 3 No. 4 No. 5 No. 6 No. 7

Figure 11. Surface profiles of coatings that were applied to the blades of a rotor tested in industrial conditions.

A macroscopic inspection of the rotor blades was also carried out after the industrial test. Tests of the rotor blades after operation in industrial conditions showed that the least wear was observed for the following coatings: No. 2—WC-Co (HVOF), $HV_{0.2}$ = 1240; No. 4—WC-Co-CrC-Ni (HVOF), $HV_{0.2}$ = 1195; and No. 7—Cr_3C_2-NiCr (HVOF), $HV_{0.2}$ = 863.

Among the mentioned coatings, the surface that was least damaged and destroyed by the centrifuged medium was coating No. 4, made of tungsten carbide and also containing chromium carbide WC-Co-CrC-Ni, deposited using the HVOF method, with a microhardness value of $HV_{0.2}$ = 1195 (Figure 12).

The microhardness of coating No. 4 is 12% lower than the value of the hardest WC-Co-Cr coating deposited from nanopowders using the HVOF method ($HV_{0.2}$ = 1369) (Table 4). Nevertheless, it shows a greater resistance to impact erosion by hard corundum particles.

Test rotor blade with WC Co CrC Ni coating (HVOF) $HV_{0.2}$ = 1195

Test rotor blade with WC Co Cr coating (HVOF) $HV_{0.2}$ = 1369

Figure 12. Macroscopic view of the rotor blades after an industrial test.

4. Discussion

Tungsten is an excellent conductor of electricity and heat, and it becomes a superconductor at low temperatures. It is the element with the highest melting point of all metals. This value is 2870 °C. The density of tungsten carbide is 15.5 g/cm^3. This structure significantly increases the strength of the material and improves its thermal conductivity. Another important parameter of tungsten carbide is its tensile and compressive strength, which is directly responsible for the strength of the material during its use. The Young's modulus in this case is 530–700 GPa. This means that tungsten carbide subjected to stresses of the indicated value will not undergo permanent deformation.

Chromium carbides are also characterized by a high resistance to erosive and corrosive wear [58]. The hardness of chromium carbide coatings is in the range of 770–1600 HV, depending on the production method [59,60]. The Cr_3C_2 compound has exhibited the highest hardness value and the maximum Cr-C bond strength. The influence of Cr-C bond strength and the density of Cr_3C_2 play a remarkable role. The Cr_3C_2 exhibited the highest strength under the applied pressure. It has also exhibited the best mechanical properties compared to Cr_7C_3 and $Cr_{23}C_6$ [61]. The presence of the metal binder (Co, Ni and Cr) and its different content in the coatings affects the hardness of the coatings.

In the tests carried out on the susceptibility to the impact erosion of particles, the highest hardness was obtained by coating No. 1—WC-Co-Cr, HVOF ($HV_{0.2}$ = 1369), and the lowest was obtained by coating No. 7—Cr_3C_2—NiCr, HVOF ($HV_{0.2}$ = 863). It was also found that coating No. 4—WC-Co-CrC-Ni, HVOF ($HV_{0.2}$ = 1195)—has the best wear resistance.

This confirms the results of the study [62], which showed that the WC-based coating has a better resistance to abrasive wear compared to the CrC-based coating. In addition, the carbide-based coatings applied via the HVOF method have an excellent wear resistance compared to other application methods. The results are also confirmed by the research work [41], which suggests that there is no simple relationship between the resistance to erosive wear and the hardness level. Overall, the high hardness provides wear resistance. However, depending on the method of impact of erosive particles, the speed of their impact on the substrate, the impact angle, the particle size and other parameters, a complex nature of the relationship between abrasive wear resistance and the material and strength parameters of the coating can be expected.

The good thermal conductivity of tungsten causes a very rapid reduction in the particle temperature during the production of coatings. This may contribute to an increase in the fraction of unmelted particles. The carbide composition may also be transformed as a result of decarbonization. In this kind of cermet coating, some brittle phases unavoidably appear (e.g., W_2C, W and $Co_xW_yC_z$) due to the decomposition and decarburization of WC in the high-temperature thermal spraying flame [63]. The melting of carbide particles and their plastic deformation in thermal spraying processes, caused by the strong impact of particles on the substrate, affect the hardness of the coating. The particles decarbonize and also break down [64]. As a result of decarbonization, the following phases may appear in the structure: WC, W_2C, W, Co_3W_3C or Co [65]. In the case of chromium carbides, depending on the C/N ratio, the following phases are formed: Cr_3C_2, Cr_7C_3, CrO_2, $Cr_{6.2}C_{3.5}$ or $Cr_3C_{1.5}$ [66]. Changes in the phase composition affect the hardness level of the coatings and their resistance to abrasive wear resulting from interaction with particles of the centrifuged medium.

It is assumed that the reason for the almost two-fold acceleration of the wear process of the WC-17%Co coating compared to WC-12%Co may be the reduction in the WC carbide share at the expense of W_2C and $(W,Co)_6$ C. The thermal dissolution of WC is primarily responsible for the occurrence of W_2C. The semicarbide (W_2C) crystals are observed to grow epitaxially on the affected WC grain and extend radially with progressively thinning cross-sections [67].

The results of the erosive wear tests conducted allow for us to conclude that the chemical composition of multi-component coatings based on WC-Co has a significant impact on their resistance to erosive wear. The synergistic interaction of tungsten carbide and chromium carbide results in the best resistance to abrasive wear by centrifuged particles (Figure 12). It reduces the wear parameter almost ten times compared to other chemical compositions of coatings. Abrasion tests indicate that, for tungsten carbide-based coatings, sand impact craters increase with increasing coating hardness. This is probably related to a greater susceptibility to cracking of hard WC grains and a lower plasticity of the coating [68]. The micro-cracks at grain boundaries significantly influence both the sliding and abrasive wear rate of WC-Co coatings. This paper hypothesized that the crack propagation for smaller carbides proceeds through the interlamellar (i.e., intergranular) boundary, whereas for larger carbides, cracks propagate through the carbide (i.e., transgranular). Yuan et al. [69] concluded that submicron-sized WC particles uniformly distributed at the interfaces could cause deflection or impede fracture. This indicates that the more homogeneous the coating structure, the more resistant it should be to erosive influences.

The work [70] concluded that the average friction coefficient depends strongly on the cermet fraction, exhibiting a maximum value of 0.64 at 50% WC-Co/Cu content. The wear coefficient values decrease almost linearly with increasing cermet fraction.

Tests of cermet coatings based on WC and Cr_3C_2 do not reveal simple relationships between wear resistance and the hardness and structure of the coating. The coatings contain numerous defects in the form of discontinuities, unmelted grains and other artifacts that affect the hardness of the coating. The size of powders and coating production conditions also affect their properties [71–73].

The tests carried out showed that coatings based on WC powder deposited using the PS method showed lower hardness than coatings deposited using the HVOF method. The WC-Co coating deposited using the HVOF method had a hardness of $HV_{0.2}$ = 1240, and the one deposited using the PS method had a hardness of $HV_{0.2}$ = 830. Coatings based on NiCrSiB powder with various admixtures had a comparable hardness level both in the case of deposition using the HVOF method and in the case of deposition using the PS method. It was a level in the range of $HV_{0.2}$ = 400–500. Coatings based on Cr_3C_2 powder showed a hardness of $HV_{0.2}$ = 950 when deposited using the PS method, and $HV_{0.2}$ = 863 when deposited using the HVOF method. In the industrial test, four coatings were selected that were most resistant to the erosive impact of dynamic particle streams. These were coatings deposited using the HVOF method, No. 2—WC-Co ($HV_{0.2}$ = 1240), No. 4—WC-Co-CrC-Ni ($HV_{0.2}$ = 1195) and No. 7—Cr_3C_2-NiCr ($HV_{0.2}$ = 863), and a coating deposited using the PS Cr_3C_2NiCr method ($HV_{0.2}$ = 950). Among the four coatings most resistant to erosion in industrial conditions, one coating was selected that also showed the highest resistance in the sandblasting test. It was coating No. 4—WC-Co-CrC-Ni.

In this case, coating No. 4 also turned out to be the most resistant to damage of centrifuged corundum particles (No. 4—WC-Co-CrC-Ni—sprayed using the HVOF method

($HV_{0,2}$ = 1195)). It was not the coating with the highest hardness, but it belonged to the group of the hardest WC-based carbide coatings sprayed onto the surface of the rotor blades. The qualitative assessment of the damage to the blades caused by the corundum particles as well as the similar level of hardness of the tested coatings and the discontinuities present in the coatings are reasons to consider this coating as potentially offering the best anti-erosion properties. The performed technical test indicates the suitability of coating No. 4 for operation in difficult conditions of exposure to impacts of hard corundum particles found in the centrifuged environment.

Coating No. 4—WC-Co-CrC-Ni—had a composition similar to the WC-Co-Cr coating tested in the work by Jonda and Łatka [74]. The hardness of both coatings was practically the same. The hardness of the WC-Co-CrC-Ni coating was $HV_{0,2}$ = 1195, and the hardness of the coating tested in the work of Jonda and Łatka, WC-Co-Cr, was HV = 1198. In the work of Jonda and Łatka, it was shown that WC-Co-Cr is the most promising candidate for the further dry sliding, erosion and cavitation resistance coating. It was also characterized by a considerable hardness, a relatively good fracture toughness and a high value of elastic modulus.

The convergence of the results confirms the data obtained and indicates that the chemical composition of coating No. 4 is favorable for practical applications. Studies of carbide coatings produced by the HVOF method also showed that, in the case of the WC-CrC-Ni coating, the hardening of the matrix caused by the dissolution of Cr is accompanied by an increase in the matrix's plasticity, which prevents carbides from detaching from the matrix [75]. According to Korobov et al. [76], the fine structure of the WC-CrC-Ni coating increases the specific surface area of carbide particles and, consequently, the required crack propagation energy. This improves their resistance to chipping and, consequently, to cavitation compared, for example, to the coarser WC-CoCr coating. Tests on the properties of WC−Co, WC−Co−Cr_3C_2 and WC−Co−TaC showed that the addition of Cr_3C_2 or TaC improves the hardness and fracture toughness of the obtained sintered compacts [76].

The increasing use of surface protection is associated with reduced operating costs of protected devices [77]. In particular, thermal spraying methods, which have multiple varieties and enable the spraying of protective coatings in various conditions, present a favorable level of costs and high effectiveness of protective layers. The hardness achieved in the case of carbide coatings based on WC and Cr_3C_2 is approximately 1000–1200HV, and in many cases, it is sufficient to protect the surface against particle impact erosion.

5. Conclusions

Based on the research conducted, the following final conclusions were made.

1. It was found that the coatings most resistant to particle impact erosion were WC-Co, $HV_{0.2}$ = 1240; WC-Co-CrC-Ni, $HV_{0.2}$ = 1195; and Cr_3C_2-NiCr, $HV_{0.2}$ = 863 deposited by the HVOF method and coating Cr_3C_2NiCr deposited by the PS method.
2. The WC-Co-CrC-Ni coating showed the highest, simultaneous resistance to the impact of a stream of corundum particles and SiO_2 particles.
3. The greatest resistance of the WC-Co-CrC-Ni coating to the erosive impact of dynamic particle streams results from the synergistic effect of tungsten and chromium carbides.

Funding: This research received no external funding.

Informed Consent Statement: Not applicable.

Data Availability Statement: The original contributions presented in the study are included in the article, further inquiries can be directed to the corresponding author.

Acknowledgments: The work was performed as part of the project INNOTECH-K2/IN2/9/181851/NCBR/13. I would like to thank all the contractors of the INNOTECH-K2/IN2/9/181851/NCBR/13 project I manage.

Conflicts of Interest: The author declares no conflict of interest.

References

1. Kania, H. Corrosion and Anticorrosion of Alloys/Metals: The Important Global Issue. *Coatings* **2023**, *13*, 216. [CrossRef]
2. Hubicki, R.; Richert, M.; Łebkowski, P. Influence of Operating Temperature on the Service Life of Aluminum Extrusion Dies. *Materials* **2022**, *15*, 6656. [CrossRef] [PubMed]
3. Li, H.; Li, L.; Li, D.; Tang, L.; Luo, Y.; Li, G.; Wu, Y.; Li, G.; Xu, Y.; Han, M.; et al. Erosion Performance of TiAlSiN Coatings Prepared by High-Power Pulsed Magnetron Sputtering. *Metals* **2023**, *13*, 1306. [CrossRef]
4. Mishnaevsky, L., Jr.; Tempelis, A.; Kuthe, N.; Mahajan, P. Recent developments in the protection of wind turbine blades against leading erosion: Materials solutions and predictive modelling, Edge. *Renew. Energy* **2023**, *215*, 118966. [CrossRef]
5. Maurer, C.; Shulz, U. Erosion resistant titanium based PVD coatings on CFRP. *Wear* **2013**, *302*, 937–945. [CrossRef]
6. Zhang, Z.; Yang, M.; He, G. Structure, mechanical, and sand erosion behavior of TiN/Ti coating deposited at various temperature. *Ceram. Int.* **2023**, *49*, 16786–16795. [CrossRef]
7. Dobrzańska-Danikiewicz, A.D.; Gołombek, K.; Pakuła, D.; Mikuła, J.; Staszuk, M.; Żukowska, L.W. Long-term development directions of PVD/CVD coatings deposited onto sintered tool materials. *Mater. Surf. Eng.* **2011**, *6*, 278–330. Available online: http://openaccesslibrary.com/vol06/8.pdf (accessed on 5 July 2024).
8. Daniel, J.; Grossman, J.; Houdková, Š.; Bystrianský, M. Impact Wear of the Protective Cr3C2-Based HVOF-Sprayed Coatings. *Materials* **2020**, *13*, 2132. [CrossRef]
9. Yanjun, Z.; Shiqi, Z.; Qiuyu, Y.; Biyu, H. An Approach to Simulate the Manufacturing Process of Thermal Spray Coatings. In Proceedings of the 5th International Conference on Mechanical Engineering, Materials and Energy (ICMEME 2016), Hong Kong, China, 10–11 December 2016; Atlantis Press: Amsterdam, The Netherlands, 2016.
10. Szczucka-Lasota, B.; Węgrzyn, T.; Stanik, Z.; Piwnik, J.; Sidun, P. Selected parameters of microjet cooling gases in hybrid spraying process. *Arch. Metall. Mater.* **2016**, *61*, 621–624. [CrossRef]
11. Feltrin, A.C.; Xing, Q.; Akinwekomi, A.D.; Waseem, O.A.; Akhtar, F. Review of Novel High-Entropy Protective Materials: Wear, Irradiation, and Erosion Resistance Properties, Review. *Entropy* **2023**, *25*, 73. [CrossRef]
12. Keshavamurthy, R.; Naveena, B.E.; Sekhar, N. Thermal Spray Coatings for Erosion-Corrosion Protection. In *Production, Properties, and Applications of High Temperature Coatings*; Book Chapture in Advances in Chemical and Materials Engineering (ACME); IGI Global: Hershey, PA, USA, 2018; pp. 246–267. [CrossRef]
13. Amin, S.; Panchal, H. A Review on Thermal Spray Coating Processes. *Int. J. Curr. Trends Eng. Res. IJCTER* **2016**, *2*, 556–563. Available online: https://www.ijcter.com (accessed on 5 July 2024).
14. Łatka, L.; Pawłowski, L.; Winnicki, M.; Sokołowski, P.; Małachowska, A.; Kozerski, S. Review of Functionally Graded Thermal Sprayed Coatings. *Appl. Sci.* **2020**, *10*, 5153. [CrossRef]
15. Siao, A.; Ming Ang, M.; Berndt, C.C. A review of testing methods for thermal spray coatings. *Int. Mater. Rev.* **2014**, *59*, 179–223. [CrossRef]
16. Tejero-Martin, D.; Rezvani Rad, M.; McDonald, A.; Hussain, T. Beyond Traditional Coatings: A Review on Thermal-Sprayed Functional and Smart Coatings. *J. Therm. Spray Technol.* **2019**, *28*, 598–644. [CrossRef]
17. Shuo Yin, S.; Cizek, J.; Suo, X.; Li, W.; Liao, H. Thermal Spray Technology. *Adv. Mater. Sci. Eng.* **2019**, *2019*, 8654764. [CrossRef]
18. Gaur, U.P.; Kamari, E. Applications of Thermal Spray Coatings: A Review. *J. Therm. Spray Eng.* **2024**, *4*, 106–114. [CrossRef] [PubMed]
19. Fauchais, P.L.; Heberlein, J.V.R.; Boulos, M.I. *Thermal Spray Fundamentals, From Powder to Part*; Springer Science + Business Media: New York, NY, USA, 2021. [CrossRef]
20. Venkatachalapathy, V.; Katiyar, N.K.; Matthews, A.; Endrino, J.L.; Goel, S.A. Guiding Framework for Process Parameter Optimisation of Thermal Spraying. *Coatings* **2023**, *13*, 713. [CrossRef]
21. Guilemany, J.M.; Nin, J. Thermal spraying methods for protection against wear. In *Surface Coatings for Protection Against Wear*; Mellor, B.G., Ed.; Woodhead Publishing in Material: Cambridge, UK, 2006.
22. Hu, C.; He, G.; Chen, J.; Fang, Z.; Yang, Z.; Zhang, Z. Research on Cleaning Mechanism of Anti-Erosion Coating Based on Thermal and Force Efects of Laser Shock. *Coatings* **2020**, *10*, 683. [CrossRef]
23. Kay, C.M. ASB Industries, Inc. Thermal Spray Applications in the Steel Industry. In *ASM Handbook*; Tucker, R.C., Jr., Ed.; Thermal Spray Technology; ASM International: Detroit, MI, USA, 2013; Volume 5A. Available online: https://www.asbindustries.com/uploads/Handbook%205A%20Thermal%20Spray%20Apps%20in%20Steel%20Industry.pdf (accessed on 5 July 2024).
24. Hardwicke, C.U.; Lau, Y.C. Advances in Thermal Spray Coatings for Gas Turbines and Energy Generation: A Review. *J. Therm. Spray Technol.* **2013**, *22*, 564–576. [CrossRef]
25. Baiamonte, L.; Pulci, G.; Hlede, E.; Marra, F.; Bartuli, C. Corrosione Thermal spray coatings for corrosion and wear protection of naval Diesel engines components. *Metall. Ital.* **2014**, *6*, 9. Available online: http://www.aimnet.it/allpdf/pdf_pubbli/giu14/Bartuli.pdf (accessed on 5 July 2024).
26. Singh, S.; Berndt, C.C.; Singh Raman, R.K.; Singh, H.; Ang, A.S.M. Applications and Developments of Thermal Spray Coatings for the Iron and Steel Industry. *Materials* **2023**, *16*, 516. [CrossRef] [PubMed]
27. Fauchais, P.; Vardelle, A. Thermal Sprayed Coatings Used Against Corrosion and Corrosive Wear, Advanced Plasma Spray Applications. 2012. Available online: https://cdn.intechopen.com/pdfs/32974/InTech-Thermal_sprayed_coatings_used_against_corrosion_and_corrosive_wear.pdf (accessed on 5 July 2024).

28. Miller, R.A. History of Thermal Barrier Coatings for Gas Turbine Engines Emphasizing NASA's Role from 1942 to 1990, NASA/TM—2009-215459. 2009. Available online: https://ntrs.nasa.gov/api/citations/20090018047/downloads/20090018047.pdf (accessed on 5 July 2024).
29. Brezinová, J.; Guzanová, A.; Draganovská, D.; Maruschak, P.O.; Landová, M. Study of Selected Properties of Thermally Sprayed Coatings Containing WC and WB Hard Particles. *Acta Mech. Et Autom.* **2016**, *10*, 4. [CrossRef]
30. Özbek, Y.Y.; Canikoğlu, N.; Ipek, M. The Mechanical Properties and Wear Resistance of HVOF Sprayed WC–Co Coatings. *Acta Phys. Pol. A* **2016**, *129*, 600–603. [CrossRef]
31. Myalska, H.; Szymański, J.K.; Moskal, G. Microstructure and properties of WC-Co HVOF coatings obtained from standard, superfine and modified by sub-micrometric carbide powders. *Arch. Metall. Mater.* **2014**, *60*, 759–766. [CrossRef]
32. Abderrahmane, A.; Gaceb, M.; Cheikh, M.; Le Roux, S. Wear Behavior and Microstructure of Thermally Sprayed NiCrBSiFeC and Composite NiCrBSiFeC-WC(Co) Coatings. 2021. Available online: https://imt-mines-albi.hal.science/hal-03149468/file/Wear-Behavior-and-Microstructure-of-Thermally-Sprayed.pdf (accessed on 5 July 2024).
33. Rachidi, R.; El Kihel1, B.; Delaunois, F.; Vitry, V.; Deschuyteneer, D. Wear Performance of Thermally Sprayed NiCrBSi and NiCrBSi-WC Coatings Under Two Different Wear Modes. *J. Mater. Environ. Sci.* **2017**, *8*, 4550–4559. [CrossRef]
34. Sharmaa, M.; Goyalb, D.K.; Kaushal, G. Tribological investigation of HVOF-spray Cr3C2-25NiCr and WC-10Co-4Cr coated turbine steel under varied slurry erosion conditions. *Indian J. Eng. Mater. Sci.* **2020**, *27*, 150–167.
35. Chang, L.; Wang, W.; Ma, D.; Douqin, M.; Jingpei, X. Deposition effects and interface structure of HVOF-sprayed multimodal WC-CoCr coatings. *J. Mater. Res.* **2023**, *38*, 4345–4356. [CrossRef]
36. Wang, Q.; Zhong, Y.; Li, H.; Wang, S.; Liu, J.; Wang, Y.; Ramachandran, C.S. Effect of Cobalt and Chromium Content on Microstructure and Properties of WC-Co-Cr Coatings Prepared by High-Velocity Oxy-Fuel Spraying. *Materials* **2023**, *16*, 7003. [CrossRef] [PubMed]
37. Jiang, H.; Zhao, X.; Song, H.; Li, C. The Effect of High-Velocity Air-Fuel WC-12Co Coatings on the Wear and Corrosion Resistance of TC18 Titanium Alloy. *Coatings* **2023**, *13*, 755. [CrossRef]
38. Maiti, A.K.; Mukhopadhyay, N.; Raman, R. Effect of adding WC powder to the feedstock of WC–Co–Cr based HVOF coating and its impact on erosion and abrasion resistance. *Surf.* **2007**, *201*, 7781–7788. [CrossRef]
39. Szala, M.; Walczak, M.; Świetlicki, A. Effect of Microstructure and Hardness on Cavitation Erosion and Dry Sliding Wear of HVOF Deposited CoNiCrAlY, NiCoCrAlY and NiCrMoNbTa Coatings. *Materials* **2022**, *15*, 93. [CrossRef]
40. Bulnes, A.G.; Fuentes, V.A.; Cano, I.G.; Dosta, S. Understanding the Influence of High Velocity Thermal Spray Techniques on the Properties of Different Anti-Wear WC-Based Coatings. *Coatings* **2020**, *10*, 1157. [CrossRef]
41. Cesanek, Z.; Schubert, J. Abrasive resistance of HVOF thermally sprayed coatings based on alloys and cermets compared to nitrided stainless steel. *Arch. Mater. Sci. Eng.* **2014**, *67*, 18–23.
42. Sundararajan, G. The differential effect of the hardness of metallic materials on their erosion and abrasion resistance. *Wear* **1993**, *162–164*, 773–781. [CrossRef]
43. Oka, Y.; Hutchings, J.M. Dependence of Material Hardness on Erosion by Solid Particle Impact. *Corros. Eng.* **1990**, *39*, 610–616. [CrossRef]
44. Sheldon, G.L. Effects of Surface Hardness and Other Material Properties on Erosive Wear of Metals by Solid Particles. *J. Eng. Mater. Technol.* **1977**, *99*, 133–137. [CrossRef]
45. Wang, S.; Liu, G.; Mao, J.; He, Q.; Feng, Z. Effects of Coating Thickness, Test Temperature, and Coating Hardness on the Erosion Resistance of Steam Turbine Blades. *J. Eng. Gas Turbines Power* **2010**, *132*, 022102. [CrossRef]
46. Levy, A.V.; Wang, B. Erosion of hard material coating systems. *Wear* **1988**, *121*, 325–346. [CrossRef]
47. Wheeler, D.W.; Wood, R.J.K. Erosion of hard surface coatings for use in offshore gate valves. *Wear* **2005**, *258*, 526–536. [CrossRef]
48. Wen, D.C. Influence of Material Surface Hardness and Particle Size on the Erosion of DC 11 Tool Steel. *J. China Inst. Technol.* **2007**, *36*, 157–170. Available online: https://www.yoke.net/Pic/PublicFile/X0KFKHUAWAJN0036.pdf (accessed on 5 July 2024).
49. Saha, G.C.; Khan, T.I.; Zhang, G.A. Erosion–corrosion resistance of microcrystalline and near-nanocrystalline WC–17Co high velocity oxy-fuel thermal spray coatings. *Corros. Sci.* **2011**, *53*, 2106–2114. [CrossRef]
50. Mehta, J.; Mittal, V.K.; Gupta, P. Role of Thermal Spray Coatings on Wear, Erosion and Corrosion Behavior: A Review. *J. Appl. Sci. Eng.* **2017**, *20*, 445–452. [CrossRef]
51. Kosa, E.; Göksenli, A. Influence of Material Hardness and Particle Velocity on Erosive Wear Rate. *J. Mech. Eng.* **2017**, *47*, 1–14. [CrossRef]
52. Malik, J.; Toor, I.H.; Ahmed, W.H.; Gasem, Z.M.; Habib, M.A.; Mansour, R.B.; Badr, H.M. Evaluating the Effect of Hardness on Erosion Characteristics of Aluminum and Steels. *J. Mater. Eng. Perform.* **2016**, *23*, 2274–2282. [CrossRef]
53. Stack, M.M.; Jana, B.D.; Abdelrahman, S.M. Erosion-Corrosion Mechanisms and Maps. 2011. Strathprints. Available online: https://strathprints.strath.ac.uk (accessed on 5 July 2024).
54. Govande, A.R.; Chandak, A.; Sunil, B.R.; Dumpala, R. Carbide-based thermal spray coatings: A review on performance characteristics and post-treatment. *Int. J. Refract. Met. Hard Mater.* **2022**, *103*, 105772. [CrossRef]
55. Bhatia, A. Thermal Spraying Technology and Applications. 2024. Available online: https://www.cedengineering.com/userfiles/T04-002%20-%20Thermal%20Spraying%20Technology%20and%20Applications%20-%20US.pdf (accessed on 5 July 2024).
56. Brezinová, J.; Guzanová, A.; Draganovská, D.; Brezina, J. Characterization of selected properties of WC–WB–Co and WC–FeCrAl coatings. *Výzkumné Články Koroze A Ochr. Mater.* **2019**, *63*, 167–173. [CrossRef]

57. Richert, M.; Mikułowski, B.; Pałka, P.; Hotloś, A.; Perek-Nowak, M. The Effect of Chemical Composition and Thermal Sprayed Method on the Chromium and Tungsten Carbides Coatings Microstructure. *J. Surf. Eng. Mater. Adv. Technol.* **2013**, *3*, 1–5. [CrossRef]
58. Blum, M.; Krieg, P.; Killinger, A.; Gadow, R.; Luth, J.; Trenkle, F. High Velocity Suspension Flame Spraying (HVSFS)of Metal Suspensions. *Materials* **2020**, *13*, 621. [CrossRef]
59. Najari, M.R.; Sajjadi, S.A.; Ganji, O. Microstructural evolution and wear properties of chromium carbide coating formed by thermo-reactive diffusion (TRD) process on a cold-work tool steel. *Results Surf. Interfaces* **2022**, *8*, 1. [CrossRef]
60. Zong, X.; Jiang, W.; Fan, Z. Evaluation of Chromium Carbide Coatings on AISI 52100 Steel Obtained by Thermo-Reactive Diffusion Technique. *Mater. Sci. Medžiagotyra* **2018**, *25*, 1. [CrossRef]
61. Sun, L.; Ji, X.; Zhao, L.; Zhai, W.; Xu, L.; Dong, H.; Liu, Y.; Peng, J. First Principles Investigation of Binary Chromium Carbides Cr_7C_3, Cr_3C_2 and $Cr_{23}C_6$: Electronic Structures, Mechanical Properties and Thermodynamic Properties under Pressure. *Materials* **2022**, *15*, 558. [CrossRef]
62. Marques, A.S.; Dalcin, R.L.; Oliveira, L.F.; Vitor da Silva, L.A.; dos Santos, G.R.; da Silva Rocha, A. Comparative Analysis of the Friction and Microstructural Properties of WC-10Co-4Cr and Cr_3C_2-25NiCr Coatings Sprayed by High-Velocity Oxy-fuel (HVO). *Am. J. Mater. Sci.* **2018**, *8*, 51–57.
63. Wang, H.; Li, Y.; Gee, M.; Zhang, H.; Liu, X.; Sang, X. Sliding wear resistance enhancement by controlling W_2C precipitation in HVOF sprayed WC-based cermet coating. *Surf. Coat. Technol.* **2020**, *387*, 125533. [CrossRef]
64. Fan, L.; Peng Ou, P.; Rong, J.; Yu, X. Preparation of high wear resistance nickel based WC coating by carefully adjusting interface structure. *Mater. Res. Express* **2022**, *9*, 086402. [CrossRef]
65. Méndez-Medrano, K.O.; Martínez-González, C.J.; Alvarado-Hernández, F.; Jiménez, O.; Baltazar-Hernández, V.H.; Ruiz-Luna, H. Microstructure and Properties Characterization of WC-Co-Cr Thermal Spray Coatings. *J. Miner. Mater. Charact. Eng.* **2018**, *6*, 4. [CrossRef]
66. Staia, M.H.; Valente, T.; Bartuli, C.; Lewis, D.B.; Constable, C.P. Part I: Characterization of Cr_3C_2-25% NiCr reactive plasma sprayed coatings produced at different pressures. *Surf. Coat. Technol.* **2001**, *146-147*, 553–562. [CrossRef]
67. Katranidis, V.; Gu, S.; Cox, D.C.; Whiting, M.J.; Kamnis, S. FIB-SEM Sectioning Study of Decarburization Products in the Microstructure of HVOF-Sprayed WC-Co Coatings. *J. Therm. Spray Technol.* **2018**, *27*, 898–908. [CrossRef]
68. Ahmed, R.; Ali, O.; Berndt, C.C.; Fardan, A. Sliding Wear of Conventional and Suspension Sprayed Nanocomposite WC-Co Coatings: An Invited Review. *J. Therm. Spray Technol.* **2021**, *30*, 800–861. [CrossRef]
69. Yuan, J.; Ma, C.; Yang, S.; Yu, Z.; Li, H. Improving the Wear Resistance of HVOF Sprayed WC-Co Coatings by Adding Submicron-Sized WC Particles at the Splats' Interfaces. *Surf. Coat. Technol.* **2016**, *285*, 17–23. [CrossRef]
70. Kekes, D.; Psyllaki, P.; Vardavoulias, M. Wear Micro-Mechanisms of Composite WC-Co/Cr -NiCrFeBSiC Coatings. Part I: Dry Sliding. *Tribol. Ind.* **2014**, *36*, 361–374. Available online: https://www.tribology.rs/journals/2014/2014-4/4.pdf (accessed on 5 July 2024).
71. Wielage, B.; Pokhmurska, H.; Wank, A.; Reisel, G.; Steinhaeuser, S.; Woezel, M. Influence of thermal spraying method on the properties of tungsten carbide coatings. In Proceedings of the Conference on Modern Wear and Corrosion Resistant Coatings Obtained by Thermal Spraying, Warsaw, Poland, 20–21 November 2003.
72. Berger, L.-M. Binary WC– and Cr_3C_2–containing hardmetal compositions for thermally sprayed coatings. *IOP Conf. Ser. Mater. Sci. Eng.* **2016**, *118*, 012010. [CrossRef]
73. Fan, K.; Jiang, W.; Luzin, V.; Gong, T.; Feng, W.; Ruiz-Hervias, J.; Yao, P. Influence of WC Particle Size on the Mechanical Properties and Residual Stress of HVOF Thermally Sprayed WC-10Co-4Cr Coatings. *Materials* **2022**, *15*, 5537. [CrossRef] [PubMed]
74. Jonda, E.; Łatka, L. Comparative Analysis of Mechanical Properties of WC-Based Cermet Coatings Sprayed by HVOF onto AZ31 Magnesium Alloy Substrates. *Adv. Sci. Technol. Res. J.* **2021**, *15*, 57–64. [CrossRef]
75. Avcu, Y.Y.; Guney, M.; Avcu, E. High-velocity air fuel coatings for steel for erosion-resistant applications. *J. Electrochem. Sci. Eng.* **2023**, *13*, 407–420. [CrossRef]
76. Korobov, H.; Alwan, N.; Soboleva, A.; Makarov, N.; Lezhnin, V.; Shumyakov, M.; Antonov, M. Deviatiarov. *J. Therm. Spray Technol.* **2021**, *31*, 234–246. [CrossRef]
77. Siwak, P.; Garbiec, D. Microstructure and mechanical properties of WC−Co, WC−Co−Cr_3C_2 and WC−Co−TaC cermets fabricated by spark plasma sintering. *Trans. Nonferrous Met. Soc. China* **2016**, *26*, 2641–2646. [CrossRef]

materials

MDPI

Article

Physico-Chemical Properties of Copper-Doped Hydroxyapatite Coatings Obtained by Vacuum Deposition Technique

Yassine Benali [1], Daniela Predoi [2,*], Krzysztof Rokosz [3,*], Carmen Steluta Ciobanu [2], Simona Liliana Iconaru [2], Steinar Raaen [4], Catalin Constantin Negrila [2], Carmen Cimpeanu [5], Roxana Trusca [6], Liliana Ghegoiu [2], Coralia Bleotu [7,8,9], Ioana Cristina Marinas [8,10], Miruna Stan [11] and Khaled Boughzala [12]

1 Faculty of Sciences, University de Gafsa, Route de Tozeur, Gafsa 2112, Tunisia; yassin.bn.ali27@gmail.com
2 National Institute of Materials Physics, Atomistilor Street, No. 405A, 077125 Magurele, Romania; ciobanucs@gmail.com (C.S.C.); simonaiconaru@gmail.com (S.L.I.); catalin.negrila@infim.ro (C.C.N.); ghegoiuliliana@gmail.com (L.G.)
3 Faculty of Electronics and Computer Science, Koszalin University of Technology, Śniadeckich 2, PL 75-453 Koszalin, Poland
4 Department of Physics, Norwegian University of Science and Technology (NTNU), Realfagbygget E3-124 Høgskoleringen 5, NO 7491 Trondheim, Norway; steinar.raaen@ntnu.no
5 Faculty of Land Reclamation and Environmental Engineering, University of Agronomic Sciences and Veterinary Medicine of Bucharest, 59 Marasti Blvd, 011464 Bucharest, Romania; carmencimpeanu@yahoo.com
6 National Centre for Micro and Nanomaterials, University Politehnica of Bucharest, 060042 Bucharest, Romania; truscaroxana@yahoo.com
7 Department of Cellular and Molecular Pathology, Stefan S. Nicolau Institute of Virology, 030304 Bucharest, Romania; cbleotu@yahoo.com
8 Research Institute of the University of Bucharest (ICUB), University of Bucharest, 060023 Bucharest, Romania; ioana.cristina.marinas@gmail.com
9 The Academy of Romanian Scientist, 050711 Bucharest, Romania
10 Department of Microbiology, Faculty of Biology, University of Bucharest, 1-3 Aleea Portocalelor Str., District 5, 060101 Bucharest, Romania
11 Department of Biochemistry and Molecular Biology, Faculty of Biology, University of Bucharest, 91-95 Splaiul Independentei, 050095 Bucharest, Romania; miruna.stan@bio.unibuc.ro
12 Higher Institute of Technological Studies of Ksar Hellal, Ksar-Hellal 5070, Tunisia; khaledboughzala@gmail.com
* Correspondence: dpredoi@gmail.com (D.P.); rokosz@tu.koszalin.pl (K.R.)

Abstract: The hydroxyapatite and copper-doped hydroxyapatite coatings ($Ca_{10-x}Cu_x(PO_4)_6(OH)_2$; x_{Cu} = 0, 0.03; HAp and 3CuHAp) were obtained by the vacuum deposition technique. Then, both coatings were analyzed by the X-ray diffraction (XRD), scanning electron microscopy (SEM), atomic force microscopy (AFM), X-ray photoelectron spectroscopy (XPS), Fourier transform infrared spectroscopy (FTIR) and water contact angle techniques. Information regarding the in vitro antibacterial activity and biological evaluation were obtained. The XRD studies confirmed that the obtained thin films consist of a single phase associated with hydroxyapatite (HAp). The obtained 2D and 3D SEM images did not show cracks or other types of surface defects. The FTIR studies' results proved the presence of vibrational bands characteristic of the hydroxyapatite structure in the studied coating. Moreover, information regarding the HAp and 3CuHAp surface wettability was obtained by water contact angle measurements. The biocompatibility of the HAp and 3CuHAp coatings was evaluated using the HeLa and MG63 cell lines. The cytotoxicity evaluation of the coatings was performed by assessing the cell viability through the MTT assay after incubation with the HAp and 3CuHAp coatings for 24, 48, and 72 h. The results proved that the 3CuHAp coatings exhibited good biocompatible activity for all the tested intervals. The ability of *Pseudomonas aeruginosa* 27853 ATCC (*P. aeruginosa*) cells to adhere to and develop on the surface of the HAp and 3CuHAp coatings was investigated using AFM studies. The AFM studies revealed that the 3CuHAp coatings inhibited the formation of *P. aeruginosa* biofilms. The AFM data indicated that *P. aeruginosa*'s attachment and development on the 3CuHAp coatings were significantly inhibited within the first 24 h. Both the 2D and 3D topographies showed a rapid decrease in attached bacterial cells over time, with a significant reduction observed after 72 h of exposure. Our studies suggest that 3CuHAp coatings could be suitable candidates for biomedical uses such as the development of new antimicrobial agents.

Keywords: copper-doped hydroxyapatite; coatings; vacuum deposition; biocompatibility; antimicrobial activity

1. Introduction

In the last few years, significant efforts have been undertaken by the scientific community in order to enhance the biocompatibility of commonly used implant materials in orthopedics/stomatology. One potential solution purposed involves the deposition of bioactive coatings, such as hydroxyapatite (HAp), on the surface of the implants [1]. Hydroxyapatite is a bioceramic that has attracted researchers' attention because of its structural and chemical resemblance to the main inorganic component of bone tissue. Hydroxyapatite has been used as a coating material for metallic implants due to its excellent cytocompatibility, ability to stimulate cellular functions, and good osteoconductivity [2]. Nonetheless, previous studies have shown that natural hydroxyapatite, which is found in the mineral phases of bone, dentin, and enamel, possesses a chemical composition that is different from synthetic HAp [1–3]. Typically, natural HAp encompasses a wide range of trace elements, such as silicon (Si), fluorine (F), magnesium (Mg), strontium (Sr), zinc (Zn), etc., each possessing distinct biological features [4]. The appearance of post-surgical infections can prevent proper bone integration and cause tissue necrosis, posing serious health risks to patients and altering their quality of life [5–7]. The increase in antibiotic-resistant bacteria is driving research toward new solutions, including the use of metallic ions for their antibacterial properties [5,8]. It is well known that copper (Cu) is a crucial trace element for all organisms and that it is vital for a wide range of physiological functions, including energy production, respiration and tissue development [9,10]. On the other hand, it serves as an enzyme cofactor in many metabolic processes. Furthermore, copper is crucial for bone mineralization and for osteoblast activity [9–15]. Insufficient levels of copper are associated with a range of medical conditions, such as myeloneuropathy, leucopenia, anemia, and Menke's disease [10,13]. Moreover, high levels of Cu are harmful for the human body and lead to Wilson's disease [10,14]. Moreover, copper possesses excellent antimicrobial activity, having been used from ancient times by the Egyptians for maintaining the purity of water [10,15]. Therefore, nowadays, the copper ion is regarded as a promising doping agent due to its pronounced antibacterial activity and low cytotoxicity.

According to the study conducted by Hidalgo-Robatto and coworkers [16], infections related to prosthetic implants and medical devices are mainly caused by certain bacteria, including *Staphylococcus epidermidis* and *Staphylococcus aureus* (Gram-positive), as well as *Escherichia coli* and *Pseudomonas aeruginosa* (Gram-negative). Therefore, the enriching of biomaterials that cover implants/medical devices with antimicrobial agents could represent a promising alternative as a systemic treatment with antibiotics [16]. For example, *Pseudomonas aeruginosa* (*P. aeruginosa*), a Gram-negative bacterium, is known for its multidrug-resistant and extensively drug-resistant strains, which frequently cause severe infections [17]. This fact presents significant challenges in selecting effective antimicrobial treatments due to their resistance [17]. The study reported by Jacobs et al. [5], which was conducted on the copper-doped biphasic calcium phosphate materials, underlined their good antibacterial activity against Gram-positive (methicillin-resistant *Staphylococcus aureus* and methicillin-sensitive) and Gram-negative (*Pseudomonas aeruginosa* and *Escherichia coli*) strains [5]. Other similar studies that have previously been reported revealed that the antibacterial efficiency of copper-doped biomaterials against *Pseudomonas aeruginosa* is dose- and incubation time-dependent [10]. Information about copper-doped hydroxyapatite was provided in various studies conducted by Bazin et al. [18] and Noori et al. [19]. Thus, the in vitro biological studies conducted by Bazin et al. [18] on the MC3T3-E1 cell line revealed that the incorporation of copper into the sintered hydroxyapatite ceramics did not alter the proliferation and cell adhesion, behavior that confirms the sample's biocompatibility [18]. Furthermore, the results of the studies performed by Noori et al. [19]

underlined that copper-doped hydroxyapatite does not disrupt the differentiation of stem cells into osteoblasts, which is a crucial process for the regeneration of bone tissue [19]. Also, the biocompatibility of copper-doped hydroxyapatite is influenced by the copper content [18–20]. Previous studies [20] showed that concentrations of copper up to 5% typically promote cellular growth and functionality, whereas higher levels of copper can lead to toxic effects on cells [20]. The groundbreaking aspect of this work lies in the development of new antimicrobial and biocompatible coatings using 3CuHAp powders through a cost-efficient method: vacuum deposition. Moreover, another aspect underlining the novelty and originality of this work is the achievement (for the first time) of continuous and pure 3CuHAp layers through vacuum deposition.

This study aimed at the development for the first time of copper-doped hydroxyapatite coatings ($Ca_{10-x}Cu_x(PO_4)_6(OH)_2$; $x_{Cu} = 0.03$; 3CuHAp) by vacuum deposition. Also, this study explored the physico-chemical properties of 3CuHAp coatings, together with their toxic effects on cells and their antibacterial activity. Techniques such as X-ray diffraction (XRD), X-ray photoelectron spectroscopy (XPS), atomic force microscopy (AFM), scanning electron microscopy (SEM), Fourier transform infrared spectroscopy (FTIR) and water contact angle were employed in order to obtain valuable information about the features of the HAp and 3CuHAp coatings. In summary, this work pioneers a novel approach by combining superior physico-chemical and biological properties and cost efficiency in the development of coatings based on HAp and 3CuHAp. These coatings have the potential to improve medical devices/implants, enhancing overall health.

2. Materials and Methods

2.1. Materials

2.1.1. Synthesis of Hydroxyapatite (HAp) and Copper-Doped Hydroxyapatite (3CuHAp)

For the synthesis of undoped hydroxyapatite coatings ($Ca_{10}Cu(PO_4)_6(OH)_2$; Ca/P ratio equal with 1.67; referred to as HAp), the steps described in our previous work [21] were followed. Firstly, two solutions were obtained, containing 0.1 mol of $Ca(NO_3)_2{\cdot}4H_2O$ (Sigma Aldrich, St. Louis, MO, USA) and 0.06 mol of $(NH_4)_2HPO_4$ (Alfa Aesar, Karlsruhe, Germany), respectively. Then, the solution containing $Ca(NO_3)_2{\cdot}4H_2O$ was slowly added into the $(NH_4)_2HPO_4$ solution. During the synthesis, the pH value was maintained at 11. The resulting mixture was stirred at 100 °C for 4 h, then centrifuged and washed for 5 times. Finally, the HAp precipitate was dried at 100 °C and used for the vacuum deposition of the HAp coatings.

A similar procedure was used for obtaining of the copper-doped hydroxyapatite powders (with the chemical formula: $Ca_{10-x}Cu_x(PO_4)_6(OH)_2$; where $x_{Cu} = 0.03$; referred to as 3CuHAp). During the synthesis, the [Ca + Cu]/P ratio was maintained at 1.67. Briefly, the $Cu(NO_3)_2{\cdot}3H_2O$ (Alfa Aesar, Karlsruhe, Germany) was dissolved with the $Ca(NO_3)_2{\cdot}4H_2O$. The next steps were carried out identically to the those described for the hydroxyapatite powders' synthesis. Finally, the 3CuHAp precipitate was dried at 100 °C and used for the vacuum deposition of the 3CuHAp coatings.

2.1.2. Development of Hydroxyapatite (HAp) and Copper-Doped Hydroxyapatite (3CuHAp) Coatings

The deposition of the HAp and 3CuHAp coatings was performed on silicon (Si) substrates. Prior to the vacuum deposition process, the substrates underwent multiple cleanings with acetone and were air-dried at 40 °C. The deposition parameters for the coatings were consistent with the ones presented by Predoi et al. [22]. Briefly, the deposition process took place in a high vacuum environment (~10^{-6} mbar). The powders were thermally evaporated from a W boat, slowly increasing the current through them until the melting state became visible. Afterward, the melt was completely evaporated in about 120 s. Finally, the HAp and 3CuHAp coatings were thermally treated at 500 °C in air. Figure 1 presents a schematic representation of the synthesis, characterization techniques and key findings of the studies conducted on the HAp and 3CuHAp coatings.

Furthermore, the vacuum deposition technique used for the development of the 3CuHAp coatings is a versatile and cost-efficient method. While the setup cost is moderate, the ongoing operation and maintenance costs are manageable. Compared to other deposition techniques, vacuum deposition is cost-effective due to its simplicity and high deposition rate. Other deposition techniques use expensive equipment that requires considerable operation and maintenance costs.

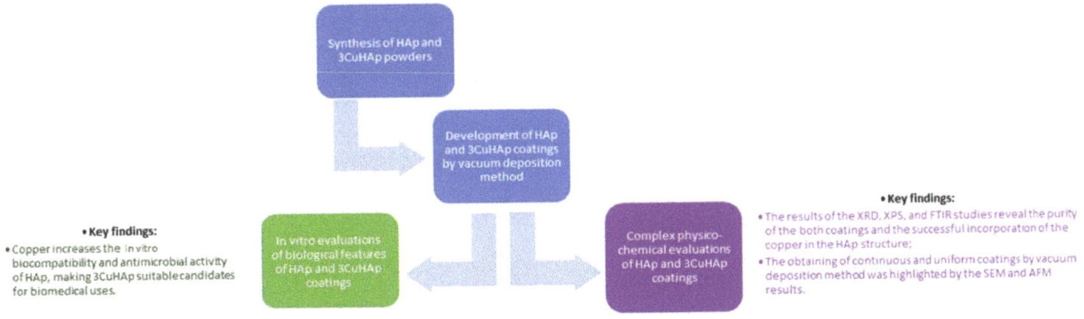

Figure 1. Schematic representation of the synthesis, characterization techniques and key findings.

2.2. *Characterizations of 3CuHAp Coatings*

2.2.1. X-ray Diffraction (XRD)

The X-ray diffraction (XRD) studies were conducted using a Bruker D8 Advance X-ray diffractometer (manufactured by Bruker in Karlsruhe, Germany). The XRD experimental data were acquired within the 20° to 60° (2θ) range, with a step size of 0.02°, using Cu Kα radiation (wavelength λ = 1.5418 Å).

2.2.2. Scanning Electron Microscopy (SEM)

The FEI Quanta Inspect F scanning electron microscope (manufactured by FEI Company, Hillsboro, OR, USA) was used in order to evaluate the surface morphology of both coatings. Additionally, the microscope is equipped with an energy-dispersive X-ray (EDS) attachment, which enabled the evaluation of the chemical composition of the HAp and 3CuHAp coatings. More than that, the thickness of both coatings was evaluated using SEM transversal cross-section images. The 3D SEM images were obtained with the aid of ImageJ 1.51j8 software [23].

2.2.3. Atomic Force Microscopy (AFM)

For this study, an atomic force microscope (AFM, NT-MDT NTEGRA Probe NanoLaboratory system, Moscow, Russia) operated in non-contact mode was used to obtain information about the surface features of the HAp and 3CuHAp coatings [24]. The AFM studies were performed using a silicon NT-MDT NSG01 cantilever. The AFM images were recorded on a surface area of 3×3 μm^2 and Gwyddion 2.55 software was used for their analysis [25].

2.2.4. X-ray Photoelectron Spectroscopy (XPS)

A multimethod SPECS surface analysis system (SPECS GmbH, Berlin, Germany) operating with Al Kα monochromatic radiation (1486.6 eV) was used in order to perform the X-ray photoelectron spectroscopy (XPS) studies. The experimental conditions were reported in previous research conducted by Iconaru et al. [26]. The XPS data were processed using Spectral Data Processor v. 2.3 (SDP) software. The binding energy of the calibration using C1s was 284.8 eV.

2.2.5. Fourier Transform Infrared Spectroscopy (FTIR)

The PerkinElmer Spectrum 100 FT-IR spectrometer (Waltham, MA, USA) is an instrument widely used for establishing the presence of functional groups in the structure of coatings. Experimental data were collected in the spectral range of 450–1200 cm^{-1}. Additionally, following the procedure detailed in [27], the second derivative spectra of the HAp and 3CuHAp coatings were obtained. The FTIR data were analyzed using the OriginPro 2021b software (OriginLab Corporation, Northampton, MA, USA).

2.2.6. Water Contact Angle Studies

Water contact angle studies were conducted under ambient conditions with the aid of a contact angle goniometer (DSA30 Kruess GmbH, Hamburg, Germany). The sessile drop method was used for the experiments. The HAp and 3CuHAp coatings underwent three repetitions of the contact angle measurement. The mean contact angle values are reported, along with the standard deviation (SD).

2.2.7. In Vitro Antibacterial Activity

The antibacterial properties of the HAp and 3CuHAp coatings were evaluated against the standard *Pseudomonas aeruginosa* 27853 ATCC bacterial strain. The in vitro antibacterial assays were conducted following the protocol outlined in [28]. The antibacterial activity of the 3CuHAp coatings was assessed after 24, 48, and 72 h of incubation with the *P. aeruginosa* 27853 ATCC bacterial suspensions. The quantitative measurements of the bacterial cell survival were recorded at these intervals. For this purpose, *P. aeruginosa* suspensions with a bacterial density of approximately 5×10^6 colony-forming units (CFUs)/mL, derived from 18–24 h cultures, were used in the study. The 3CuHAp coatings were incubated at 37 °C for 24, 48, and 72 h with these bacterial suspensions, and the bacterial growth was monitored over time. For each incubation period, the bacterial suspensions were collected and incubated on LB agar medium. A free bacterial suspension served as a positive control (C+). The CFU count per milliliter (CFU/mL) was determined and graphically represented as the log CFU/mL over time. The antibacterial experiments were conducted in triplicate, and the results are expressed as the mean $\pm$ standard deviation (SD). The qualitative evaluation of the bacterial cell adherence and proliferation on the surface of the 3CuHAp coatings was also studied using the atomic force microscopy (AFM) technique. For these assays, *P. aeruginosa* bacterial cells were cultured on the surface of the 3CuHAp coatings for three different time intervals. After each incubation period, the 3CuHAp coatings were removed from the culture medium, washed using sterile saline solution to remove the unattached bacterial cells, fixed with cold methanol, and prepared for visualization.

2.2.8. In Vitro Biological Evaluation

The cytotoxicity of the HAp and 3CuHAp coatings was assessed using both HeLa (ATCC CRM-CCL2) and osteosarcoma MG63 (ATCC CRL-1427) cells, following a similar methodology to that described by Iconaru et al. [29]. The cells were cultured in Dulbecco's modified Eagle's medium (DMEM) enriched with heat-inactivated fetal bovine serum at 37 °C in an atmosphere containing 5% CO_2. The HeLa and MG63 cells were seeded at a concentration of 1×10^5 cells/well in complete medium and incubated with the HAp and 3CuHAp coatings for 24, 48, and 72 h. The cytotoxicity was evaluated by measuring the cell viability using the MTT [3-(4,5-dimethylthiazolyl)-2,5-diphenyltetrazolium bromide] reduction assay. After the incubation periods, the cells were washed with phosphate-buffered saline (PBS) and incubated with 1 mg/mL MTT solution for 2.5 h. The cell viability was determined by measuring the optical density of the medium at 595 nm using a TECAN spectrophotometer. The percentage of viable HeLa and MG63 cells was calculated relative to a control sample, which was set to 100% viability. The HeLa and MG63 cells were seeded at a concentration of 1×10^5 cells/well in complete medium and maintained at 37 °C for 72 h. The morphology of the cells maintained together with the HAp and 3CuHAp coatings was observed under visible light using an Axio Observer Inverted microscope (Carl Zeiss, GMBH, Munich, Germany). In addition, the adhesion of the cells on the material was highlighted by fixing them with 70% ethanol, staining them with propidium iodide and photographing them with an Axio Observer D Inverted microscope equipped with a fluorescence module (Carl Zeiss, GMBH, Munich, Germany).

Furthermore, the cells were also visualized with the aid of an inversed trinocular metallographic microscope OX.2153-PLM (Euromex, Arnhem, The Netherlands). The metallographic microscope was equipped with an CMEX digital camera and the images were acquired with ImageFocusAlpha software (v 1.3.7.19728, Euromex, Arnhem, The Netherlands) using the 10× magnification objective. These studies were performed under ambient conditions.

3. Results and Discussion

The samples consisting of the HAp and 3CuHAp coatings obtained by the vacuum deposition process were analyzed from a structural point of view by X-ray diffraction (Figure 2).

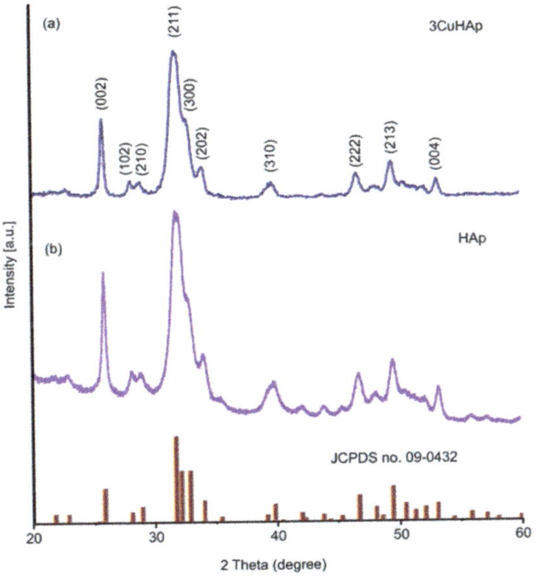

Figure 2. XRD pattern of the HAp (**b**) and 3CuHAp (**a**) coatings and JCPDS no. 09-0432.

The X-ray diffraction analysis of the deposited HAp and 3CuHAp coatings revealed the presence of hydroxyapatite (Figure 2a,b). This was highlighted by the diffraction pattern, which exhibited maxima associated with pure hexagonal hydroxyapatite (JCPDS no. 09-0432).

The diffraction peaks corresponded to specific crystallographic planes of the hexagonal HAp structure. These planes were (002), (210), (211), (300), (202), (310), (222), (213), (004), and (322). There were no peaks associated with impurities, indicating that the coatings were composed of a single phase of hydroxyapatite. The calculated lattice parameters were a = 9.393 Å and b = 6.869 Å for the 3CuHAp sample, while for the HAp coatings, the values were a = 9.416 Å and b = 6.881 Å. The values of the lattice parameters were lower compared to the values of pure hydroxyapatite (a = 9.418 Å and b = 6.884 Å). The cell volume in the case of the 3CuHAp sample was equal of 524.72 Å^3, while the cell volume of the HAp coatings was 528.32 Å^3. The cell volume of pure hexagonal HAp is 528.80 Å^3. For the HAp (x_{Cu} = 0.00) and 3CuHAp (x_{Cu} = 0.03) samples, the crystallite size was determined using Scherrer's equation [30–32]. The calculated crystallite size for the HAp sample was

19.45 nm, while for the 3CuHAp sample, it was 12.18 nm. The substitution of calcium (Ca) for copper (Cu) at the Ca (II) sites in the hydroxyapatite (HAp) lattice leads to a decrease in the lattice parameters and a contraction of the cell volume. This behavior aligns with findings from prior studies reported by Mariappan et al. [33].

Figure 3 shows the 2D and 3D SEM micrographs, together with the SEM particle size distribution histogram and SEM transversal cross-section image, obtained for HAp. The SEM images (both the 2D and 3D representations) reveal the presence of a continuous coating with no evidence of cracks. Furthermore, the mean particle size obtained by SEM for HAp was 22.1 nm. The transversal cross-section image obtained for the HAp coatings deposited by vacuum deposition on the Si substrate shows that their thickness was around 159.2 nm.

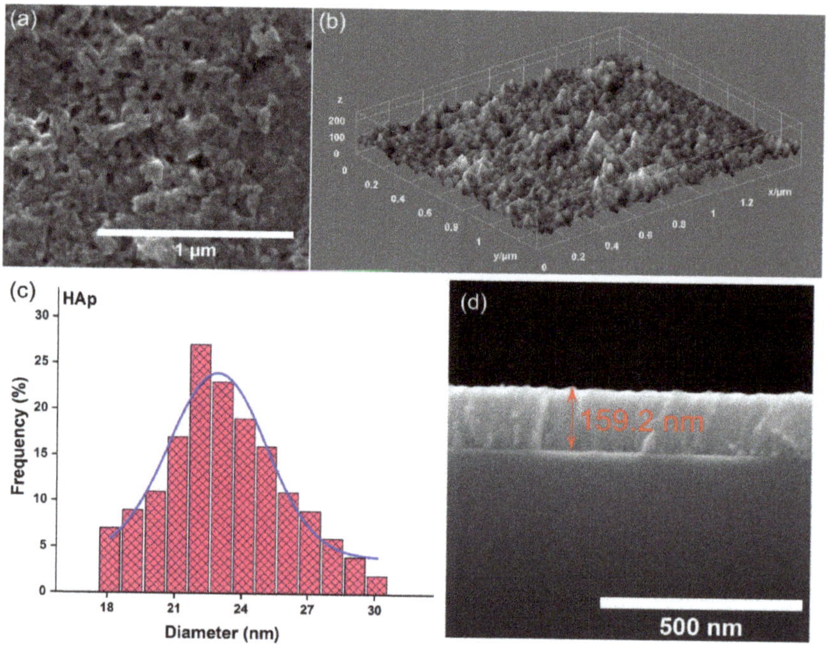

Figure 3. (**a**) SEM image (2D) of the HAp coatings; (**b**) SEM image (3D) of the HAp coatings; (**c**) SEM particle size distribution of the HAp coatings; and (**d**) SEM transversal cross-section image of the HAp coatings.

Both the 2D and 3D scanning electron microscope images acquired from the 3CuHAp coatings are presented (Figure 4a,b). Notably, the 3CuHAp coatings' surface exhibits an absence of cracks and fissures. Additionally, the SEM images highlight the continuous and homogeneous surface morphology of the 3CuHAp coatings. Importantly, no other surface defects are noticeable in the obtained 2D and 3D SEM images (Figure 4a,b). The mean particle size determined via SEM was 15.8 nm for the 3CuHAp. Moreover, the SEM transversal cross-section image suggests that the thickness of the 3CuHAp coatings was around 158.1 nm.

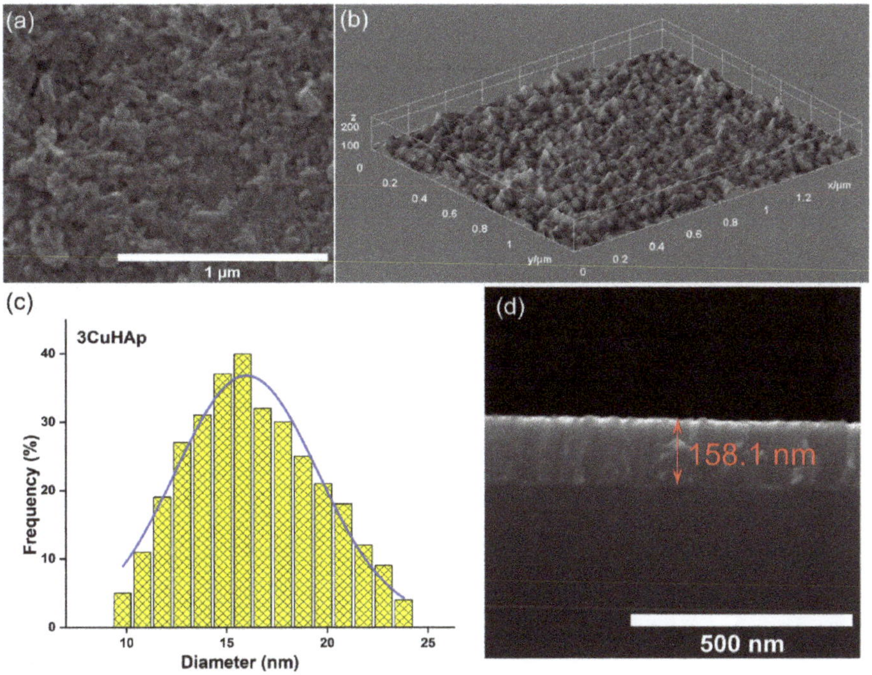

Figure 4. (**a**) SEM image (2D) of the 3CuHAp coatings; (**b**) SEM image (3D) of the 3CuHAp coatings; (**c**) SEM particle size distribution of the 3CuHAp coatings; and (**d**) SEM transversal cross-section image of the 3CuHAp coatings.

The EDS spectra obtained for the HAp and 3CuHAp coatings deposited on a silicon (Si) substrate by the vacuum deposition coating process are presented in Figure 5a,b. The results indicate that both coatings are pure. This conclusion is based on the absence of maxima in the EDS spectra that could be attributed to contaminants. In the EDS spectra specific to the HAp coatings (Figure 5b), only the lines associated with the next chemical elements that belong to the HAp structure are present: calcium (Ca), oxygen (O) and phosphorus (P). The EDS spectra of the 3CuHAp coatings revealed the presence of calcium (Ca), oxygen (O), phosphorus (P), and copper (Cu) chemicals. These elements collectively contribute to the composition of the $Ca_{10-x}Cu_x(PO_4)_6(OH)_2$ structure. Additionally, both EDS spectra showed an Si line. This Si signal originates from the silicon substrate on which the HAp and 3CuHAp coatings were deposited.

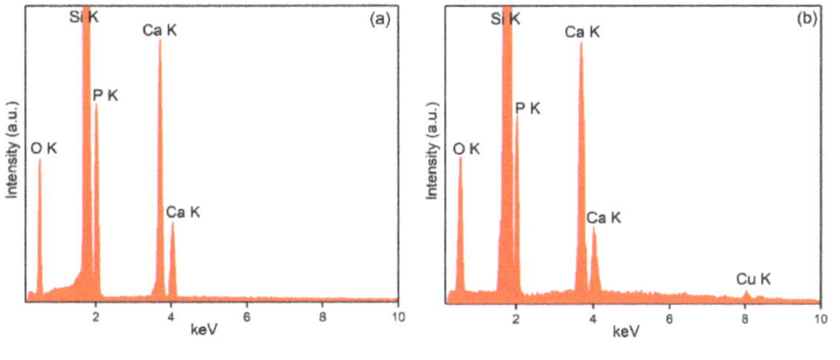

Figure 5. The EDS spectra obtained for the HAp (**a**) and 3CuHAp (**b**) coatings.

Atomic force microscopy (AFM) was used to examine the surface topography of the HAp and 3CuHAp coatings. The findings from the AFM studies (2D, together with their 3D representations) are illustrated in Figure 6a–d. The results of the AFM studies obtained for the hydroxyapatite coatings are presented in Figure 6a,b and indicate the presence of a uniform coating consisting of nanoagregates that are well distributed on the surface. The AFM images (2D and their 3D representations) reveal that no important cracks or fissures could be noticed on the HAp coatings' surface. The value obtained for the roughness parameter (R_{RMS}) by AFM for the HAp coatings was 19.02 nm.

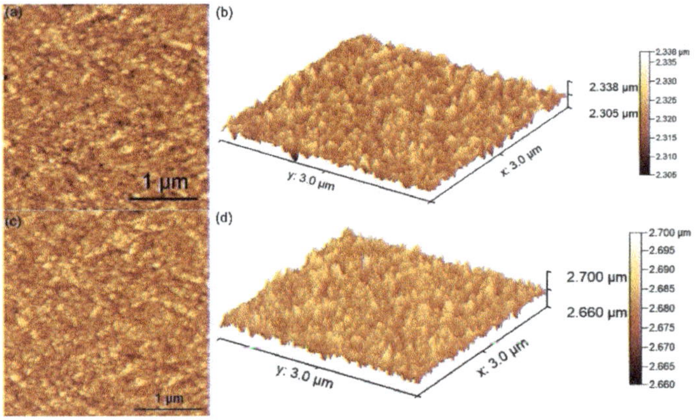

Figure 6. (**a**) The 2D and (**b**) 3D AFM images obtained for the HAp coatings; and (**c**) the 2D and (**d**) 3D AFM images obtained for the 3CuHAp coatings.

Both the 2D and 3D AFM micrographs reveal the surface topography of the 3CuHAp coatings (Figure 6c,d). The AFM analysis confirms that the studied coating displays a uniform and continuous morphology. Additionally, the 2D surface images clearly show an absence of cracks or fissures, indicating that the nanoaggregates are evenly distributed across the surface of the 3CuHAp coatings. Furthermore, the value obtained from the AFM studies for the roughness parameter (R_{RMS}) was equal to 18.73 nm. As can be seen, the value of R_{RMS} for the 3CuHAp coatings is smaller compared with the value of R_{RMS} obtained for the HAp coatings. This behavior indicates a decrease in roughness in the case of 3CuHAp and could be attributed to the HAp lattice distortion induced by the copper substitution.

Supplementary information about the roughness of the HAp and 3CuHAp coatings was obtained from the AFM images. The values of the roughness parameters (R_a and R_q), were obtained from the AFM images and are presented in Table 1. The value obtained for the roughness average (R_a) parameter was 15.43 nm for HAp and 15.16 nm for 3CuHAp. On the other hand, the values obtained for the root mean square (RMS) roughness (R_q) were nearly equal, with 19.02 nm for HAp and 18.73 nm for 3CuHAp.

Table 1. Roughness parameters of the HAp and 3CuHAp coatings' surface obtained by AFM.

Sample	R_q (nm)	R_a (nm)
HAp	19.02	15.43
3CuHAp	18.73	15.16

Table 2 depicts the values obtained for the roughness parameters (R_a and R_q), as quantified from the SEM images. For the determination of the roughness parameters from the SEM images, the images were first preprocessed and the Z height was correlated from the AFM images. The results obtained from the SEM images concerning the roughness parameters of the HAp and 3CuHAp coatings are similar and could be correlated with the results determined from the AFM surface topographies.

Table 2. Roughness parameters of the HAp and 3CuHAp coatings' surface obtained by SEM.

Sample	R_q (nm)	R_a (nm)
HAp	19.87	15.45
3CuHAp	18.79	14.78

For the HAp coatings, the value obtained for the roughness average (R_a) parameter was 15.45 nm. Meanwhile for the 3CuHAp, the roughness average (R_a) parameter was 14.78 nm. The root mean square (RMS) roughness (R_q) values were 19.87 nm for HAp and 18.79 nm for 3CuHAp.

The values obtained for the R_a and R_q parameters (from AFM and SEM images) indicate that the studied samples exhibit a low roughness. Moreover, in the case of the 3CuHAp sample, a slight decrease in the roughness parameters could be noticed, a fact that could be attributed to the presence of copper ions in the HAp structure. Furthermore, according to the study reported by Osman et al. [34], a surface with a lower roughness could promote superior biological properties.

To characterize the surface of the HAp and 3CuHAp coatings, X-ray photoelectron spectroscopy was used. Figure 7 presents the XPS survey spectra of the of the HAp ($x_{Cu} = 0$) coatings. The qualitative analysis revealed the presence of key constituents of hydroxyapatite (Figure 7a), including calcium (Ca), phosphorus (P) and oxygen (O). The presence of copper (Cu) was observed in the general spectrum of 3CuHAp (Figure 7b). The results are in agreement with previous studies [35]. These results align with those achieved from the energy-dispersive X-ray spectroscopy (EDS) analysis, attesting to the presence of copper in the examined samples.

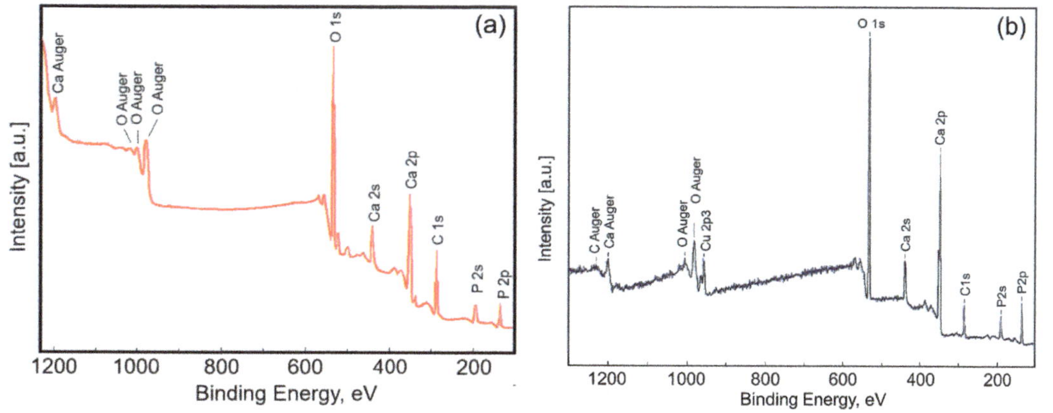

Figure 7. XPS survey scan of the HAp (**a**) and 3CuHAp (**b**) coatings obtained by the vacuum deposition process.

High-energy resolution analysis of the individual peaks was performed. Thus, the high-resolution spectra of HAp for carbon, oxygen, calcium and phosphorus are presented in Figure 8. The measured binding energy (EB) scale was referenced to a C 1s at the EB value of 284.8 eV [36]. Figure 8 reveals the high-resolution C 1s spectra for the pure HAp ($x_{Cu} = 0$) coatings. As can be seen, we have only one component that was highlighted at 284.8 eV. This peak is associated with residual or accidental carbon. Serra et al. [37] associated the 284.8 eV peak with C–C and C–H bonds. The high-resolution XPS spectrum of O 1s for the HAp coatings shows one peak located at 531.4 eV (Figure 8b). The peak at

531.4 eV is associated with the hydroxyl groups resulting from the chemisorption of water or oxygen. The results are in line with previous studies by Moulder et al. [38]. The Ca 2p high-resolution spectrum of the HAp coatings exhibited a well-resolved doublet with two components assigned to Ca 2p3/2 and Ca 2p1/2 (Figure 8c). The peak located at about 347.2 eV shows that the calcium atoms are bound to a phosphate group (PO_4^{3-}). Following the processing of the deconvolution data, the P 2p photoelectron line consists of one single peak at a binding energy of 133.1 eV. In accordance with past XPS studies [39], the binding energy of the photoelectron peaks for Ca and P is characteristic of their full oxidation states (Ca^{2+} and P^{5+}) for hydroxyapatite.

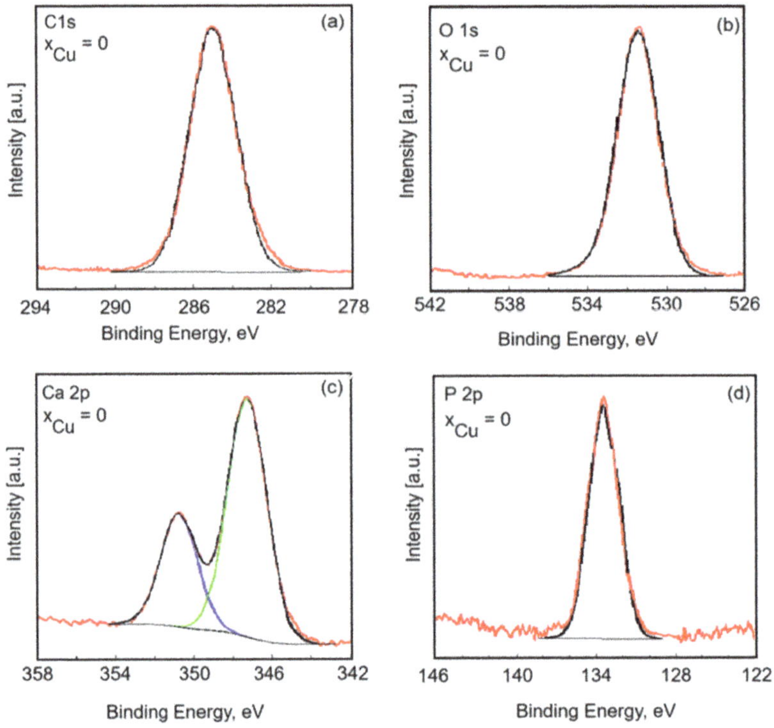

Figure 8. High-resolution XPS spectra of C1s (**a**), O 1s (**b**), Ca 2p (**c**) and P 2p (**d**) of the HAp coatings obtained by the vacuum deposition process.

The high-resolution spectra of the 3CuHAp sample are presented in Figure 9. The high-resolution XPS spectrum of C 1s of the 3CuHAp sample shows three peaks identified at binding energies of 284.8, 286.4 and 289.2 eV. The peak at 284.8 eV is attributed to C–C and C–H single bonds [40,41]. The peaks at 286.4 eV and 289.2 eV are assigned to C–O–C bonds and O–C=O bonds [40,41]. The high-resolution XPS spectrum of O 1s for the 3CuHAp thin film shows three peaks located at 530.9 eV, 532.3 eV and 533.9 eV (Figure 9b). The peak observed at 530.9 eV is attributed to HAp (the lattice oxygen of the P O species). The peak identified at 532.3eV is associated with oxygen adsorbed on the HAp surface [42,43]. The peak located at the binding energy of 533.9 eV could be due to possible traces of water. The high-resolution spectrum for calcium (Figure 9c) shows two peaks, Ca 2p 3/2 and Ca 2p ½, which are identified at 347.24 eV and 350.79 eV, being associated with the tetravalent state (Ca^{2+}), in agreement with previous studies [42]. The high-resolution spectrum of P2p for the 3CuHAp thin film is presented in Figure 9d. The P 2p 3/2 and P 2p 1/2 peaks were located at 133.20 eV and 134.19 eV, respectively. The given information states that the ratio of the areas is 2:1, and the separation occurs at an energy of 0.9 eV. The bond-binding energy is specific to PO_4 in hydroxyapatite. In Figure 9e, the high-resolution XPS spectra of copper (Cu) reveal three distinct maxima resulting from deconvolution. These peaks occur at specific binding energies of 936.16 eV, 941.79 eV and 945.05 eV. The peak identified at 936.16 eV corresponds to the Cu2p3/2 state. The peaks noticed at 941.79 eV and 945.05 eV are shake-up peaks, suggesting the presence of Cu^{2+} species [44–46]. The peaks associated with Cu^{1+} [47] and CuO at binding energies of 952.3 eV and 933.3 eV [48] were not identified.

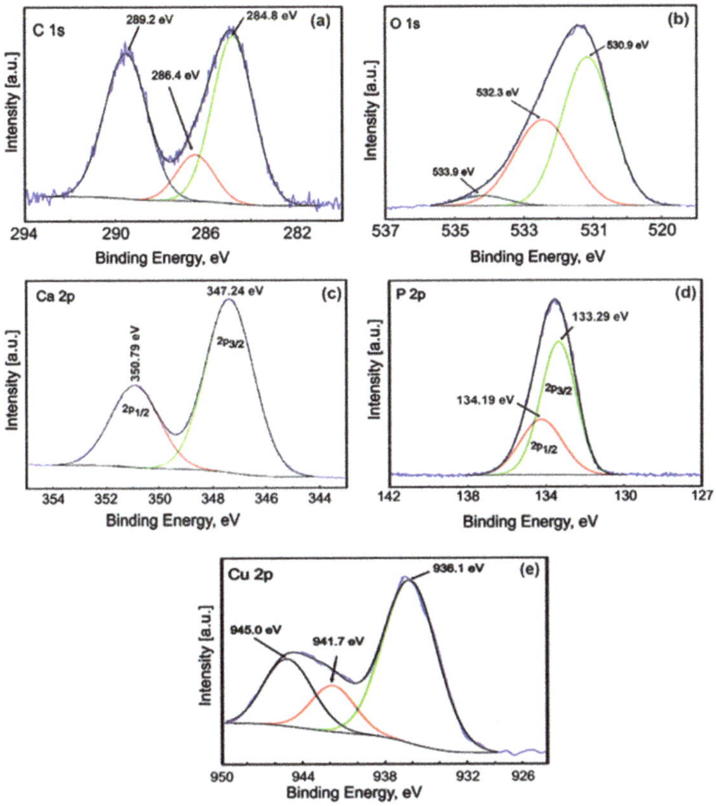

Figure 9. High-resolution XPS spectra of C1s (**a**), O 1s (**b**), Ca 2p (**c**), P 2p (**d**) and Cu 2p (**e**) of the 3CuHAp coatings obtained by the vacuum deposition process.

Figure 10 presents the absorption FTIR spectra obtained for the HAp and 3CuHAp coatings. The FTIR spectra of the HAp coatings reveal the presence of the main characteristic vibrational bands of the phosphate and hydroxyl groups from the hydroxyapatite structure. Due to the similarity, in the following, only the maxima noticed in the FTIR spectra of 3CuHAp will be discussed.

The FTIR spectra are dominated by two intense vibrational bands that are centered at ~563 cm^{-1} (ν_4) and ~1030 cm^{-1} (ν_3) and are characteristics of phosphate groups from HAp. Other vibrational bands that are attributed to phosphate groups' vibrations are observed at ~460 cm^{-1} (ν_2), ~605 cm^{-1} (ν_4), ~962 cm^{-1} (ν_1) and ~1096 cm^{-1} (ν_3) [16,27,49–55]. The maxima observed at ~962 cm^{-1} (ν_1) clearly indicate the presence of HAp in the 3CuHAp coatings [16,27,49–55]. On the other hand, the vibration peak specific to OH^- groups from HAp appears at ~631 cm^{-1}. Furthermore, in the FTIR spectra obtained for the undoped HAp coatings presented in Figure 10, the presence of maxima could be observed that are associated with the same vibrational groups as the one observed for the 3CuHAp coatings. Also, the doping of HAp with copper led to a decrease in the FTIR maxima, together with a slight displacement of their position. No other significant vibrational bands are noticed in the FTIR spectra, a fact that indicates the coatings' purity. Therefore, the FTIR results are in concordance with the XRD results.

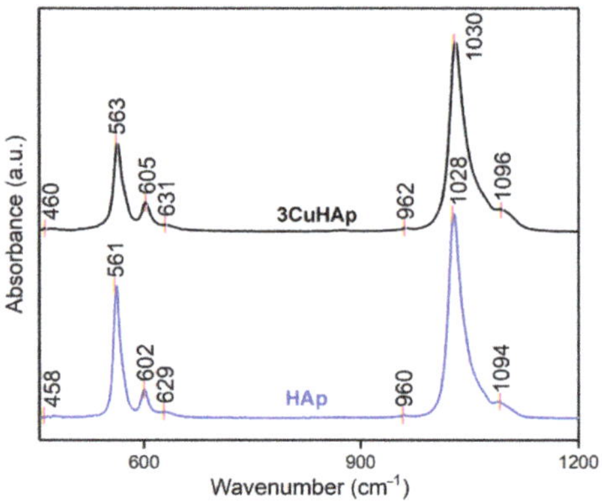

Figure 10. The FTIR spectra of the HAp and 3CuHAp coatings.

To acquire detailed information about the vibrational bands present in the HAp and 3CuHAp samples, we conducted an FTIR second-derivative analysis. The obtained second-derivative spectra (Figure 11), which pertain to the spectral regions characteristic of the ν_4, ν_3, ν_2, and ν_1 phosphate group vibrations, are illustrated in Figure 11. The ν_1 vibration band of the phosphate group could be observed at around 962 cm^{-1} in the second-derivative spectra [16,27,49–55]. Another vibrational band observed in the Figure 11 belongs to the ν_4 (observed at around 563 cm^{-1}, 574 cm^{-1}, 587 cm^{-1} and 605 cm^{-1}) and ν_3 (noticed at about 1030 cm^{-1}, 1041 cm^{-1}, 1059 cm^{-1} and 1096 cm^{-1}) vibrations of the phosphate group [16,27,49–55]. Furthermore, the presence of the vibrational peak associated with the vibration of the hydroxyl group (librational mode) could be noticed in the spectra. Also, the band specific to the ν_2 vibration could be observed in the spectra at around 460 cm^{-1} [16,27,49–55]. Table 3 summarizes the position of the vibrational bands observed in the second-derivative spectra obtained for the HAp and 3CuHAp coatings that are similar to the one observed in the FTIR spectra.

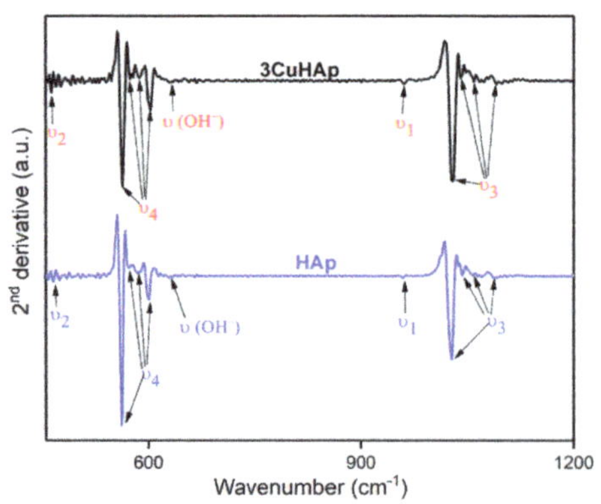

Figure 11. FTIR second-derivative spectra of the HAp and 3CuHAp coatings.

Table 3. IR wavenumber positions (cm^{-1}) specific to the ν_1, ν_2, ν_3, and ν_4 phosphate bands, as identified in the FTIR second-derivative spectra of the HAp and 3CuHAp coatings.

Assignments	Position (cm^{-1})	
	HAp	3CuHAp
Hydroxyl group	629	631
(ν_2) phosphate groups	458	460
(ν_4) phosphate groups	561, 573, 586, 602	563, 574, 587, 605;
(ν_1) phosphate groups	960	962
(ν_3) phosphate groups	1028, 1039, 1057, 1094	1030, 1041, 1059, 1096;

As can be seen in both second-derivative spectra, no intense vibrational bands were observed that could be associated with the presence of a supplementary phase or impurities. These results are consistent with those obtained by the XRD studies.

Wettability is a key property of implant surfaces that affects the cell–material interactions. Research has shown that hydrophilic surfaces promote enhanced mineral deposition and osteoblast cell growth in comparison to hydrophobic surfaces [52,56]. Therefore, in order to obtain valuable information about the wettability properties of the HAp and 3CuHAp coatings, water contact angle studies were performed (Figure 12). The mean value of the contact angle obtained for HAp was equal to 18.30 ± 5.7°. Meanwhile, the mean value obtained for the contact angle of the 3CuHAp coatings was 19.84 ± 6.6°. These values indicate the presence of a surface with a hydrophilic nature. As can be seen in Figure 12, by doping the hydroxyapatite with copper, a slight increase in the contact angle occurs. This behavior is consistent with the Wenzel's relation for hydrophilic surfaces, which states that increasing the surface roughness will decrease the water contact angle value [57,58]. Such hydrophilic surfaces are known to enhance the biological properties of the materials, including the bioactivity and bone-bonding behavior [52,56]. Moreover, these results suggest that copper-doped hydroxyapatite coatings, when deposited through the vacuum deposition method, hold significant potential for improving the performance and safety of biomedical implants.

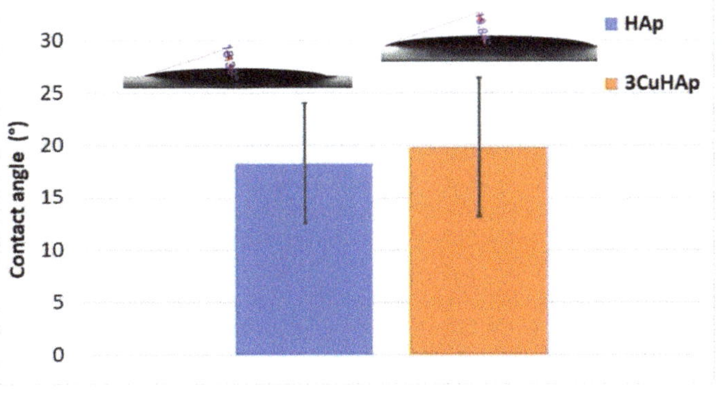

Figure 12. Water contact angle of the HAp and 3CuHAp coatings.

The cytotoxicity of the HAp and 3CuHAp coatings was assessed using both the HeLa and MG63 cell lines by determining the cell viability with the aid of an MTT assay after 24, 48, and 72 h of incubation with the 3CuHAp coatings. The results of the MTT assays, as illustrated in Figure 13, are presented as the mean ± standard deviation (SD) from three experiments, with the cell viability expressed as a percentage of the control. The control was considered 100%. Statistical significance was determined using the ANOVA single-factor test, with $p \leq 0.05$ considered to be significant. The data indicated that the viability of the HeLa and MG63 cells was not significantly different from the control after 24, 48, and 72 h of incubation with the HAp and 3CuHAp coatings, demonstrating good biocompatibility.

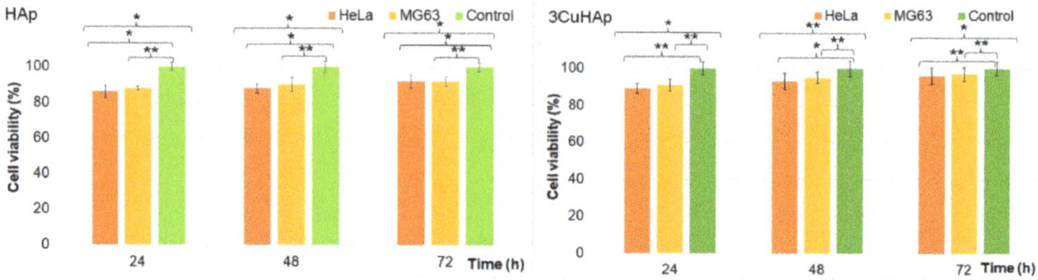

Figure 13. Graphical representation of the cell viability of HeLa and MG63 cells exposed to the HAp and 3CuHAp coatings for 24, 48 and 72 h. The results are depicted as the mean ± standard deviation (SD) and quantified as percentages of the control (100% viability). The ANOVA single-factor test was used for the statistical analysis and $p \leq 0.05$ was accepted as statistically significant (* $p < 0.03$, ** $p < 0.001$).

The viability of the HeLa and MG63 cells incubated with the HAp coatings was above 86% in the first 24 h and increased with the exposure time, reaching 92% after 72 h of exposure.

In the case of the 3CuHAp coatings, the cell viability remained above 88% after the initial 24 h and increased to 93% after 48 h and 95% after 72 h in the case of the HeLa cells. The result regarding the cell viability obtained for the MG63 cells was above 90% in the first 24 h and increased to 94% after 48 h and 96% after 72 h. These results are consistent with other reported studies regarding the biocompatibility of hydroxyapatite and hydroxyapatite composite coatings [16,18,59–65]. Furthermore, the cell viability of the HeLa and MG63 cells after exposure to the 3CuHAp coatings was consistently above 88%, exceeding the 70% cytotoxicity threshold set by ISO standard 10993-5 [59]. These findings suggest that the 3CuHAp coatings do not present a cytotoxic effect against HeLa and MG63 cells and could be promising candidates for the development of advanced implant materials with good biocompatibility.

Additional information regarding the biocompatibility of the HAp and 3CuHAp was achieved by visual observation of the morphology of the cells exposed to the HAp and 3CuHAp coatings for 72 h. The results revealed that the morphology of the HeLa and MG63 cells remained unchanged when cultured in the presence of both the HAp and 3CuHAp coatings (Figure 14). Moreover, the results also emphasized that the HeLa and MG63 cells seeded on the HAp and 3CuHAp coatings were allowed to proliferate for 72 h without altering their morphology. This observation supports the notion that the HAp and 3CuHAp coatings are biocompatible and could facilitate cell proliferation and tissue regeneration. Moreover, the results revealed that the integration of the implant into the surrounding tissue is possible. These biocompatible properties are attributed to both HAp, which is renowned for its osteoconductive nature and ability to mimic the mineral properties of bones, as well as the presence of the copper ions that form the HAp matrix, which plays an important role in bone growth [1,10,16,18,60–67].

Additional information regarding the adhesion and proliferation of the HeLa and MG63 cells on the surface of the HAp and 3CuHAp coatings was obtained by fluorescence microscopy evaluation. For this purpose, the cells were allowed to grow on the surface of the HAp and 3CuHAp coatings and stained before observation. The images from the fluorescence microscopy studies are shown in Figure 15a–d. Furthermore, metallographic microscopy was used to assess the adherence and proliferation of the cells on the HAp and 3CuHAp coatings. The results of the metallographic microscopy are presented in Figure 15e–h.

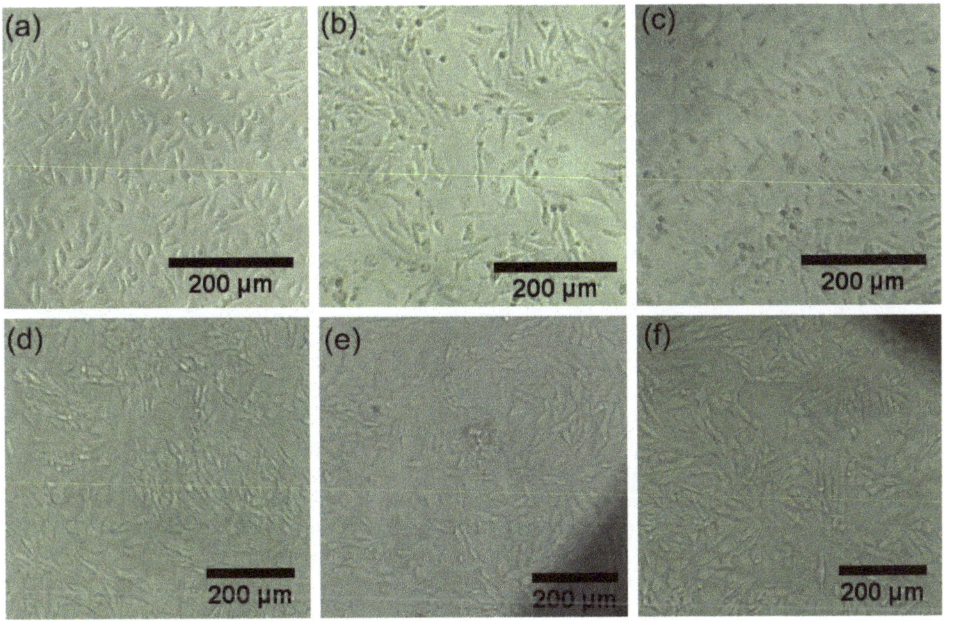

Figure 14. The morphology of HeLa and MG63 cells exposed to the HAp (**b**,**e**) and 3CuHAp (**c**,**f**) coatings for 72 h. HeLa control cells (**a**) and MG63 control cells (**d**).

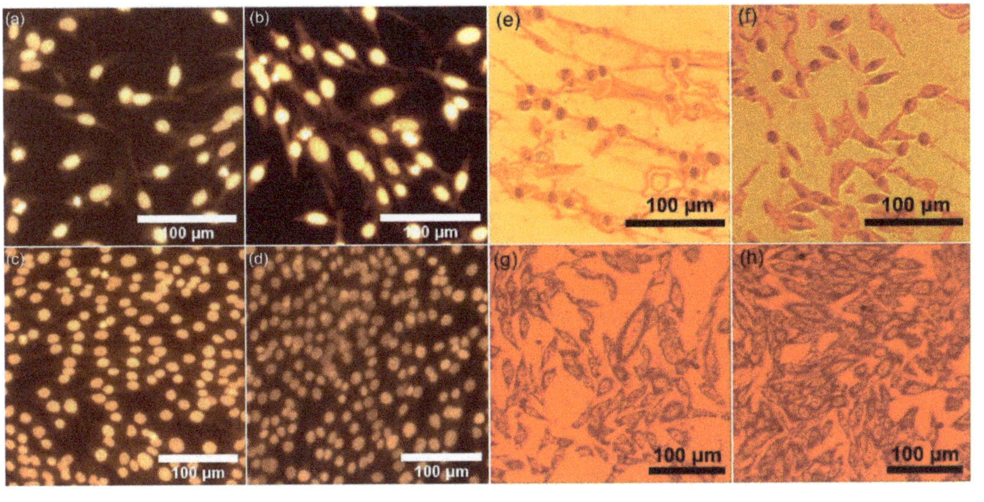

Figure 15. The morphology of HeLa and MG63 cells grown on the HAp coatings (**a**,**c**) and 3CuHAp coatings (**b**,**d**) visualized by fluorescence microscopy evaluation and the morphology of HeLa and MG63 cells grown on the HAp coatings (**e**,**g**) and 3CuHAp coatings (**f**,**h**) visualized by metallographic microscopy.

The results of the visual observation revealed that the surfaces of both the HAp and 3CuHAp coatings allowed the studied cells to adhere and to develop. Furthermore, no morphological abnormalities were detected in the HeLa and MG63 cells that adhered to the surface of the coatings. Our studies reveal that the HeLa and MG63 cell viability increases with the time of exposure (from 24 h to 72 h) to the HAp and 3CuHAp coatings. The cell viability increases over time in the presence of the HAp and 3CuHAp coatings due to their biocompatibility, bioactivity, and ability to support cell adhesion, proliferation, and differentiation. Even though there are several specific mechanisms that are involved in cell adhesion and proliferation, depending on the cell type and the conditions of the experiment, the overall positive interaction between the cells and the HAp and 3CuHAp coatings is a consistent theme. Hydroxyapatite closely resembles the mineral component of bone, making HAp coatings well tolerated by various cells and tissues. They are bioactive, promoting cell adhesion, proliferation, and differentiation, and serve as a source of essential calcium and phosphate ions, enhancing cellular processes. The surface properties of HAp also support cell adhesion and protein adsorption, further enhancing the cell viability. HAp is also known to promote osteogenic differentiation in certain cell types, leading to increased cell function. Additionally, the low toxicity and ability to stabilize cell membranes reduces cell death and improves overall cell health. On the other hand, copper ions could increase the cellular viability by acting as essential cofactors for enzymes involved in various essential biological processes, such as cellular respiration, antioxidant defense, and collagen synthesis. Copper ions have the ability to protect cells from oxidative stress, support ATP production via the mitochondrial electron transport chain, and facilitate collagen and elastin cross-linking, which is crucial for tissue integrity. Furthermore, copper is also reported to be responsible for promoting angiogenesis, enhancing nutrient and oxygen supply. More than that, copper ions support immune function and regulate both the gene expression and signal transduction pathways, all of which collectively enhances cell growth, differentiation, and survival [10–13,68]. In this context, the increase in the HeLa and MG63 cell viability with the incubation time in the presence of the HAp and 3CuHAp coatings could be attributed to several factors: enhanced cell adhesion, improved nutrient exchange, and increased bioactivity, protection against cytotoxic agents, gradual ion release, and reduced oxidative stress. Together, these factors help to create a supportive environment for cells' growth and survival.

Furthermore, to obtain a more comprehensive image of the biological properties of the 3CuHAp coatings, their antibacterial activity was also assessed against *Pseudomonas aeruginosa*, a common Gram-negative bacterial strain that is responsible for infections that are tough to treat in the blood, lungs, or other parts of the body.

The ability of *Pseudomonas aeruginosa* cells to adhere to and develop on the surface of 3CuHAp coatings was investigated with the aid of AFM studies, with the aim of emphasizing the role of copper ions in combating bacterial growth. For this purpose, the AFM topographies of the 3CuHAp coatings were recorded after incubation with a *P. aeruginosa* bacterial suspension at various time intervals (24, 48, and 72 h), under ambient conditions and at room temperature. The 2D surface topographies of the 3CuHAp coatings exposed to the *P. aeruginosa* bacterial suspensions were captured in non-contact mode over an area of $10 \times 10\ \mu m^2$. The 2D AFM images of the 3CuHAp coatings, incubated with *P. aeruginosa* bacterial suspensions for 24, 48, and 72 h, along with their 3D representations, are displayed in Figure 16a–f.

The AFM recordings of the 2D surface topography of the 3CuHAp coatings demonstrated that these coatings effectively inhibited *P. aeruginosa* bacterial cell adherence and growth, even in the early stages of development. The AFM studies further highlighted the coatings' ability to prevent *P. aeruginosa* biofilm formation on their surfaces. The adhered bacterial cells exhibited the characteristic rod-shaped morphology of *P. aeruginosa*, with lengths ranging from 1.03 to 2.57 μm and widths from 0.54 to 0.75 μm. The AFM data indicated that *P. aeruginosa* attachment and development on the 3CuHAp coating surfaces were significantly inhibited even within the first 24 h of incubation. Additionally, both the 2D topography and 3D representations suggested that the antibacterial activity of the coatings is correlated with the incubation time, showing a rapid decrease in the number of attached bacterial cells over time. After 72 h of exposure to the bacterial suspensions, the number of attached bacterial cells decreased significantly. The cell count performed on the AFM 2D images revealed that after 24 h of exposure, only 16 bacterial cells remained attached to the 3CuHAp surface, while after 48 h, only 9 bacterial cells were still adhered on the surface. After 72 h of incubation, both the 2D AFM topography as well as the 3D representation emphasized the presence of only a few isolated *P. aeruginosa* bacterial cells on the 3CuHAp coatings' surface (the cell count revealed the presence of approximately three isolated bacterial cells). In addition, the AFM studies revealed that the 3CuHAp coatings inhibited *P. aeruginosa* biofilm formation.

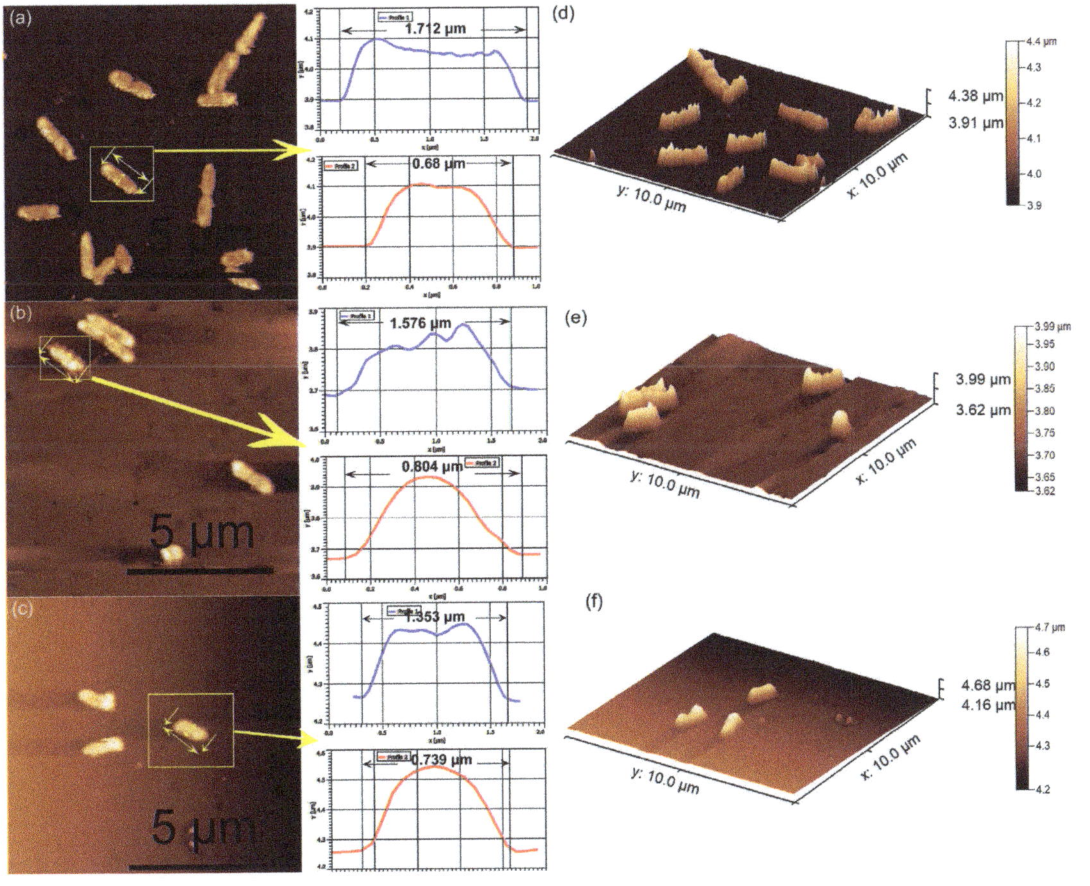

Figure 16. Two-dimensional AFM topography of *Pseudomonas aeruginosa* 27853 ATCC cells attached to the surface of the 3CuHAp coatings after a 24 (**a**), 48 (**b**) and 72 h (**c**) incubation period and their 3D representation (**d**–**f**). Individual bacterial cells chosen and their measured profile in width and in length, where the measurement is pointed out by the yellow arrow.

The antibacterial activity of the HAp coatings against *P. aeruginosa* was also evaluated. The results of the AFM topography recorded on the surface of the HAp coatings demonstrated that these coatings promoted the adhesion and development of *P. aeruginosa* on their surface, even from the early stage of the cell's development. Furthermore, the data also showed that the HAp coatings provided the *P. aeruginosa* bacterial cells with an adhesive surface and allowed the development of bacterial biofilm. Also, the data suggested that the development of the bacterial cells on the surface of the HAp coatings increased with the incubation time. The data depicting the 2D topographies as well as the 3D representations of the HAp coatings' surfaces after incubation with *P. aeruginosa* bacterial suspensions for 24, 48 and 72 are presented in Figure 17a–f. The results emphasized that since the HAp coatings did not exhibit any inhibitory activity against the *P. aeruginosa* bacterial cells, the antibacterial activity of the 3CuHAp could be attributed to the copper ions present from the HAp lattice. Furthermore, the data emphasized that the cells adhered to the entire HAp coating's surface.

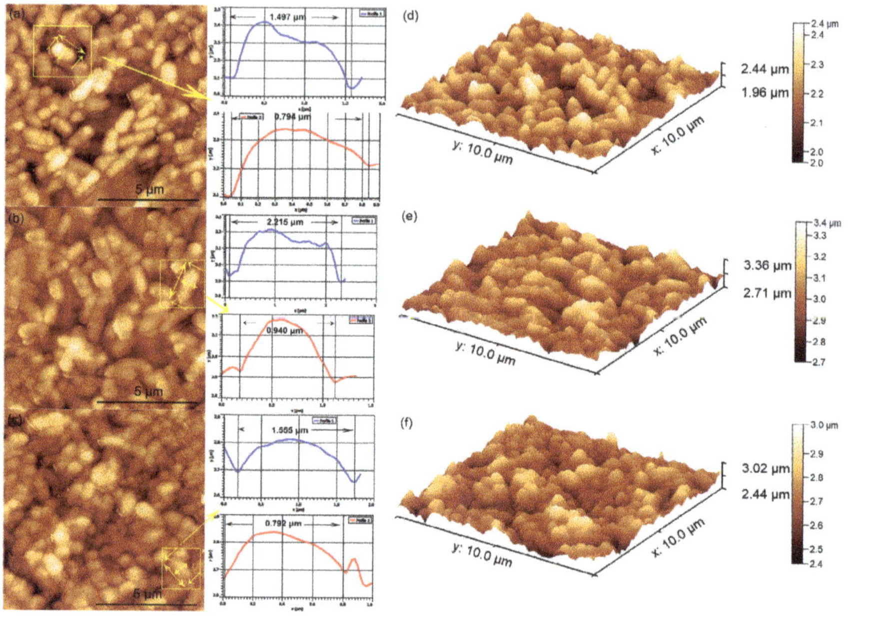

Figure 17. Two-dimensional AFM topography of *Pseudomonas aeruginosa* 27853 ATCC cells attached to the surface of the HAp coatings after a 24 (**a**), 48 (**b**) and 72 h (**c**) incubation period and their 3D representation (**d–f**). Individual bacterial cells chosen and their measured profile in width and in length, where the measurement is pointed out by the yellow arrow.

An individual analysis was also performed from the AFM data, which provided important quantitative measurements regarding the attached bacterial cell dimensions and surface roughness. The cross-section measurements along the lines depicted in Figures 16 and 17, indicated by the yellow arrows, show information about the individual bacterial cell sizes. The results indicated that the adhered cells were well structured and integrated on the surface of the HAp coatings, and also that they were able to colonize the entire surface of the coatings. On the other hand, the results also emphasized that the bacterial cells were reduced in numbers and that the formation of biofilm was inhibited by the 3CuHAp coatings. From the profile images, the bacterial cells were observed to present the distinctive morphological features of *P. aeruginosa* bacterial cells, being relatively smooth and having lengths between 1.35 µm and 2.25 µm and diameters between 0.68 µm to 0.94 µm.

Additional information regarding the antibacterial activity of the HAp and 3CuHAp coatings was obtained by employing a quantitative in vitro antibacterial assay. The results of the quantitative studies are graphically represented in Figure 18. The quantitative assays depicted in Figure 18 highlighted the antibacterial activity of the 3CuHAp coatings on the adherence to and development of *P. aeruginosa* colony-forming units (CFUs) after exposure periods of 24, 48, and 72 h.

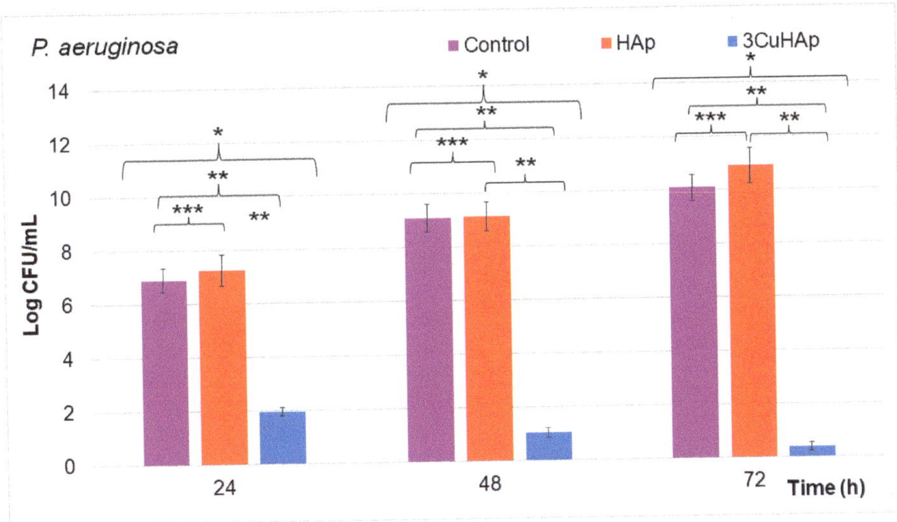

Figure 18. Graphical representation of the log colony forming units (CFUs)/mL of the HAp and 3CuHAp coatings incubated with *Pseudomonas aeruginosa* cells for 24, 48 and 72 h. The ANOVA single-factor test was used for the statistical analysis and $p \leq 0.05$ was accepted as statistically significant (* $p < 0.001$, ** $p < 0.002$, and *** $p < 0.007$).

The antibacterial properties of the HAp and 3CuHAp coatings were examined, using a free *P. aeruginosa* culture as the positive control (C+). The results showed that the HAp coatings promoted the *P. aeruginosa* bacterial cell proliferation and positively influenced their development and adherence across all the tested time intervals. These findings are in good agreement with previous studies emphasizing that hydroxyapatite did not exhibit antibacterial activity [24]. The results of the quantitative antibacterial assays depicted in Figure 18 demonstrated that the development of *P. aeruginosa* CFUs was inhibited early within the first 24 h of exposure to the 3CuHAp coatings compared to the control. The in vitro antibacterial studies indicated a significant reduction in the *P. aeruginosa* colonies within the first 24 h compared to the positive control culture (C+) [69].

The results highlighted the crucial role of copper ions in the antibacterial properties of the 3CuHAp coatings, which were enhanced over time. The data indicated that the 3CuHAp coatings exhibited the strongest antibacterial effect after 72 h of incubation. Over this 72 h exposure period, the coatings, leveraging the antibacterial efficacy of copper ions, effectively inhibited the development of *P. aeruginosa* bacterial cells, significantly reducing the number of bacterial colonies. Even though the exact mechanisms responsible for the antibacterial behavior of the material are still unelucidated, there are several hypotheses. In this case, the most reported mechanism states that the exposure of bacteria to copper ions can interfere and modify the permeability of the cell membrane, leading to the leakage of cellular contents, and eventually, to cell lysis [70].

Another proposed mechanism affirms that copper ions could interfere with protein synthesis and with the DNA replication processes. The antimicrobial activity of copper ions has been attributed to their ability to generate reactive oxygen species (ROS) to inflict damage on bacteria [66]. Gram-negative bacteria have a negatively charged outer layer (lipopolysaccharides), which attracts cations and causes membrane damage. According to various studies [71–74], the anionic surface cell membrane promotes the binding of high concentrations of metal ions, enhancing toxicity. Copper ions can also interact with DNA by interfering with nucleic acid strands and disturb biochemical processes [75]. Copper ions can also generate intracellular hydroxyl radicals, which harm essential proteins [76].

The main mechanism of antimicrobial activity for metal ions is the generation of ROS, whether dependent or independent of Fenton reactions, and mainly leads to damage to cell membranes. Metals such as copper act as catalysts in Fenton-like reactions to generate ROS such as O2•, OH•, etc. [77], effects observed in the *E. coli* cell model [78]. Also, the free radicals created can cause damage to the mitochondrial membrane, resulting in cell degradation and ultimately cell death [71]. In addition, copper ions have been found to cause multiple toxic effects; in addition to generating reactive oxygen species, they can also lead to lipid peroxidation and protein oxidation [71], which ultimately lead to the stopping of cell division [79].

Even though the exact mechanisms responsible for the antimicrobial activity are still uncertain, it was hypothesized that the principal and most important of copper's antibacterial mechanisms is attributed to the generation of reactive oxygen species (ROS), which induce oxidative stress that could damage bacterial cells. ROS damage various cellular components, including lipids, proteins, and DNA, leading to bacterial cell dysfunction and eventually death. Furthermore, copper ions have the ability to disrupt the bacterial cell membrane by binding and destabilizing their lipid bilayers, thus increasing the membrane permeability and causing leakage of its cellular contents. It was also reported that copper could interfere with the essential enzyme functions by binding to their active sites, further inhibiting bacterial growth and survival. This multifaceted strategy renders copper highly effective in preventing bacterial attachment, inhibiting biofilm formation, and ultimately, eliminating bacterial infections.

The data from the antibacterial assays are in good agreement with the present data from existing studies [66,67,80]. Furthermore, the results of the biological studies conducted emphasized that 3CuHAp possesses good biocompatibility and also exhibits strong antibacterial activity, making the coating potential candidates for future development in biomedical applications (coatings for various types of implants used in orthopedy and stomatology, medical devices, etc.).

The biocompatibility of the 3CuHAp coatings was thoroughly assessed using the HeLa and MG63 cell lines. The viability of the HeLa and MG63 cells following exposure to the 3CuHAp coatings was consistently above 88%, surpassing the 70% cytotoxicity threshold specified by ISO standard 10993-5. These results indicate that the 3CuHAp coatings exhibit no cytotoxic effects on HeLa and MG63 cells, thereby indicating their potential as promising candidates for the development of advanced implant materials with favorable biocompatibility. Furthermore, the microscopic observation showed that the morphology of the HeLa and MG63 cells remained unchanged when cultured in the presence of the HAp and 3CuHAp coatings. The antibacterial assays against *P. aeruginosa* also emphasized that the 3CuHAp exhibited significant antibacterial activity, even from the first 24 h of incubation. The findings of the AFM studies revealed that the 3CuHAp coatings effectively inhibited the formation of *P. aeruginosa* biofilms. Analysis of the AFM data highlighted a significant inhibition of *P. aeruginosa* attachment and development on the coating's surface within the initial 24 h. Furthermore, both the 2D AFM topography and its 3D representation demonstrated the rapid decline of the attached bacterial cells over time, with a notable reduction observed after 72 h of exposure. The development of the 3CuHAp coatings represents a notable advancement in biomedical applications. The evident suppression of *Pseudomonas aeruginosa* biofilm formation and the notable decrease in bacterial cell attachment demonstrate the potential of these coatings to address infections linked with implantable medical devices.

In summary, our results suggest that the coatings based on HAp and 3CuHAp ($Ca_{10-x}Cu_x(PO_4)_6(OH)_2$; $x_{Cu} = 0, 0.03$), as obtained by the vacuum deposition technique, exhibit structural, chemical, and morphological features that enhance their biological activity. The presence of copper induces good antibacterial activity to HAp. The coatings' surface topography promotes enhanced in vitro biocompatibility. These findings underline that these types of coatings could represent promising candidates for uses in the biomedical field.

On the other hand, this study on copper-doped hydroxyapatite has some limitations, including a narrow range of doping concentrations, use of a single deposition technique and controlled laboratory conditions. Further research is needed and should explore a broader range of copper-doping levels, investigate alternative deposition methods, test long-term stability and analyze the mechanical properties, including in vitro and in vivo behaviors. These improvements will enhance the understanding and practical use of biomaterials based on copper-doped hydroxyapatite.

4. Conclusions

This study reported for the first time the development of copper-doped hydroxyapatite coatings by the vacuum deposition technique. Various techniques were used for the coatings' complex characterization. In summary, the XRD studies confirmed the successful deposition of the HAp and 3CuHAp coatings, both containing pure hexagonal hydroxyapatite without impurities. The results of the AFM and SEM studies confirmed the continuous and homogeneous surface morphology of the coatings. Furthermore, the presence of hydroxyapatite in the studied coatings was proved by the FTIR results. The chemical composition of the HAp and 3CuHAp coatings was analyzed by XPS and EDS measurements. The results highlighted that the 3CuHAp coatings are highly biocompatible with the HeLa and MG63 cell lines, showing over 88% cell viability and no cytotoxic effects, making them suitable for advanced implant materials. Additionally, the in vitro antibacterial assays emphasized that the 3CuHAp coatings exhibit significant antibacterial activity against *Pseudomonas aeruginosa*, effectively inhibiting biofilm formation and bacterial attachment, highlighting their potential to prevent infections in implantable medical devices. Moreover, their proven compatibility with human cells indicates their potential integration into biomedical devices, thereby creating opportunities for the development of safer and more effective medical implants. In conclusion, copper-doped hydroxyapatite represents a promising candidate as a covering biomaterial for implants due to the superior biocompatibility and improved antimicrobial activity.

Author Contributions: Conceptualization, K.R. and D.P.; methodology, K.R., D.P. and S.R.; software, K.R., D.P. and S.R.; validation, Y.B., D.P., K.R., C.S.C., S.L.I., S.R., R.T., L.G., C.B. and K.B.; formal analysis, Y.B., D.P., K.R., C.S.C., S.L.I., S.R., R.T., C.C., I.C.M., M.S., C.C.N., L.G., C.B. and K.B.; investigation, Y.B., D.P., K.R., C.S.C., S.L.I., S.R., R.T., C.C., I.C.M., M.S., C.C.N., L.G., C.B. and K.B.; resources, Y.B., D.P., K.R., C.S.C., S.L.I., S.R., R.T., L.G., C.B. and K.B.; data curation, K.R. and D.P.; writing—original draft preparation, K.R., D.P., S.R., S.L.I., C.S.C. and Y.B.; writing—review and editing, Y.B., D.P., K.R., C.S.C., S.L.I., S.R., R.T., C.C., I.C.M., M.S., C.C.N., L.G., C.B. and K.B.; visualization, Y.B., D.P., K.R., C.S.C., S.L.I., S.R., R.T., C.C., I.C.M., M.S., C.C.N., L.G., C.B. and K.B.; supervision, K.R., D.P. and S.R.; project administration, K.R.; funding acquisition, K.R. and D.P. All authors have read and agreed to the published version of the manuscript.

Funding: This research was funded by the Core Program of the National Institute of Materials Physics, granted by the Romanian Ministry of Research, Innovation and Digitalization through the Project PC1-PN23080101.

Institutional Review Board Statement: Not applicable.

Informed Consent Statement: Not applicable.

Data Availability Statement: The original contributions presented in the study are included in the article, further inquiries can be directed to the corresponding authors.

Conflicts of Interest: The authors declare no conflicts of interest. The funders had no role in the design of the study; in the collection, analyses, or interpretation of data; in the writing of the manuscript; or in the decision to publish the results.

References

1. Furko, M.; Havasi, V.; Kónya, Z.; Grünewald, A.; Detsch, R.; Boccaccini, A.R.; Balázsi, C. Development and characterization of multi-element doped hydroxyapatite bioceramic coatings on metallic implants for orthopedic applications. *Bol. Soc. Esp. Ceram. Vidr.* **2018**, *57*, 55–65. [CrossRef]
2. Huang, Y.; Zhang, X.; Zhao, R.; Mao, H.; Yan, Y.; Pang, X. Antibacterial efficacy, corrosion resistance, and cytotoxicity studies of copper-substituted carbonated hydroxyapatite coating on titanium substrate. *J. Mater. Sci.* **2015**, *50*, 1688–1700. [CrossRef]
3. San Thian, E.; Huang, J.; Barber, Z.H.; Best, S.M.; Bonfield, W. Surface modification of magnetron-sputtered hydroxyapatite thin films via silicon substitution for orthopaedic and dental applications. *Surf. Coat. Technol.* **2011**, *205*, 3472–3477. [CrossRef]
4. Gomes, S.; Vichery, C.; Descamps, S.; Martinez, H.; Kaur, A.; Jacobs, A.; Nedelec, J.M.; Renaudin, G. Cu-doping of calcium phosphate bioceramics: From mechanism to the control of cytotoxicity. *Acta Biomater.* **2018**, *65*, 462–474. [CrossRef] [PubMed]
5. Jacobs, A.; Renaudin, G.; Charbonnel, N.; Nedelec, J.-M.; Forestier, C.; Descamps, S. Copper-Doped Biphasic Calcium Phosphate Powders: Dopant Release, Cytotoxicity and Antibacterial Properties. *Materials* **2021**, *14*, 2393. [CrossRef] [PubMed]
6. Calhoun, J.H.; Manring, M.; Shirtliff, M. Osteomyelitis of the Long Bones. *Semin. Plast. Surg.* **2009**, *23*, 59–72. [CrossRef]
7. ter Boo, G.-J.A.; Grijpma, D.W.; Moriarty, T.F.; Richards, R.G.; Eglin, D. Antimicrobial delivery systems for local infection prophylaxis in orthopedic- and trauma surgery. *Biomaterials* **2015**, *52*, 113–125. [CrossRef] [PubMed]
8. Tenover, F.C. Mechanisms of antimicrobial resistance in bacteria. *Am. J. Infect. Control* **2006**, *34*, S3–S10. [CrossRef]
9. Vincent, M.; Duval, R.E.; Hartemann, P.; Engels-Deutsch, M. Contact killing and antimicrobial properties of copper. *J. Appl. Microbiol.* **2018**, *124*, 1032–1046. [CrossRef]
10. Jacobs, A.; Renaudin, G.; Forestier, C.; Nedelec, J.M.; Descamps, S. Biological properties of copper-doped biomaterials for orthopedic applications: A review of antibacterial, angiogenic and osteogenic aspects. *Acta Biomater.* **2020**, *117*, 21–39. [CrossRef]
11. Gérard, C.; Bordeleau, L.J.; Barralet, J.; Doillon, C.J. The stimulation of angiogenesis and collagen deposition by copper. *Biomaterials* **2010**, *31*, 824–831. [CrossRef] [PubMed]
12. Heidenau, F.; Mittelmeier, W.; Detsch, R.; Haenle, M.; Stenzel, F.; Ziegler, G.; Gollwitzer, H. A novel antibacterial titania coating: Metal ion toxicity and in vitro surface colonization. *J. Mater. Sci. Mater. Med.* **2005**, *16*, 883–888. [CrossRef] [PubMed]
13. Hordyjewska, A.; Popiołek, Ł.; Kocot, J. The many "faces" of copper in medicine and treatment. *BioMetals* **2014**, *27*, 611–621. [CrossRef] [PubMed]
14. Lorincz, M.T. Neurologic Wilson's disease. *Ann. N. Y. Acad. Sci.* **2010**, *1184*, 173–187. [CrossRef] [PubMed]
15. Grass, G.; Rensing, C.; Solioz, M. Metallic copper as an antimicrobial surface. *Appl. Environ. Microbiol.* **2011**, *77*, 1541–1547. [CrossRef] [PubMed]
16. Hidalgo-Robatto, B.M.; López-Álvarez, M.; Azevedo, A.S.; Dorado, J.; Serra, J.; Azevedo, N.F.; González, P. Pulsed laser deposition of copper and zinc doped hydroxyapatite coatings for biomedical applications. *Surf. Coat. Technol.* **2018**, *333*, 168–177. [CrossRef]
17. Kunz Coyne, A.J.; El Ghali, A.; Holger, D.; Rebold, N.; Rybak, M.J. Therapeutic Strategies for Emerging Multidrug-Resistant *Pseudomonas aeruginosa*. *Infect. Dis. Ther.* **2022**, *11*, 661–682. [CrossRef] [PubMed]
18. Bazin, T.; Magnaudeix, A.; Mayet, R.; Carles, P.; Julien, I.; Demourgues, A.; Gaudon, M.; Champion, E. Sintering and biocompatibility of copper-doped hydroxyapatite bioceramics. *Ceram. Int.* **2021**, *47*, 13644–13654. [CrossRef]
19. Noori, A.; Hoseinpour, M.; Kolivand, S.; Lotfibakhshaiesh, N.; Ebrahimi-Barough, S.; Ai, J.; Azami, M. Exploring the various effects of Cu doping in hydroxyapatite nanoparticle. *Sci. Rep.* **2024**, *14*, 3421. [CrossRef]
20. Li, Q.; Song, S.; Li, J.; Yang, J.; Zhang, R.; Niinomi, M.; Nakano, T. Antibacterial properties and biocompatibility of hydroxyapatite coating doped with various Cu contents on titanium. *Mater. Trans.* **2022**, *63*, 1072–1079. [CrossRef]
21. Motelica-Heino, M.; Predoi, M.V.; Ciobanu, S.C.; Iconaru, S.L.; Predoi, D. Studies of New Layer Formation on the Surface of Zinc Doped Hydroxyapatite/Chitosan Composite Coatings in Biological Medium. *Coatings* **2023**, *13*, 472. [CrossRef]
22. Predoi, D.; Ciobanu, C.S.; Iconaru, S.L.; Raaen, S.; Badea, M.L.; Rokosz, K. Physicochemical and Biological Evaluation of Chitosan-Coated Magnesium-Doped Hydroxyapatite Composite Layers Obtained by Vacuum Deposition. *Coatings* **2022**, *12*, 702. [CrossRef]
23. Image, J. Available online: http://imagej.nih.gov/ij (accessed on 28 June 2024).
24. Predoi, D.; Ciobanu, C.S.; Iconaru, S.L.; Predoi, S.A.; Chifiriuc, M.C.; Raaen, S.; Badea, M.L.; Rokosz, K. Impact of Gamma Irradiation on the Properties of Magnesium-Doped Hydroxyapatite in Chitosan Matrix. *Materials* **2022**, *15*, 5372. [CrossRef]
25. Gwyddion. Available online: http://gwyddion.net/ (accessed on 20 January 2024).
26. Iconaru, S.L.; Ciobanu, C.S.; Predoi, D.; Motelica-Heino, M.; Negrilă, C.C.; Badea, M.L.; Predoi, M.V.; Chifiriuc, C.M.; Popa, M. Nitrogen and Bromide Co-Doped Hydroxyapatite Thin Films with Antimicrobial Properties. *Coatings* **2021**, *11*, 1505. [CrossRef]
27. Iconaru, S.-L.; Motelica-Heino, M.; Predoi, D. Study on Europium-Doped Hydroxyapatite Nanoparticles by Fourier Transform Infrared Spectroscopy and Their Antimicrobial Properties. *J. Spectrosc.* **2013**, *2013*, 284285. [CrossRef]
28. Ciobanu, C.S.; Iconaru, S.L.; Predoi, D.; Trușcă, R.-D.; Prodan, A.M.; Groza, A.; Chifiriuc, M.C.; Beuran, M. Fabrication of Novel Chitosan–Hydroxyapatite Nanostructured Coatings for Biomedical Applications. *Coatings* **2021**, *11*, 1561. [CrossRef]
29. Iconaru, S.L.; Predoi, D.; Ciobanu, C.S.; Motelica-Heino, M.; Guegan, R.; Bleotu, C. Development of Silver Doped Hydroxyapatite Coatings for Biomedical Applications. *Coatings* **2022**, *12*, 341. [CrossRef]
30. Mansour, S.F.; El-dek, S.I.; Ahmed, M.K. Tailoring the structure of biphasic calcium phosphate via synthesis procedure. *Mater. Res. Express* **2017**, *4*, 125015. [CrossRef]

31. Ismail, A.M.; Menazea, A.A.; Kabary, H.A.; El-Sherbiny, A.E.; Samy, A. The influence of calcination temperature on structural and antimicrobial characteristics of zinc oxide nanoparticles synthesized by Sol–Gel method. *J. Mol. Struct.* **2019**, *1196*, 332–337. [CrossRef]
32. Elashmawi, I.S.; Menazea, A.A. Different time's Nd:YAG laser-irradiated PVA/Ag nanocomposites: Structural, optical, and electrical characterization. *J. Mater. Res. Technol.* **2019**, *8*, 1944–1951. [CrossRef]
33. Mariappan, A.; Pandi, P.; Rani, K.B.; Neyvasagam, K. Study of the photocatalytic and antibacterial effect of Zn and Cu doped hydroxyapatite. *Inorg. Chem. Commun.* **2022**, *136*, 109128. [CrossRef]
34. Osman, M.A.; Alamoush, R.A.; Kushnerev, E.; Seymour, K.G.; Watts, D.C.; Yates, J.M. Biological response of epithelial and connective tissue cells to titanium surfaces with different ranges of roughness: An in-vitro study. *Dent. Mater. Off. Publ. Acad. Dent. Mater.* **2022**, *38*, 1777–1788. [CrossRef] [PubMed]
35. Mavropoulos, E.; Hausen, M.; Costa, A.M.; Alves, G.; Mello, A.; Ospina, C.A.; Mir, M.; Granjeiro, J.M.; Rossi, A.M. The impact of the RGD peptide on osteoblast adhesion and spreading on zinc-substituted hydroxyapatite surface. *J. Mater. Sci. Mater. Med.* **2013**, *24*, 1271–1283. [CrossRef]
36. Lopez, D.A.; Simison, S.N.; De Sanchez, S.R. The influence of steel microstructure on CO2 corrosion. EIS studies on the inhibition efficiency of benzimidazole. *Electrochim. Acta* **2003**, *48*, 845–854. [CrossRef]
37. Serra, J.; González, P.; Liste, S.; Serra, C.; Chiussi, S.; León, B.; Pérez-Amor, M.; Ylänen, H.O.; Hupa, M. FTIR and XPS studies of bioactive silica based glasses. *J. Non-Cryst. Solids* **2003**, *332*, 20–27. [CrossRef]
38. Moulder, J.F.; Stickle, W.F.; Sobol, P.E.; Bomben, K.D. *Handbook of X-ray Photoelectron Spectroscopy*; Physical Electronics Inc.: Chanhassen, MN, USA, 1995.
39. Battistoni, C.; Casaletto, M.P.; Ingo, G.M.; Kaciulis, S.; Mattogno, G.; Pandolfi, L. Surface characterization of biocompatible hydroxyapatite coatings. *Surf. Interface Anal.* **2000**, *29*, 773–781. [CrossRef]
40. Greczynski, G.; Hultman, L. X-ray photoelectron spectroscopy: Towards reliable binding energy referencing. *Prog. Mater. Sci.* **2020**, *107*, 100591. [CrossRef]
41. Available online: https://www.thermofisher.com/ro/en/home/materials-science/learning-center/periodic-table/non-metal/carbon.html (accessed on 30 January 2024).
42. Nagakane, K.; Yoshida, Y.; Hirata, I.; Fukuda, R.; Nakayama, Y.; Shirai, K.; Ogawa, T.; Suzuki, K.; Van Meerbeek, B.; Okazaki, M. Analysis of chemical interaction of 4-MET with hydroxyapatite using XPS. *Dent. Mater. J.* **2006**, *25*, 645–649. [CrossRef] [PubMed]
43. Stranick, M.A.; Root, M.J. Influence of strontium on monofluorophosphate uptake by hydroxyapatite XPS characterization of the hydroxyapatite surface. *Colloids Surf.* **1991**, *55*, 137–147. [CrossRef]
44. Chusuei, C.C.; Brookshier, M.A.; Goodman, D.W. Correlation of relative X-ray photoelectron spectroscopy shake-up intensity with CuO particle size. *Langmuir* **1999**, *15*, 2806–2808. [CrossRef]
45. Zhang, Y.C.; Tang, J.Y.; Wang, G.L.; Zhang, M.; Hu, X.Y. Facile synthesis of submicron Cu2O and CuO crystallites from a solid metallorganic molecular precursor. *J. Cryst. Growth* **2006**, *294*, 278–282. [CrossRef]
46. Schon, G. ESCA studies of Cu, Cu_2O and CuO. *Surf. Sci.* **1973**, *35*, 96–108. [CrossRef]
47. Kumar, M.; Bhatt, V.; Nayal, O.S.; Sharma, S.; Kumar, V.; Thakur, M.S.; Kumar, N.; Bal, R.; Singh, B.; Sharma, U. CuI nanoparticles as recyclable heterogeneous catalysts for C–N bond formation reactions. *Catal. Sci. Technol.* **2017**, *7*, 2857–2864. [CrossRef]
48. Chen, S.; Brown, L.; Levendorf, M.; Cai, W.; Ju, S.Y.; Edgeworth, J.; Li, X.; Magnuson, C.W.; Velamakanni, A.; Piner, R.D.; et al. Oxidation resistance of graphene-coated Cu and Cu/Ni alloy. *ACS Nano* **2011**, *5*, 1321–1327. [CrossRef] [PubMed]
49. Bai, X.; More, K.; Rouleau, C.M.; Rabiei, A. Functionally graded hydroxyapatite coatings doped with antibacterial components. *Acta Biomater.* **2010**, *6*, 2264–2273. [CrossRef] [PubMed]
50. Shanmugam, S.; Gopal, B. Copper substituted hydroxyapatite and fluorapatite: Synthesis, characterization and antimicrobial properties. *Ceram. Int.* **2014**, *40*, 15655–15662. [CrossRef]
51. Ciobanu, S.C.; Predoi, D.; Chifiriuc, M.C.; Iconaru, S.L.; Predoi, M.V.; Popa, M.; Rokosz, K.; Raaen, S.; Marinas, I.C. *Salvia officinalis*–Hydroxyapatite Nanocomposites with Antibacterial Properties. *Polymers* **2023**, *15*, 4484. [CrossRef]
52. Huang, Y.; Hao, M.; Nian, X.; Qiao, H.; Zhang, X.; Zhang, X.; Song, G.; Guo, J.; Pang, X.; Zhang, H. Strontium and copper co-substituted hydroxyapatite-based coatings with improved antibacterial activity and cytocompatibility fabricated by electrodeposition. *Ceram. Int.* **2016**, *42*, 11876–11888. [CrossRef]
53. Mounika, S.; Ramakrishnan, P. Synthesis and Comparison of Chemical Changes Using FTIR Spectroscop for Copper Substituted Hydroxyapatite. *E3S Web Conf.* **2024**, *477*, 00083. [CrossRef]
54. Lukaviciute, L.; Karciauskaite, J.; Grigoraviciute, I.; Vasiliauskiene, D.; Sokol, D.; Kareiva, A. Calcium Hydroxyapatite Coatings: Low-Temperature Synthesis and Investigation of Antibacterial Properties. *Coatings* **2023**, *13*, 1991. [CrossRef]
55. Nam, P.T.; Thom, N.T.; Phuong, N.T.; Xuyen, N.T.; Hai, N.S.; Anh, N.T.; Dung, P.T.; Thanh, D.T.M. Synthesis, characterization and antimicrobial activity of copper doped hydroxyapatite. *Vietnam. J. Chem.* **2018**, *56*, 672–678. [CrossRef]
56. Eliaz, N.; Shmueli, S.; Shur, I.; Benayahu, D.; Aronov, D.; Rosenman, G. The effect of surface treatment on the surface texture and contact angle of electrochemically deposited hydroxyapatite coating and on its interaction with bone-forming cells. *Acta Biomater.* **2009**, *5*, 3178–3191. [CrossRef]
57. Ryan, B.J.; Poduska, K.M. Roughness effects on contact angle measurements. *Am. J. Phys.* **2008**, *76*, 1074–1077. [CrossRef]
58. Danish, M. Contact Angle Studies of Hydrophobic and Hydrophilic Surfaces. In *Handbook of Magnetic Hybrid Nanoalloys and their Nanocomposites*; Springer: Cham, Switzerland, 2022. [CrossRef]

59. *ISO Standard 10993-5*; Evaluation Biologique des Dispositifs Medicaux—Partie 5, Essais Concernant la Cytotoxicite' In Vitro. ISO: Geneva, Switzerland, 2009; p. 34.
60. Rodríguez-Valencia, C.; López-Álvarez, M.; Cochón-Cores, B.; Pereiro, I.; Serra, J.; González, P. Novel selenium-doped hydroxyapatite coatings for biomedical applications. *J. Biomed. Mater. Res. A* **2013**, *101*, 853–861. [CrossRef] [PubMed]
61. Marques, C.F.; Olhero, S.; Abrantes, J.C.; Marote, A.; Ferreira, S.; Vieira, S.I.; Ferreira, J.M. Biocompatibility and antimicrobial activity of biphasic calcium phosphate powders doped with metal ions for regenerative medicine. *Ceram. Int.* **2017**, *43*, 15719–15728. [CrossRef]
62. Li, K.; Xia, C.; Qiao, Y.; Liu, X. Dose-response relationships between copper and its biocompatibility/antibacterial activities. *J. Trace Elements Med. Biol.* **2019**, *55*, 127–135. [CrossRef] [PubMed]
63. Gritsch, L.; Lovell, C.; Goldmann, W.H.; Boccaccini, A.R. Fabrication and characterization of copper(II)-chitosan complexes as antibiotic-free antibacterial biomaterial. *Carbohydr. Polym.* **2018**, *179*, 370–378. [CrossRef]
64. Unabia, R.B.; Candidato, R.T., Jr.; Pawłowski, L.; Salvatori, R.; Bellucci, D.; Cannillo, V. In vitro studies of solution precursor plasma-sprayed copper-doped hydroxyapatite coatings with increasing copper content. *J. Biomed. Mater. Res. Part B Appl. Biomater.* **2020**, *108*, 2579–2589. [CrossRef]
65. Tuntun, S.M.; Hossain, M.S.; Uddin, M.N.; Shaikh, M.A.; Bahadur, N.M.; Ahmed, S. Crystallographic characterization and application of copper doped hydroxyapatite as a biomaterial. *New J. Chem.* **2023**, *47*, 2874–2885. [CrossRef]
66. Shen, Q.; Qi, Y.; Kong, Y.; Bao, H.; Wang, Y.; Dong, A.; Wu, H.; Xu, Y. Advances in Copper-Based Biomaterials with Antibacterial and Osteogenic Properties for Bone Tissue Engineering. *Front. Bioeng. Biotechnol.* **2022**, *20*, 795425. [CrossRef]
67. Sahithi, K.; Swetha, M.; Prabaharan, M.; Moorthi, A.; Saranya, N.; Ramasamy, K.; Srinivasan, N.; Partridge, N.C.; Selvamurugan, N. Synthesis and Characterization of NanoscaleHydroxyapatite-Copper for Antimicrobial Activity Towards Bone Tissue Engineering Applications. *J. Biomed. Nanotechnol.* **2010**, *6*, 333–339. [CrossRef] [PubMed]
68. Hoveidaei, A.H.; Sadat-Shojai, M.; Mosalamiaghili, S.; Salarikia, S.R.; Roghani-Shahraki, H.; Ghaderpanah, R.; Ersi, M.H.; Conway, J.D. Nano-hydroxyapatite structures for bone regenerative medicine: Cell-material interaction. *Bone* **2024**, *179*, 116956. [CrossRef] [PubMed]
69. Mone, N.S.; Syed, S.; Ravichandiran, P.; Kamble, E.E.; Pardesi, K.R.; Salunke-Gawali, S.; Rai, M.; Vikram Singh, A.; Prasad Dakua, S.; Park, B.H.; et al. Synergistic and Additive Effects of Menadione in Combination with Antibiotics on Multidrug-Resistant Staphylococcus aureus: Insights from Structure-Function Analysis of Naphthoquinones. *ChemMedChem* **2023**, *18*, e202300328. [CrossRef] [PubMed]
70. Singh, A.V.; Baylan, S.; Park, B.W.; Richter, G.; Sitti, M. Hydrophobic pinning with copper nanowhiskers leads to bactericidal properties. *PLoS ONE* **2017**, *12*, e0175428. [CrossRef] [PubMed]
71. Chatterjee, A.K.; Chakraborty, R.; Basu, T. Mechanism of antibacterial activity of copper nanoparticles. *Nanotechnology* **2014**, *25*, 135101. [CrossRef]
72. Vasiliev, G.; Kubo, A.L.; Vija, H.; Kahru, A.; Bondar, D.; Karpichev, Y.; Bondarenko, O. Synergistic antibacterial effect of copper and silver nanoparticles and their mechanism of action. *Sci. Rep.* **2023**, *13*, 9202. [CrossRef] [PubMed]
73. Nisar, P.; Ali, N.; Rahman, L.; Ali, M.; Shinwari, Z.K. Antimicrobial activities of biologically synthesized metal nanoparticles: An insight into the mechanism of action. *J. Biol. Inorg. Chem.* **2019**, *24*, 929–941. [CrossRef]
74. Ballo, M.K.; Rtimi, S.; Mancini, S.; Kiwi, J.; Pulgarin, C.; Entenza, J.M.; Bizzini, A. Bactericidal activity and mechanism of action of copper-sputtered flexible surfaces against multidrug-resistant pathogens. *Appl. Microbiol. Biotechnol.* **2016**, *100*, 5945–5953. [CrossRef]
75. Slavin, Y.N.; Asnis, J.; Hńfeli, U.O.; Bach, H. Metal nanoparticles: Understanding the mechanisms behind antibacterial activity. *J. Nanobiotechnol* **2017**, *15*, 65. [CrossRef]
76. Crisan, M.C.; Teodora, M.; Lucian, M. Copper Nanoparticles: Synthesis and Characterization, Physiology, Toxicity and Antimicrobial Applications. *Appl. Sci.* **2022**, *12*, 141. [CrossRef]
77. Salah, I.; Parkin, I.P.; Allan, E. Copper as an antimicrobial agent: Recent advances. *RSC Adv.* **2021**, *11*, 18179–18186. [CrossRef] [PubMed]
78. Gyawali, R.; Ibrahim, S.A.; Abu Hasfa, S.H.; Smqadri, S.Q.; Haik, Y. Antimicrobial Activity of Copper Alone and in Combination with Lactic Acid against Escherichia coli O157:H7 in Laboratory Medium and on the Surface of Lettuce and Tomatoes. *J. Pathog.* **2011**, *2011*, 650968. [CrossRef] [PubMed]
79. Gomaa, E.Z.; Housseiny, M.M.; Omran, A.A.A.K. Fungicidal Efficiency of Silver and Copper Nanoparticles Produced by Pseudomonas fluorescens ATCC 17397 against Four Aspergillus Species: A Molecular Study. *J. Clust. Sci.* **2019**, *30*, 181–196. [CrossRef]
80. Liu, R.; Memarzadeh, K.; Chang, B.; Zhang, Y.; Ma, Z.; Allaker, R.P.; Ren, L.; Yang, K. Antibacterial Effect of Copper-Bearing Titanium alloy (Ti-Cu) against Streptococcus Mutans and Porphyromonas Gingivalis. *Sci. Rep.* **2016**, *6*, 29985. [CrossRef] [PubMed]

materials

MDPI

Article

Hexakis[p-(hydroxymethyl)phenoxy]cyclotriphosphazene as an Environmentally Friendly Modifier for Polyurethane Powder Coatings with Increased Thermal Stability and Corrosion Resistance

Barbara Pilch-Pitera [1,*], Dominika Czachor-Jadacka [1], Łukasz Byczyński [1], Michał Dutkiewicz [2], Rafał Januszewski [3,4], Krzysztof Kowalczyk [5], Wojciech J. Nowak [6] and Katarzyna Pojnar [7]

1 Faculty of Chemistry, Department of Polymers and Biopolymers, Rzeszow University of Technology, al. Powstańców Warszawy 6, 35-959 Rzeszow, Poland; d.czachor94@gmail.com (D.C.-J.); lbyczynski@prz.edu.pl (Ł.B.)
2 Poznan Science and Technology Park, Adam Mickiewicz University Foundation, Rubież 46, 61-612 Poznan, Poland; midu@ppnt.poznan.pl
3 Faculty of Chemistry, Adam Mickiewicz University in Poznan, Umultowska 89B, ul. Uniwersytetu Poznańskiego 8, 61-614 Poznan, Poland; r.janusz@amu.edu.pl
4 Center for Advanced Technology, Adam Mickiewicz University in Poznan, ul. Uniwersytetu Poznańskiego 10, 61-614 Poznan, Poland
5 Faculty of Chemical Technology and Engineering, Polymer Institute, West Pomeranian University of Technology in Szczecin, ul. Pułaskiego 10, 70-322 Szczecin, Poland; krzysztof.kowalczyk@zut.edu.pl
6 Faculty of Mechanical Engineering and Aeronautics, Department of Materials Science, Rzeszow University of Technology, al. Powstańców Warszawy 8, 35-959 Rzeszow, Poland; wjnowak@prz.edu.pl
7 Doctoral School of Engineering and Technical Sciences at the Rzeszow University of Technology, al. Powstańców Warszawy 12, 35-959 Rzeszow, Poland; d521@stud.prz.edu.pl
* Correspondence: barbpi@prz.edu.pl; Tel.: +48-17-856-1700

Abstract: Protection against fire and the corrosion of metals is necessary to ensure human safety. Most of the fire and corrosion inhibitors do not meet the ecological requirements. Therefore, effective and ecological methods of protecting metals are currently a challenge for researchers. In this work, the influence of hexakis(4-(hydroxymethyl)phenoxy)cyclotriphosphazene (HHPCP) on the characteristics of powder coatings was examined. The coatings' properties were investigated by measuring the roughness, hardness, adhesion to the steel surface, cupping, gloss, scratch resistance, and water contact angle. The thermal stability was studied by furnace test and TGA analysis. The corrosion resistance test was carried out in a 3.5% NaCl solution. The distribution of phosphazene-derived segments in the coating was examined by GD-EOS analysis. Modified coatings show better corrosion and thermal resistance and can be used for the protection of the steel surface. Their better corrosion resistance is due to the electroactive properties of the phosphazene ring and its higher concentration at the coating surface, confirmed by GD-EOS analysis. The increase in thermal resistance is due to the effect of the formation of phosphoric metaphosphoric and polyphosphoric acids during the decomposition of HHCPC, which remain in the condensed char phase and play a crucial role in surface protection.

Keywords: hexakis(4-(hydroxymethyl)phenoxy)cyclotriphosphazene; powder coatings; anticorrosion protection; thermal stability

1. Introduction

In recent years, there has been a growing interest in environmentally friendly coatings that still provide various functionalities and satisfy consumer requirements [1,2]. Powder coatings (PCs) are a better alternative to conventional liquid products because of the multitude of advantages resulting from the lack of organic solvents in their composition. For these reasons, PCs meet requirements related to the minimization of volatile organic compound (VOC) emissions. Compared to liquid paints, powder coatings are defined by, e.g., a short time to obtain the finished painting (as opposed to liquid paints, the PC is ready for use after taking it out of the oven with no need to wait for it to dry), lower costs of waste disposal (less waste due to the possibility of reusing the powder falling from the detail during spray), and durability (excellent properties of the final product). Powder coatings offer a wide range of modifications. According to the latest reports, the conductive properties [3–6], antibacterial [7], amphiphobic [8], and wear resistance [9,10] are described.

This work focuses on the application of cyclotriphosphazene to modify polyurethane powder coatings in order to improve their properties, such as fire resistance or corrosion protection. Flame-retardant coatings are in high demand in the automotive, aviation, and construction industries. Flame retardancy of powder coating is particularly desirable to reduce fire damage and also provide time for the safe evacuation of people [11–13]. Phosphorus compounds, together with boron compounds, are commonly reported types of coating modifiers with decreased flammability [14]. The general mechanism of anti-flammable materials depends on the flame retardant given and the type of substrate [15]. Flame-retardant additives can delay combustion in both the gas phase through flame inhibition and condensed phases (e.g., polymer matrix) by enhancing charring. Efficiency depends not only on the flame retardant itself but also on its interaction with the coating binder and additives. The coating surface can undergo charring, which creates a thermal insulation barrier between the burned and unburned parts. The created barrier slows the heat transfer to the unburned part. Active additives in the gas phase also provide an increase in the protection of flammable materials, but the release of hydrogen chloride or hydrogen bromide is damaging to our health and the environment, and its usage has been restricted [16]. In addition, a fire retardant is a substance that acts by physical or chemical action, reducing the flammability of materials or causing delays in their combustion. In the physical mechanism, a protective coating can be created to lower the ignition temperature. There are also many ways to increase the flame retardancy of polyurethane. Chattopadhyay et al. described reports on the use of such compounds as flame retardants, including halogenated paraffin, chlorofluorocarbons, inorganic oxides, and hydroxides (e.g., antimony oxide, magnesium hydroxide, aluminum hydroxide), inorganic carbonates (e.g., potassium, calcium, magnesium carbonate), inorganic boron compounds (e.g., boric acid, zinc borate borax), phosphorus compounds, expanded graphite, melamine, mica, and organic compounds of boron and silicon. The addition of expanded graphite dyes the coating black, which limits its dyeing to bright colors. Compounds containing halogen atoms, during a fire, release poisonous and asphyxiating hydrogen halides that destroy the ozone layer. Furthermore, when they are burning, some additives can release poisonous carbon monoxide (CO) [17]. The application of fire-retardant coatings is among the simplest and most effective ways to prevent fire in materials. This approach presents several advantages, allowing easy processing and no modification of the mechanical properties of the protected material. Among the most promising halogen-free solutions are intumescent flame-retardant (IFR) coatings. They consist of three active ingredients: an acidic catalyst (e.g., ammonium polyphosphate), a charring agent (e.g., polyol), and a blowing agent (most commonly ammonium compounds or melamine) connected by a polymeric binder. As a result of heat exposure, the intumescent coating swells, increasing many times in its original thickness, and produces a foamed carbonaceous char, which acts as an insulating layer to protect the substrate [18]. However, an effective interaction of intumescent components requires a coating thickness of more than 300–500 μm [19]. In the case of powder coatings with such high thickness, their properties are significantly deteriorated. According to technical requirements, e.g., Qualicoat [20], the optimal thickness of a single powder coating should be 60 μm.

The discovery of the cyclotrifosphazene compound by Allock and his colleagues in the mid-1950s allowed for relatively mild combustion properties, the reduced production of toxic gases, and its use as a flame retardant [21]. Due to the high reactivity of cyclotriphosphazene compounds, it is possible to introduce a wide range of substituents through nucleophilic reactions. Consequently, various side groups affect the characteristics of polyphosphazenes. In their review, Usri et al. discussed research on cyclotriphosphazene, with concentration on the synthesis, structural characteristics, dielectric characteristics, and flame retardancy of the compound hexachlorocyclotri-phosphazene.

Many polymer materials are characterized by good chemical, physical, and mechanical properties. But it is hard to find a material that can be characterized by specific properties, such as fire resistance or corrosion protection. Zhou et al. described the synthesis, curing kinetics, thermal properties, and flame retardancy of cyclotriphosphazene-containing multifunctional epoxy resin. For its modification, the author used the following commercial products: DGEBA (diglycidyl ether of bisphenol A) and hexachlorocyclotriphosphazene (HCCP). Multifunctional epoxy resin hexa-[4-(glycidyloxymethyl)phenoxy]-cyclotriphosphazene (HGPCP) cured with 4,4′-diaminodiphenyl methane (DDM) and 4,4′-diaminodiphenyl sulfone (DDS) exhibited excellent flame retardancy during the test via the limiting oxygen index (LOI) and vertical burning test (UL-94) [22]. Xu et al. also investigated flame-retardant properties based on epoxy resin. The LOI test and horizontal flame test showed that the presence of hexakis[p-(hydroxymethyl)phenoxy]cyclotriphosphazene (HHPCP) also leads to improved flame retardancy properties of epoxy resin [23]. Epoxidized hexakis-4-(2-(4-((β-methallyl)oxy)phenyl)propan-2-yl)phenoxycyclotriphosphazene by 3-chloroperbenzoic acid was cured by its treatment with isophorone diamine, which characterized improved thermal stability and glass transition temperature [24]. In addition to fire and heat resistance, epoxy resin based on arylaminocyclotriphosphazene improves chemical and corrosion resistance, high tensile strength, adhesion, and low solidification [25]. Consequently, these substances find applications in numerous cutting-edge scientific and technological domains, including coatings, adhesives, composites, extractants, metal complexes, organometallic chemistry, inorganic chemistry, aviation, aerospace, radioengineering, nuclear, and medical fields [26,27].

Improvements in fire resistance were noted even with the application of cyclotriphosphazene to combustible polypropylene. Flame-retardant polypropylene (PP) was obtained by appropriately blending with microencapsulated ammonium polyphosphate (MAPP). The modified polypropylene was characterized by an increase in the oxygen index (LOI) [28]. Mohd Taip and al. described the increase in the LOI value in the 2K polyurethane wood coating containing hexasubstituted cyclotriphosphazene compounds [29].

Generally, cyclotriphosphazene has often been used as a modifier and as a flame retardant in many polymers. Nevertheless, based on our analysis of the relevant research, there are no reports of modifications of powder coatings with the cyclotriphosphazene compound.

Corrosion is also one of the main challenges facing metal component manufacturers. The duration of corrosion protection depends on the high adhesion to the surface, tightness, and resistance of the coating to the corrosive environment. Powder coatings perfectly protect, stop, or slow down the spontaneous corrosion process. Corrosion resistance is an important element in the durability and aesthetics of objects. Currently, the most commonly used corrosion protection is metallic coatings, e.g., Zn or Cr [30–32] and polymer coatings [33]. However, the use of Cr and Zn is limited due to environmental and economic considerations. The alternative is powder coatings, which can be modified to obtain the appropriate properties, for example, excellent adhesion to the substrate surface, resistance to atmospheric environments, and hydrophobic properties [34]. The hydrophobic coating repels moisture and reduces the absorption of water and corrosive media to the surface of the coatings. However, in a highly corrosive environment, for more effective protection, the barrier made of a polymer binder may be insufficient, and the addition of corrosion inhibitors is necessary. The anticorrosion protection of powder coatings can be improved using various types of additives, e.g., ceramic powders, pigments, and fillers [35]. An example is the invention described in CN106189682A, which reveals the use of intercalated montmorillonite for the modification of epoxy resin-based powder coatings for the protection of pipeline corrosion [36]. Similarly, the addition of electroactive conductive polymers, such as polyaniline or polypyrrole, may increase the protection of steel in acidic or neutral environments by the repassivation of exposed areas [37]. Cyclophosphazene derivatives contain P heteroatoms and may also be applicable as corrosion inhibitors on metal surfaces for their treatment/protection against the corrosive environment [26]. However, polyurethane high-solid coatings show a decrease in hydrophobicity after modification with epoxyfunctional cyclotriphosphazene containing bulky aromatic substituents and silicon chains [38].

Moreover, the influence of functionalized cyclotriphosphazene on the properties of polyuretane powder coatings has not been studied so far. For this reason, the novelty of this work is the use of hexakis(4-(hydroxymethyl)phenoxy)cyclotriphosphazene as a reactive modifier of polyurethane powder coatings and the examination of its effect on coating properties.

The scope of this research was to investigate the influence of the hexakis(4-(hydroxymethyl)phenoxy)cyclotriphosphazene modifier on the properties of powder coatings based on polyester resin and the polyisocyanate hardener. The anticorrosion protection of the coatings, gloss, water contact angle, roughness, cupping, hardness, adhesion, and scratch resistance of the coatings was examined. The coating's thermal properties were tested using the thermogravimetric method (TGA). Additionally, fire protection properties were examined using the furnace test method.

2. Experimental Section

2.1. Materials

Hexachlorocyclotriphosphazene (HC) and 4-hydroxybenzaldehyde were purchased from Sigma Aldrich (Darmstadt, Germany). Potassium carbonate, tetrahydrofuran (THF), sodium tetrahydroborate ($NaBH_4$), dichloromethane, ethanol, and methanol were purchased from Merck (Darmstadt, Germany).

Sirales PE 6110—hydroxyl functionalized saturated polyester resin containing isophthalic acid as well as neopentyl glycol (acid value: 4–8 mg KOH/g, hydroxyl value: 30–45 mg KOH/g) was supplied by Sir Industriale (Macherio, Italy). A blocked Vestagon B1530 polyisocyanate, which contained 14.8–15.7% of the isocyanate group, was delivered from Evonik Industries (Marl, Germany). A Resiflow PV 88 flow control agent was provided by Worlèe Chemie GmbH (Lauenburg, Germany), and benzoin (degassing agent) was supplied from Sigma Aldrich (Darmstadt, Germany).

2.2. Synthesis of Hexakis(4-(hydroxymethyl)phenoxy)cyclotriphosphazene (HHPCP)

Hexakis(4-(hydroxymethyl)phenoxy)cyclotriphosphazene (3) was obtained in a two-step nucleophilic substitution reaction of Cl atoms in hexachlorocyclotriphosphazene (1) with 4-hydroxybenzaldehyde in the presence of potassium carbonate, leading to the formation of hexakis(4-formylphenoxy)cyclotriphosphazene (2) in the first step and a further reduction in the aldehyde groups present in (2) to the hydroxyl ones carried out in the presence of sodium borohydride in the second step.

Based on the synthetic path illustrated in Scheme 1, in the first step, hexachlorocyclotriphosphazene (1) (5 g, 14.38 mmol), K_2CO_3 (28.52 g, 206.35 mmol), and 4-hydroxybenzaldehyde (10.54 g, 86.3 mmol) in THF (150 mL) as a solvent were placed together in a 250 mL round-bottom flask with a gas funnel and condenser attached. The reaction mixture was then refluxed under an Ar atmosphere until the complete conversion of substrates was monitored with FT-IR spectroscopy. The solvent was then evaporated, and the product was extracted three times with dichloromethane from the residue. Hexakis(4-formylphenoxy)cyclotriphosphazene (2) as a crystalline solid was isolated by vacuum dichloromethane evaporation (98.60% yield) and subjected to FT-IR, ^{1}H, ^{13}C, and ^{31}P NMR analysis, which confirmed its structure [^{1}H NMR ($CDCl_3$): 7.13 (d, 12H), 7.72 (d, 12H), 9.91 (s, 6H) ppm; ^{13}C NMR ($CDCl_3$): 121.29, 131.46,

133.83, 154.56, 190.50 ppm; ^{31}P NMR ($CDCl_3$): 7.09 ppm]. The spectroscopic analysis results for the hexakis(4-formylphenoxy)cyclotriphosphazene (2) are consistent with data from the literature [23]. The FT-IR, ^{1}H, ^{13}C, and ^{31}P NMR spectra of hexakis(4-formylphenoxy)cyclotriphosphazene (2) are presented in Figures S1–S4 (see Supplementary Materials).

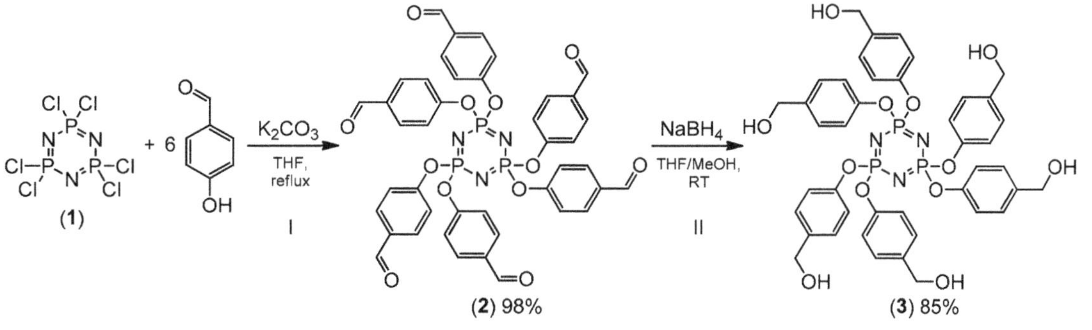

Scheme 1. Synthetic path for hexakis(4-(hydroxymethyl)phenoxy)cyclotriphosphazene (3) from hexachlorocyclotriphosphazene (1) through the formation of hexakis(4-formylphenoxy)cyclotriphosphazene (2).

In the second step, obtained hexakis(4-formylphenoxy)cyclotriphosphazene (2) (10 g, 11.6 mmol) was placed in a 250 mL round-bottom flask equipped with a condenser and gas funnel and dissolved in 200 mL of a THF/MeOH (1:1) mixture. Next, $NaBH_4$ (2.85 g, 75.31 mmol) was slowly added to the phosphazene (2) solution in RT (room temperature). The reaction was carried out in RT to complete the conversion of CHO groups (evaluated based on the band intensity at 1702 cm^{-1} on the FT-IR spectra of the reaction mixture in Figure 1). After the completion of the reaction and solvent evaporation, the obtained crude product was dissolved in EtOH, precipitated with water, filtered, and washed three times with water. Dried at 80 °C, hexakis(4-(hydroxymethyl)phenoxy)cyclotriphosphazene (3) as a white powder (85.3% yield) was subjected to the FT-IR, ^{1}H, ^{13}C, and ^{31}P-NMR analysis which confirmed its structure [^{1}H NMR (DMSO-d6): 4.47 (d, 12H), 5.23 (t, 6H), 6.80 (d, 12H) 7.20 (d, 12H) ppm; ^{13}C NMR (DMSO-d6): 62.28, 120.13, 127.70, 139.45, 148.62 ppm; ^{31}P NMR (DMSO-d6): 8.86 ppm]. The spectroscopic analysis results for hexakis(4-(hydroxymethyl)phenoxy)cyclotriphosphazene (3) are consistent with data from the literature [23]. The FT-IR, ^{1}H, ^{13}C, and ^{31}P NMR spectra of hexakis(4-(hydroxymethyl)phenoxy)cyclotriphosphazene (3) are presented in Figures S5–S8 (see Supplementary Materials).

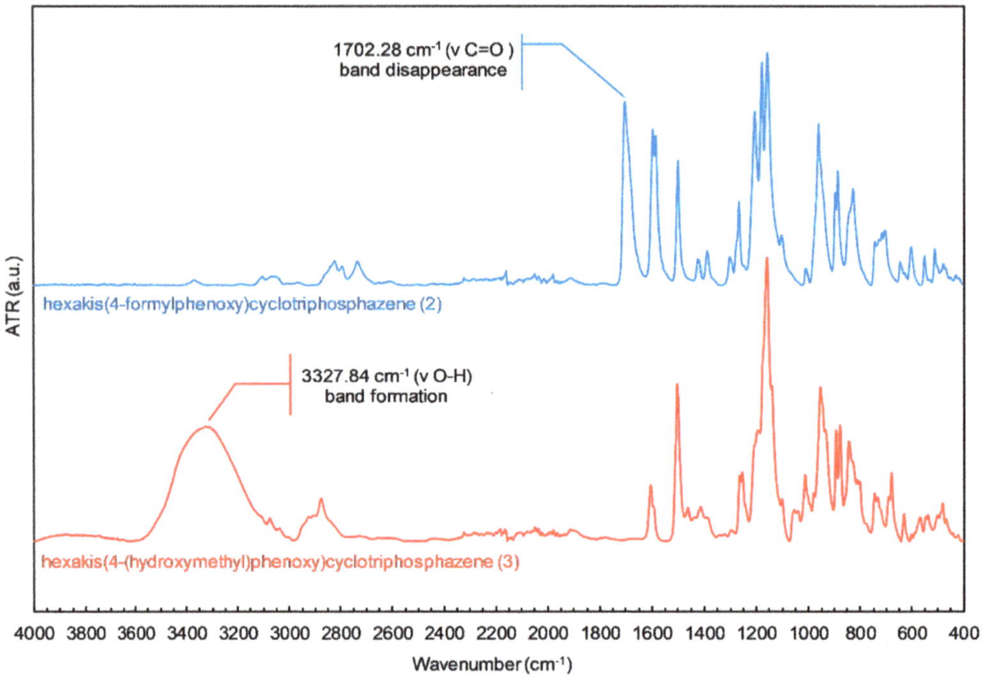

Figure 1. Stacked FT-IR spectra of hexakis(4-formylphenoxy)cyclotriphosphazene (2) as a substrate and hexakis(4-(hydroxymethyl)phenoxy)cyclotriphosphazene (3) as a product.

2.3. *Method of Preparation of Powder Coatings*

The formulations of polyurethane powder coatings included a blocked polyisocyanate Vestagon B1530 and a polyester resin PE6110 in the molar ratio of the -NCO:-OH group equal to 1:1, benzoin as degassing agents (0.25 wt%), Resiflow PV 88 as the leveling agent (0.5 wt%), which consists of a liquid acrylic polymer that has been adsorbed onto a silica filler, as well as hexakis(4-(hydroxymethyl)phenoxy)cyclotriphosphazene as a modifier (1.5 wt%, 2.5 wt% and 5 wt%). The reference sample does not contain the modifier. The PC formulation was milled and extruded in a corotating twin screw mini extruder (EHP 2×12 Sline from Zamak (Poland). The temp. distribution in the extruder was as follows: zone I—90 °C, zone II—120 °C, zone III—122 °C, and adapter—125 °C. The screw rotational speed was 120 r.p.m. After extrusion, the mixture was cooled, crushed, pulverized, and subsequently sieved using a 100 µm sieve. Using an electrostatic gun, PEM X-1, controlled by EPG Sprint X (CORONA) from Wagner (Altstätten, Switzerland), the final powder coatings were applied to the steel panels made of DC01 stamping steel, cold-rolled, 0.5 mm thick, which were previously cleaned and prepared. The electrode voltage was 30 kV. The powder particles in the gun were electrified by an electrode positioned in the nozzle after being transported from the tank by compressed air. The charged particles were then transferred to ground steel plates. To prepare the surface, the steel plates were first degreased with acetone and then immersed in a 1.5% aqueous solution of ESKAPHOR Z 2000C, pH = 5.5 at a temperature of 30 °C for 2 min to apply the zirconium phosphate conversion coating. The panels were then rinsed with distilled water and dried after the water was removed. The coatings were then cured in an oven at 220 °C for 15 min. The names given to the cured coatings were based on the modifier content; for example, PU/1.5%Ph means polyurethane coating with a 1.5% cyclophosphazene modifier (Table 1).

Table 1. Formulations of powder coatings.

Materials	Vestagon B 1530 [%]	Sirales PE 6110 [%]	Phosphazene Modifier [%]	Benzoin [%]	Resiflow PV 88 [%]
Coating Symbol					
PU/0.0%Ph	17.75	81.50	0.00	0.25	0.50
PU/1.5%Ph	19.23	78.72	1.50	0.25	0.50
PU/2.5%Ph	20.55	76.20	2.50	0.25	0.50
PU/5%Ph	20.25	74.00	5.00	0.25	0.50

2.4. Coatings Characterization

2.4.1. Polymerization Test

The polymerization test was carried out in accordance with Qualicoat specifications [20]. A cotton wool swab was saturated with the MEKO solvent. Within 30 s, the powder coating was lightly rubbed 30 times in each direction. After 30 min, the polymerization quality was evaluated in accordance with the following ratings:

1. The coating is quite soft and matt;
2. The coating is matt and can be scratched with a fingernail;
3. Gloss loss is less than 5 units;
4. No noticeable changes. The coating cannot be scratched with a fingernail.

2.4.2. Glow Discharge Optical Emission Spectrometry (GD-EOS)

Glow discharge optical emission spectroscopy analysis (GD-OES) was performed on coatings with a different amount of the phosphazene modifier on the steel substrate. The GD-OES depth profiles were achieved using a GD Profiler HR (Horiba Jobin Yvon). All analyses were conducted under the same sputtering conditions, that is, with an Ar pressure of 700 Pa and 30 W power. A copper anode of 4 mm in diameter was used. The analysis used Ar with a very high purity of 6 N. As a result of the different content of the phosphazene modifier, the phosphorus profiles of the coatings were analyzed. A higher light intensity corresponded to a higher concentration of measured elements at a certain depth. The wavelength of 178 nm was used for light detection emitted by phosphorous.

2.4.3. Performance Properties

A Mar Surf PSI profilometer (Mahr GmbH, Göttingen, Germany) was used to measure the roughness of the coating in accordance with the ISO 12085 [39] standard. The measuring needle was positioned on the cured coating, and the profilometer was leveled. The device then automatically conducted the measurement by moving the needle along the surface of the coating. Measurements were taken at multiple locations on the coating. Ten coated panels were tested, and the arithmetic mean was calculated. The roughness value was determined using the Ra parameter (arithmetic mean of the roughness profile deviated from the baseline) and the Rz parameter (arithmetic mean of the five highest-profile hills decreased by the arithmetic mean of the five lowest-profile depths).

The coating gloss and thickness were measured using a micro-TRI-gloss µ gloss meter from Byk-Gardner (Geretsried, Germany) in accordance with the ISO 2813 [40] (gloss) and ISO 2808 [41] (thickness) standard. Ten coated panels were tested, and the arithmetic mean was calculated.

The surface adhesion of the coatings was assessed by a cross-cut test in accordance with the ISO 2409 [42] standard. Six cutters knife from Byk-Gardner (Geretsried, Germany) was used to make a cut through the coating into the substrate. The surface of the incision network was examined with the naked eye and classified on a six-point scale from 0 to 5. The best surface adhesion was observed in coatings classified with a score of 0, where the edges of the incision were completely smooth. Conversely, the worst adhesion was marked with a score of 5, where the damaged surface of the incision network exceeded 65%. Measurements were taken at 3 coated panels of each composition, and the arithmetic mean was calculated.

The relative hardness of the cured powder coatings was examined using the König Pendulum tester from BYK-Gardner (Geretsried, Germany) in accordance with the ISO 1522 [43]. The relative hardness value was determined as the ratio of the damping time of an oscillating pendulum supported on the coating surface to the time noted for the pendulum supported on a glass plate. The test was conducted in three locations on the coated steel plates, with two plates of each composition examined, resulting in six measurements for each sample; then, the arithmetic mean was calculated.

The cupping of the cured powder coatings was evaluated by a manual cupping tester in accordance with the ISO 1520 [44]. Cupping is the minimum depth, expressed in millimeters, at which the coating is damaged during mechanical deformation.

The scratch resistance was determined using the Elcometer Clemen Tester (Manchester, UK) in accordance with ISO 1518-1 [45]. Three coated panels were tested for each composition. The measurement involved determining the minimum load required to scratch the coating using a needle with a semicircular tip.

2.4.4. Water Contact Angle

The water contact angle was measured by the 'sitting drop' method using an OCA15 EC optical goniometer by Data Physics according to the EN 828 [46] standard. The drop volume was 1 μL. The measurement temperature was 24 ± 1 °C. The drop images were captured with a camera. The contact angle was determined using a control program, following the prior establishment of the baseline and drop contour. The arithmetic mean of 10 measurements was taken as the final result.

2.4.5. Thermogravimetric Analysis

Thermogravimetric analysis was carried out using a Mettler Toledo TGA/DSC1 instrument (Mettler Toledo, Columbus, OH, USA). The measurements were performed in a nitrogen atmosphere in the temperature range of 25 to 700 °C at a heating rate of 10 °C/min. The sample weight was ~5 mg, the gas flow rate was 50 cm^3/min, and a 150 μL open alumina pan was used.

2.4.6. Furnace Tests

A gas flame-heated programmable laboratory furnace, equipped with two slots for steel plates coated with the tested material, was used to measure the thermal insulation properties of the prepared intumescent coatings. During the furnace test, two coated steel samples (ca. 10×10 cm) were vertically placed in a furnace. The furnace temperature was regulated according to the standard cellulosic fire curve described in ISO 834-2 [47] by a control unit of a gas burner fed with a propane–butane mixture. The temperature values, as well as the temperature of the steel substrates, were tracked using K-type thermocouples and continuously recorded. Each test continued until the steel temperature reached 350 °C (the critical temperature value measured on the backside of the tested sample). Three coated panels were tested, and the arithmetic mean was calculated.

2.4.7. Corrosion Resistance

The cuts were made on the coating in accordance with ISO 17872 [48] in the shape of a letter X, through the coating to the metal using a special knife so that the ends of the incisions were 20 mm from the edge of the tile. The samples were put in a 3.5% NaCl solution at 35 °C for 650 h and 720 h in order to test the corrosion resistance in accordance with ISO 2812-1 [49]. The degree of delamination and corrosion around the scratches was evaluated according to the ISO 4628-8 [50] standard. Three coated panels were tested, and the arithmetic mean was calculated.

3. Results and Discussion

3.1. Characteristics of the Powder Coatings Composition and Manufacturing Process

The polyurethane powder coatings containing 1.5 wt%, 2.5 wt%, and 5 wt% of the cyclophosphazene modifier were obtained as a result of the addition of hexakis(4-(hydroxymethyl)phenoxy)cyclotriphosphazene to the mixture containing polyester resin and special additives for powder systems. The composition of the mixture is presented in Table 1. Chemical modification was used to modify standard polyurethane powder coatings. The amount of cross-linking agent was chosen such that it could react with the hydroxyl groups of the resin and phosphazene. During the cross-linking process, the hydroxyl groups derived from the modifier reacted with the isocyanate groups of the polyisocyanate, which allowed them to be chemically incorporated into the coating structure. The modifier was added to the mixture at the premilling stage. During the extrusion of the modified powder mixture, no negative effect of the modifier on the homogenization of the powder coating components was observed. Due to the high melting point of the modifier, a higher curing temperature (220 °C) was used for standard polyurethane powder coatings (180 °C). After 15 min of curing in the oven, the received coatings were characterized by a smooth and transparent surface, and no defects were also observed.

3.2. Powder Coatings Performance Properties

After the curing process, a polymerization test was performed to verify whether the coating was cross-linked. Coatings were qualified at level 4 according to the polymerization test, which is the highest on this scale. After thirty double rubs, the coatings did not show a perceptible change and could not be scratched with a fingernail. According to the Qualicoat requirements, this is a very satisfactory result. The investigated powder coating exhibits perfect chemical resistance to methyl ethyl ketone. To investigate the influence of the phosphazene modifier on the mechanical properties of the powder coatings obtained, their roughness, adhesion to the steel surface, hardness, gloss, cupping, and scratch resistance were examined. The results of the mechanical properties of the cured powder coatings are shown in Table 2. The modification of polyurethane coatings with cyclophosphazene did not significantly influence the mechanical properties of the polyurethane powder coatings. The values of properties, such as adhesion to the steel substrate, hardness, cupping, and scratch resistance, remained at the same level as those of the reference sample. For example, on a scale of 0–5 (0—best, 5—worst), the obtained coatings were characterized by the highest degree in this scale. The lack of a visible effect of the cyclophosphazene modifier on the mechanical properties of the coating, such as scratch resistance, may be due to a small difference in the energy of bonds occurring in the cyclophosphazene ring (P-N = 209 kJ/mol and P-O = 351 kJ/mol) and polyurethane (N-O = 201 kJ/mol, C-N = 305 kJ/mol, C-O = 358 kJ/mol). The properties, such as hardness and cupping, can be influenced by the presence of aromatic substituents in the structure of the phosphazene modifier, which should increase these parameters. However, probably more important here is the flexible "articulated" action of the ether oxygen atom that connects these aromatic rings with the phosphazene ring. An increase in the gloss value from 101.6 GU to 107.1 GU as the modifier content in the obtained cured coatings increased. The observed increase in gloss indicates the very good compatibility of the modifier with the polyurethane matrix, which may be due to the hydrophilic–hydrophobic structure of both compounds. With the increase in the modifier content, the water contact angle decreased from 84.9 to 83.0 deg. The reduced contact angle for water is a consequence of the hydrophilic structure of the cyclophosphazene modifier.

Table 2. Specifications of the powder coating properties.

Symbol of Coating		PU/0.0%Ph	PU/1.5%Ph	PU/2.5%Ph	PU/5%Ph
Roughness ISO 12085	Ra Rz	1.20 ± 0.02 4.30 ± 0.20	1.09 ± 0.02 3.21 ± 0.18	1.12 ± 0.03 3.33 ± 0.17	0.97 ± 0.05 2.40 ± 0.14
Gloss for the angle of 60° ISO 2813	GU	101.6 ± 1.3	106.2 ± 1.5	105.9 ± 1.4	107.1 ± 1.5
Adhesion to the steel surface ISO 2409	0—best 5—worst	0	0	0	0
Hardness ISO 1522	-	0.808 ± 0.002	0.806 ± 0.003	0.806 ± 0.003	0.807 ± 0.003
Cupping ISO 1520	mm	10.66 ± 0.10	10.63 ± 0.07	10.63 ± 0.06	10.62 ± 0.09
Scratch resistance ISO 1518	g	350	350	350	350
Water contact angle EN 828	deg	84.9 ± 2.3	84.7 ± 2.5	84.0 ± 2.1	83.0 ± 2.1

3.3. Phosphorus Distribution in the Cured Coatings

The GD-OES depth profiles that illustrate the intensities of light emitted by phosphorus as a function of depth are shown in Figure 2. The concentration of phosphorus in powder coatings was qualitatively estimated. The intensity of light emission by phosphorus from the PU/0.0%Ph sample showed a value at the background level and correlated with a phosphorus concentration of zero. Coatings containing the cyclophosphazene modifier showed a higher intensity of light emission by phosphorus, which increases with the increasing cyclophosphazene content in the sample. This indicates the presence of phosphazene in these coatings, the content of which increases by the order of PU/1.5%Ph, PU/2.5%Ph and PU PU/5%Ph. The phosphorus content on the surface of the coatings slightly increases, while at a depth above 2 µm, it remains constant. The observed increase in phosphorus concentration at the coating surface may be related to the migration of phosphazene-derived segments toward the coating surface. This migration is a result of the thermodynamic incompatibility of phosphazene-derived segments with polyurethane segments and an increase in interfacial energy. As a result of the migration of these incompatible segments to the surface, the interfacial energy in the coating is reduced. This migration is inhibited by chemical bonds formed between the cyclotriphosphazene hydroxyl groups and polyisocyanate during the coating cross-linking process. A similar process was observed in our earlier research on polyurethane powder coatings modified with polysiloxanes or fluoropolyols [51].

3.4. Thermal Stability

Thermogravimetric analysis was performed to investigate the effect of cyclophosphazene on the thermal stability of fire-protective powder coatings. The TG and DTG curves of the reference PU sample and a sample that contained 5% cyclophosphazene are presented in Figure 3.

The unmodified polyurethane coating decomposes in two stages in the temperature range of 320–520 °C. The first step occurs at the temperature of the maximum mass loss rate of T_{max1} = 337 °C and is related to the scission of the urethane bond, while the second step (T_{max2} = 467 °C) is associated with the degradation of the polyester segments. The mass losses in the first and second steps amount to 23.8% and 71.2%, respectively. The temperature of 5% mass loss is 258 °C. A similar decomposition pattern is observed for the modified sample, but the thermal decomposition range is narrower (320–450 °C). The modification of the sample resulted in an increase in temperature of 5% of the mass loss by approximately 22 °C, which confirms the positive impact of cyclophosphazene on the thermal stability of the coating. That increase in thermal stability may also result from the increase in cross-link density as a result of the incorporation of hexafunctional cyclophosphazene. The T_{max} of the first degradation stage is higher (T_{max1} = 385 °C), but T_{max2} is lower, 420 °C compared to the T_{max} of the unmodified sample. This phenomenon may be related to the catalytic effect of phosphoric and polyphosphoric acids formed during the decomposition of phosphazene on the degradation of polyester segments, as reported in Ref. [52]. Moreover, the decomposition of a phosphazene-modified sample occurs in a narrower temperature range than that of an unmodified sample, and it ends at a lower temperature value. Thus, the formation of solid residue after degradation, which acts as an insulating layer and prevents further degradation, occurs earlier. Additionally, the amount of char yield, i.e., 8%, is higher for the modified sample compared to 5% for pristine polyurethane. The higher char yield enhances the fire retardancy of the polyurethane coatings.

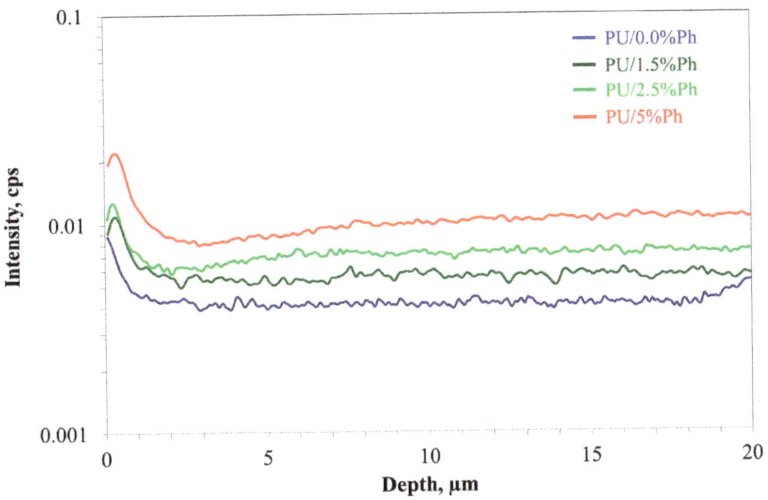

Figure 2. GD-OES depth profiles illustrating the intensity of light emitted by phosphorus as a function of depth.

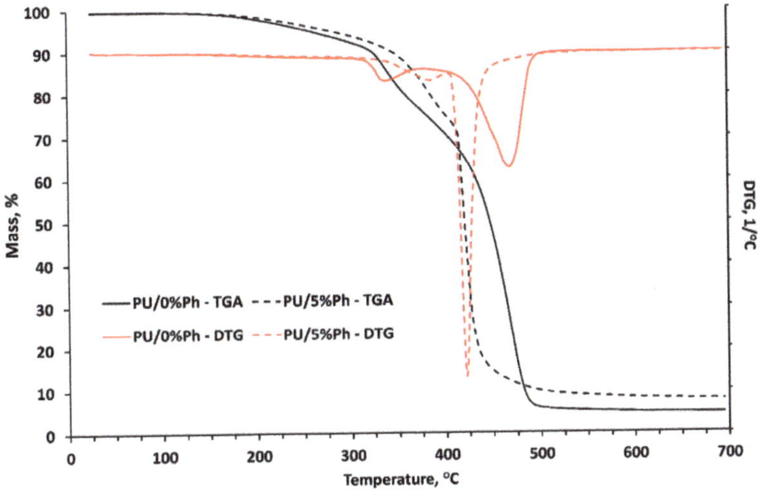

Figure 3. TG and DTG curves of samples PU/0.0%Ph and PU/5%Ph.

3.5. Thermal Protection

The furnace test was used to study the thermal-resistant character of the modified PU powder coatings. The profiles of temperature changes (of the coated steel substrates) in relation to the test duration are shown in Figure 4, while the photographs of the coatings, taken after the fire test, are presented in Figure 5. Detailed time values at which the coating reached a temperature of 350 °C are shown in Table 3.

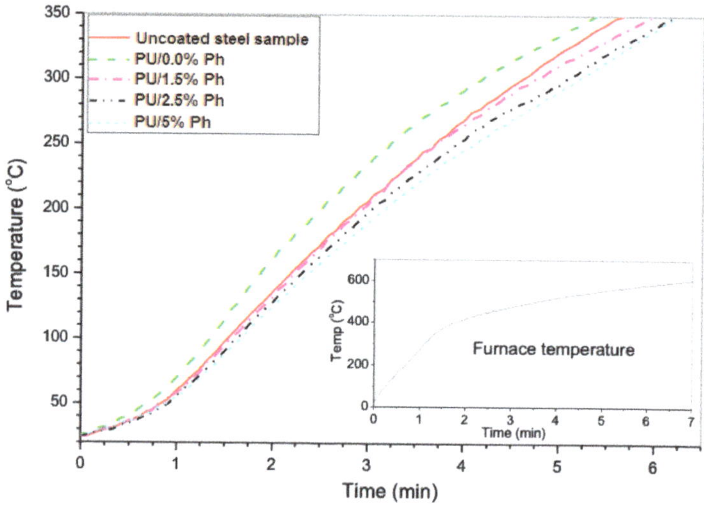

Figure 4. Temperature–time profiles for coated steel samples exposed to standard cellulose fire heating in the furnace.

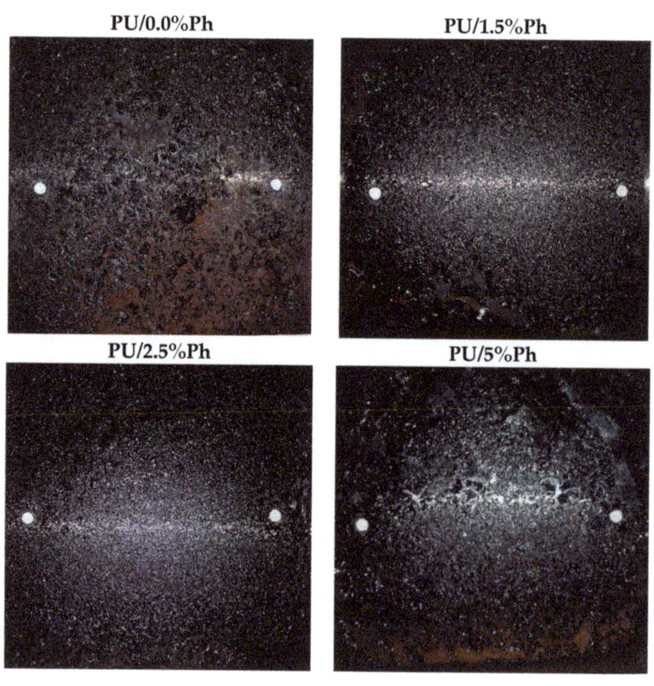

Figure 5. Photographs of powder coatings after the furnace test.

Table 3. Final time of reaching temperature of 350 °C.

Sample	Time	Coating Thickness, μm
Uncoated steel sample	5 min 40 s ± 1 s	-
PU/0.0%Ph	5 min 25 s ± 1 s	73.6 ± 1.2
PU/1.5%Ph	6 min ± 1 s	71.9 ± 1.0
PU/2.5%Ph	6 min 12 s ± 1 s	73.2 ± 1.1
PU/5%Ph	6 min 13 s ± 1 s	72.1 ± 1.0

For the uncoated sample and the sample coated with the reference paint, the increase in the steel substrate temperature was greater than for the plates painted with the cyclotriphosphazene-containing coatings. It was shown that the addition of hexakis(4-(hydroxymethyl)phenoxy)cyclotriphosphazene to the polyurethane powder coatings improved their thermal resistance. The uncoated steel sample reached 350 °C after 5 min 40 s of heating, while the PU/1.5%Ph sample reached this temperature 20 s later. Increasing the modifier amount in the coating by 1% increased the time by another 12 s (2.5%Ph sample). An additional increase in the additive concentration in the coating did not significantly increase the time required to reach the specified temperature of the steel substrate.

Differences in the time at which the coating reached a temperature of 350 °C for the tested samples were not significant due to the fact that the tested coatings had a small thickness in the range of 71.9–73.6 μm. A greater difference in the time needed to reach 350 °C is possible when the coating is thicker. However, the thickness of the tested powder coatings complies with the technical requirements of Qualicoat [28]. In the case of powder coatings, as opposed to liquid coatings, increasing their thickness results in the deterioration of mechanical properties.

Figure 5 shows the surface of the coatings after the furnace test, which had charred and created a heat insulation barrier between the unburned parts.

As confirmed by TG studies, the amount of char increases with the increasing cyclophosphazene content. Moreover, the charring of the unmodified sample is due to its loose structure, whereas the carboceneous phosphazene-modifed coatings are significantly more compact. The tighter char structure is due to the formation of phosphoric acid, metaphosphoric acid, and polyphosphoric acid during the decomposition of HHCPC, which remains in the condensed phase [23]. The liquid or semi-liquid states of these compounds cause an increase in volume and seal the char layer, which is a better protective shield to prevent the flammable volatiles from penetrating the surface of the coating into the flame zone during the process of burning. The surface morphology of the char residue after the furnace test shows an alveolate structure, which indicates that non-flammable gas, such as water, was released during combustion, which was also observed by Xu et al. during the decomposition of HHPCP [23]. Non-flammable gas and flammable volatiles, such as toluene, benzene, and the benzene fragments generated and volatilized into the gaseous phase, cannot easily penetrate the compact char layer into the flame zone during the burning process, which contributes to slowing down this process. In addition, a thicker and more compact layer of char slows down the heat transfer from the flame to the unburned part more effectively, protecting the substrate than the loose and thinner layer formed by reference coatings.

3.6. Corrosion Properties

An examination and explanation of the anticorrosion effect of HHPCP-modified coatings is a novelty of this work. To assess the protective properties of the coating against corrosion, immersion tests were performed. The coatings on the steel substrates were placed in a 3.5% NaCl solution at 35 °C for 650 and 720 h according to ISO 2812-1. After immersion tests, the samples were cleaned, the delaminated coating was removed, and coating damage was evaluated according to ISO 4628 1-10 [53]. According to ISO 4628-8, the degree of coating delamination was calculated using Formula (1), and the degree of corrosion around the scratches occurred according to Formula (2):

$$d = \frac{A_d - A_l}{2} \cdot \frac{1}{l} \quad (1)$$

where:

A_d—delamination area including scratch area, [mm^2];
A_l—scratch area, [mm^2], 30 mm^2;
l—scratch length, [mm], 100 mm.

$$c = \frac{w_c - w}{2} \quad (2)$$

where

w_c—average width of the corrosion zone, [mm];
w—width of the original crack, [mm], 0.3 mm.

The corrosion protection performance of the various coatings containing different amounts of the cyclophosphazene modifier is shown in Table 4, and photographs of modified powder coatings after 650 and 720 h of exposure to a 3.5% NaCl solution are shown in Figure 6. The first measurement was taken after 650 h.

Table 4. Characteristics of the powder coating damage after 650 and 720 h of exposure to a 3.5% NaCl solution.

Symbol of Coating	Degree of Delamination around the Scratches, mm		Degree of Corrosion, around the Scratches, mm	
	650 h	720 h	650 h	720 h
PU/0.0%Ph	6.80 ± 0.05	8.20 ± 0.04	2.63 ± 0.02	3.75 ± 0.02
PU/1.5%Ph	no damages	6.23 ± 0.05	1.08 ± 0.02	1.80 ± 0.02
PU/2.5%Ph	no damages	6.85 ± 0.03	0.52 ± 0.01	1.73 ± 0.01
PU/5%Ph	no damages	3.05 ± 0.02	0.10 ± 0.01	0.43 ± 0.01

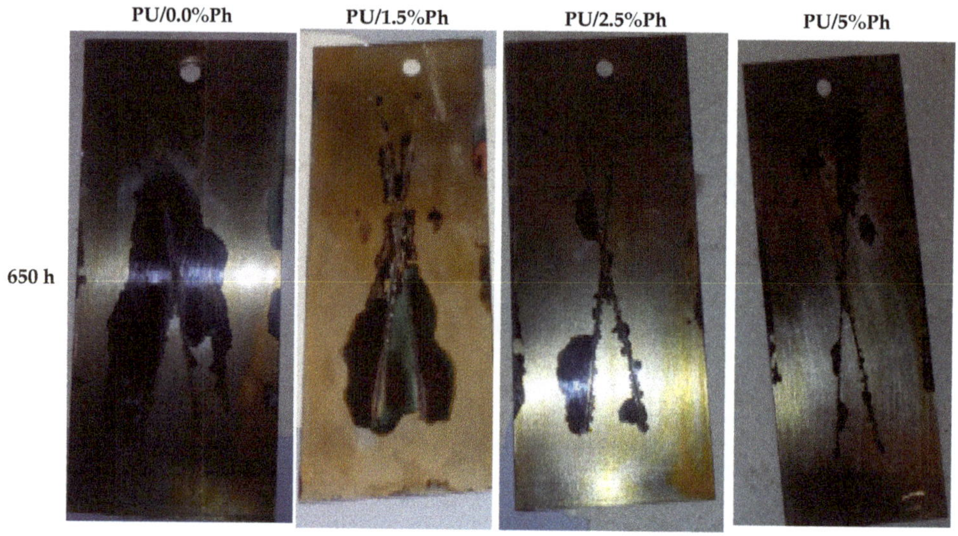

Figure 6. *Cont.*

720 h

Figure 6. Photographs of the coatings taken after 650 and 720 h of exposure to a 3.5% NaCl solution.

Based on the results obtained, it was found that the sample containing 5% cyclophosphazene revealed the best anticorrosive properties. No delamination occurred around the scratches, and the degree of corrosion around the scratches was the lowest in the case of these samples. The largest delamination and corrosion around the scratches was observed in the case of a reference sample. The subsequent measurement was made after 720 h. After this time, delamination occurred for each coating sample. However, the lowest degree of delamination around the scratches was observed in the sample containing 5% cyclophosphazene. Delamination below d < 5 mm indicates a positive test result. For each coating sample, the degree of corrosion around the scratches increased. The results confirmed that a higher amount of modifier improved the anticorrosion properties of the coating. Due to the chemical structure of the phosphazene ring (three conjugated double bonds and free electron pairs on nitrogen atoms), HHPCP can show electroactive properties, protecting the steel substrate against corrosion not only by a barrier but also by anodizing, as confirmed by Tansug et al. in the case of coatings with the addition of electroactive polymers such as polypyrrole or polyaniline [54]. The P heteroatoms contained in the phosphazene ring can adsorb on the metal surface in two ways: by physical adsorption or chemisorption, protecting it against corrosion [26]. In addition, this effect is enhanced due to the higher concentration of phosphazene-derived segments at the coating surface, which was confirmed by the GD-OES technique. The corrosion protection of the coating modified with HHPCP is related to the barrier effect of the polyurethane binder. HHCPC has a rather polar structure, as evidenced by the lack of increase in the water contact angle, which does not have a positive effect on increasing barrier protection.

4. Conclusions

As part of this study, the influence of the hexakis(4-(hydroxymethyl)phenoxy)cyclotri phosphazene modifier on the properties of PU powder coatings was investigated, which was the novelty of this work. The applied modification by the chemical incorporation of the modifier into the coating structure prevents the uncontrolled release of the modifier during exploitation. The release of the modifier is impossible due to permanent chemical bonding. When the modifier forms only a physical mixture without chemical bonds, it is easy to release the modifier into the environment during exploitation, leading to the deterioration of the coating performance.

We proved that HHPCP improved the thermal and anticorrosion resistance of the coatings. The HHPCP-modified coatings revealed increased fire protection compared to that of the unmodified sample; however, due to the technical requirements of the powder coating thickness, the improvement was not significant. The appropriate thickness of the powder coating is necessary so that such properties of the coating, such as gloss, roughness, hardness, scratch resistance, cupping, water contact angle, and adhesion to steel, do not deteriorate. These parameters were maintained at a very high level, especially adhesion to the substrate, which is crucial in ensuring adequate anticorrosion protection of the steel.

The properties of cyclotriphosphazene-modified coatings are highly advantageous when they are used to provide protection for steel or other materials against different environmental influences, especially corrosion. Due to these improved properties, modified coatings with success can be used to protect materials used outdoors, especially susceptible to corrosion, such as steel exploited in highly corrosive environments (e.g., corrosivity category C4 and C5). This property extends the service life of treated materials compared to those which are protected with unmodified coatings.

Based on this research, it can be concluded that the HHPCP modifier can be successfully used as a green alternative to obtain environmentally friendly halogen- and zinc-free powder coatings with increased thermal stability and corrosion resistance. The optimal content of the phosphazene modifier in the powder coating should be 5 wt.%.

Supplementary Materials: The following supporting information can be downloaded at: https://www.mdpi.com/article/10.3390/ma17112672/s1, Figure S1: FT-IR spectrum of Hexakis(4-formylphenoxy)cyclotriphosphazene (2); Figure S2: ^{1}H-NMR spectrum of Hexakis(4-formylphenoxy) cyclotriphosphazene (2) in $CDCl_3$; Figure S3: ^{13}C NMR spectrum of Hexakis(4-formylphenoxy)cyclotri phosphazene (2) in $CDCl_3$; Figure S4: ^{31}P NMR spectrum of Hexakis(4-formylphenoxy)cyclotriphos phazene (2) in $CDCl_3$; Figure S5: FT-IR spectrum of Hexakis(4-(hydroxymethyl)phenoxy)cyclotriphos phazene (3); Figure S6: ^{1}H NMR spectrum of Hexakis(4-(hydroxymethyl)phenoxy)cyclotriphospha zene (3) in DMSO-d6; Figure S7: ^{13}C NMR spectrum of Hexakis(4-(hydroxymethyl)phenoxy)cyclotri phosphazene (3) in DMSO-d6; Figure S8: ^{31}P NMR spectrum of Hexakis(4-(hydroxymethyl)phenoxy) cyclotriphosphazene (3) in DMSO-d6.

Author Contributions: Conceptualization, B.P.-P.; Methodology, D.C.-J., Ł.B., M.D., R.J., K.K., W.J.N. and K.P.; Investigation, D.C.-J., Ł.B., M.D., R.J., K.K., W.J.N. and K.P.; writing: B.P.-P., D.C.-J., Ł.B., M.D., R.J., K.K. and K.P. All authors have read and agreed to the published version of the manuscript.

Funding: This research was funded by Minister of Science and Higher Education Republic of Poland within the program "Regional Excellence Initiative", grant number RCD.RB.24.002. This work was supported by the Polish National Science Centre, Grant No. UMO-2020/37/B/ST5/03266.

Institutional Review Board Statement: Not applicable.

Informed Consent Statement: Not applicable.

Data Availability Statement: The original data presented in the study are included in the article. Additional raw data supporting the conclusions of this article will be made available by the authors on request.

Acknowledgments: The authors would like to thank Sir Industriale from Italy as well as Evonic Degussa GmbH and Worlèe Chemie GmbH from Germany for sending free samples of raw materials.

Conflicts of Interest: The authors declare no conflict of interest.

References

1. Thomas, J.; Patil, R. Enabling Green Manufacture of Polymer Products via Vegetable Oil Epoxides. *Ind. Eng. Chem. Res.* **2023**, *62*, 1725–1735. [CrossRef]
2. Thomas, J.; Singh, V.; Jain, R. Synthesis and characterization of solvent free acrylic copolymer for polyurethane coatings. *Prog. Org. Coat.* **2020**, *145*, 105677. [CrossRef]
3. Pilch-Pitera, B.; Czachor, D.; Kowalczyk, K.; Pavlova, E.; Wojturski, J.; Florczak, Ł.; Byczyński, Ł. Conductive polyurethane-based powder clear coatings modified with carbon nanotubes. *Prog. Org. Coat.* **2019**, *137*, 105367. [CrossRef]
4. Pourhashem, S.; Ghasemy, E.; Rashidi, A.; Vaezi, M.R. A review on application of carbon nanostructures as nanofiller in corrosion-resistant organic coatings. *J. Coat. Technol. Res.* **2020**, *17*, 19–55. [CrossRef]
5. Golgoon, A.; Aliofkhazrae, M.; Toorani, M. Nanocomposite Protective Coatings Fabricated by Electrostatic Spray Method. *Protect. Met. Phys. Chem. Surf.* **2018**, *54*, 192–221. [CrossRef]

6. Stojanović, I.; Šimunović, V.; Alar, V.; Kapor, F. Experimental evaluation of polyester and epoxy-polyester powder coatings in aggressive media. *Coatings* **2018**, *8*, 98. [CrossRef]
7. Binay, M.I.; Kirdeciler, S.K.; Akata, B. Development of antibacterial powder coatings using single and binary ion-exchanged zeolite A prepared from local kaolin. *Appl. Clay Sci.* **2019**, *182*, 105251. [CrossRef]
8. Zhang, Y.; Xiao, Z.; Liu, C.; Yu, X. Durable superamphiphobic coatings from one-step electrostatic dusting. *Soft Matter* **2019**, *15*, 7374–7380. [CrossRef] [PubMed]
9. Hu, H.; Zhao, S.; Sun, G.; Zhong, Y.; You, B. Evaluation of scratch resistance of functionalized graphene oxide/polysiloxane nanocomposite coatings. *Prog. Org. Coat.* **2018**, *117*, 118–129. [CrossRef]
10. Fernández-Álvarez, M.; Velasco, F.; Bautista, A.; Galiana, B. Functionalizing organic powder coatings with nanoparticles through ball milling for wear applications. *Appl. Surf. Sci.* **2020**, *513*, 145834. [CrossRef]
11. Ahmad, F.; Zulkurnain, E.S.B.; Ullah, S.; Al-Sehemi, A.G.; Raza, M.R. Improved fire resistance of boron nitride/epoxy intumescent coating upon minor addition of nano-alumina. *Mater. Chem. Phys.* **2020**, *256*, 123634. [CrossRef]
12. Tang, B.; Feng, W.; Guo, J.; Sun, J.; Zhang, S.; Gu, X.; Li, H.; Yang, W. Hydrophobic modification of pentaerythritol and its application in fire-retardant coatings for steel structures. *Prog. Org. Coat.* **2020**, *138*, 105391. [CrossRef]
13. Hu, X.; Zhu, X.; Sun, Z. Efficient flame-retardant and smoke-suppression properties of MgAlCO3-LDHs on the intumescent fire retardant coating for steel structures. *Prog. Org. Coat.* **2019**, *135*, 291–298. [CrossRef]
14. Xu, Z.; Liu, D.; Yan, L.; Xie, X. Synergistic effect of sepiolite and polyphosphate ester on the fire protection and smoke suppression properties of an amino transparent fire-retardant coating. *Prog. Org. Coat.* **2020**, *141*, 105572. [CrossRef]
15. Schartel, B. Phosphorus-based Flame Retardancy Mechanisms—Old Hat or a Starting Point for Future Development? *Materials* **2010**, *3*, 4710–4745. [CrossRef] [PubMed]
16. Costes, L.; Laoutid, F.; Brohez, S.; Dubois, P. Bio-based flame retardants: When nature meets fire protection. *Mater. Sci. Eng. R Rep.* **2017**, *117*, 1–25. [CrossRef]
17. Chattopadhyay, D.K.; Webster, D.C. Thermal stability and flame retardancy of polyurethanes. *Prog. Polym. Sci.* **2009**, *34*, 1068–1133. [CrossRef]
18. Puri, R.G.; Khanna, A.S. Intumescent coatings: A review on recent progress. *J. Coat. Technol. Res.* **2017**, *14*, 1–20. [CrossRef]
19. Horacek, H. Preparation and Fire Test of Intumescent Powder Coatings. *OAlib* **2014**, *1*, 1–15. [CrossRef]
20. QUALICOAT. *Specifications for a Quality Label for Liquid and Powder Organic Coatings on Aluminium for Architectural Applications*; QUALICOAT: Zurich, Switzerland, 2022.
21. Usri, S.N.K.; Jamain, Z.; Makmud, M.Z.H. A review on synthesis, structural, flame retardancy and dielectric properties of hexasubstituted cyclotriphosphazene. *Polymers* **2021**, *13*, 2916. [CrossRef]
22. Zhou, L.; Zhang, G.; Yang, S.; Yang, L.; Cao, J.; Yang, K. The synthesis, curing kinetics, thermal properties and flame rertardancy of cyclotriphosphazene-containing multifunctional epoxy resin. *Thermochim. Acta* **2019**, *680*, 178348. [CrossRef]
23. Xu, J.; He, Z.; Wu, W.; Ma, H.; Xie, J.; Qu, H.; Jiao, Y. Study of thermal properties of flame retardant epoxy resin treated with hexakis[p-(hydroxymethyl)phenoxy]cyclotriphosphazene. *J. Therm. Anal. Calorim.* **2013**, *114*, 1341–1350. [CrossRef]
24. Chistyakov, E.M.; Filatov, S.N.; Yudaev, P.A.; Kireev, V.V. Synthesis, characterization and epoxidation of hexakis-4-(2-(4-((β-methallyl)oxy)phenyl)propan-2-yl)phenoxycyclotriphosphazene. *Tetrahedron. Lett.* **2019**, *60*, 444–448. [CrossRef]
25. Rybyan, A.A.; Bilichenko, J.V.; Kireev, V.V.; Kolenchenko, A.A.; Chistyakov, E.M. Curing of DER-331 Epoxy Resin with Arylaminocyclotriphosphazenes Based on o-, m-, and p-methylanilines. *Polymers* **2022**, *14*, 5334. [CrossRef] [PubMed]
26. Dagdag, O.; Kim, H. Progress in the Field of Cyclophosphazenes: Preparation, Properties, and Applications. *Polymers* **2023**, *16*, 122. [CrossRef] [PubMed]
27. Malkappa, K.; Ray, S.S. Thermal Stability, Pyrolysis Behavior, and Fire-Retardant Performance of Melamine Cyanurate@Poly (cyclotriphosphazene-*co*-4,4′-sulfonyl diphenol) Hybrid Nanosheet-Containing Polyamide 6 Composites. *ACS Omega* **2019**, *4*, 9615–9628. [CrossRef]
28. Wen, P.; Tai, Q.; Hu, Y.; Yuen, R.K.K. Cyclotriphosphazene-Based Intumescent Flame Retardant against the Combustible Polypropylene. *Ind. Eng. Chem. Res.* **2016**, *55*, 8018–8024. [CrossRef]
29. Taip, N.A.M.; Jamain, Z.; Palle, I. Fire-Retardant Property of Hexasubstituted Cyclotriphosphazene Derivatives with Schiff Base Linking Unit Applied as an Additives in Polyurethane Coating for Wood Fabrication. *Polymers* **2022**, *14*, 3768. [CrossRef]
30. Shi, Z.-Z.; Gao, X.-X.; Chen, H.-T.; Liu, X.-F.; Li, A.; Zhang, H.-J.; Wang, L.-N. Enhancement in mechanical and corrosion resistance properties of a biodegradable Zn-Fe alloy through second phase refinement. *Mater. Sci. Eng. C* **2020**, *116*, 111197. [CrossRef]
31. Tomachuk, C.R.; Elsner, C.I.; Di Sarli, A.R.; Ferraz, O.B. Corrosion resistance of Cr(III) conversion treatments applied on electrogalvanised steel and subjected to chloride containing media. *Mater. Chem. Phys.* **2010**, *119*, 19–29. [CrossRef]
32. Chen, K.H.; Fang, H.C.; Zhang, Z.; Chen, X.; Liu, G. Effect of of Yb, Cr and Zr additions on recrystallization and corrosion resistance of Al–Zn–Mg–Cu alloys. *Mater. Sci. Eng. A* **2008**, *497*, 426–431. [CrossRef]
33. Radhakrishnan, S.; Sonawane, N.; Siju, C.R. Epoxy powder coatings containing polyaniline for enhanced corrosion protection. *Prog. Org. Coat.* **2009**, *64*, 383–386. [CrossRef]
34. Chang, K.-C.; Hsu, M.-H.; Lu, H.-I.; Lai, M.-C.; Liu, P.-J.; Hsu, C.-H.; Ji, W.-F.; Chuang, T.-L.; Wei, Y.; Yeh, J.-M.; et al. Room-temperature cured hydrophobic epoxy/graphene composites as corrosion inhibitor for cold-rolled steel. *Carbon N. Y.* **2014**, *66*, 144–153. [CrossRef]

35. Cai, S.; Wang, G.; Zhang, S.; Ni, Y.; Huang, J.; Li, Z.; Jin, R. Flexible Ceramic Wear-Resistant Heat-Proof Dual-Anticorrosive. Coating. Patent CN 1279130 C, 11 October 2016.
36. Chen, J.; Liu, X.; Zhao, Z.; Huo, B.; Li, Z.; Sun, H. A Kind of Montmorillonite Intercalation Polymeric Modification Epoxy Resin Based Powder Coating for Pipeline Corrosion Protection and Preparation Method. Thereof. Patent CN 106189682 C, 7 December 2016.
37. Sørensen, P.A.; Kiil, S.; Dam-Johansen, K.; Weinell, C.E. Anticorrosive coatings: A review. *J. Coat. Technol. Res.* **2009**, *6*, 135–176. [CrossRef]
38. Byczyński, Ł.; Dutkiewicz, M.; Januszewski, R.; Rojewski, S. Polyurethane high-solids coatings modified with silicon-containing functionalized cyclotriphosphazene. *Prog. Org. Coat.* **2022**, *172*, 107139. [CrossRef]
39. *ISO 12085*; Geometrical Product Specifications (GPS)—Surface Texture: Profile Method—Motif Parameters. Technical Committees ISO/TC 213; ISO: Geneva, Switzerland, 1996.
40. *ISO 2813*; Paints and Varnishes—Determination of Gloss Value at 20°, 60° and 85°. Technical Committee: ISO/TC 35/SC 9; ISO: Geneva, Switzerland, 2014.
41. *ISO 2808*; Paints and Varnishes—Determination of Film Thickness. Technical Committee: ISO/TC 35/SC 9; ISO: Geneva, Switzerland, 2019.
42. *ISO 2409*; Paints and Varnishes—Cross-Cut Test. Technical Committee: ISO/TC 35/SC 9; ISO: Geneva, Switzerland, 2020.
43. *ISO 1522*; Paints and Varnishes—Pendulum Damping Test. Technical Committee: ISO/TC 35/SC 9; ISO: Geneva, Switzerland, 2022.
44. *ISO 1520*; Paints and Varnishes—Cupping Test. Technical Committee: ISO/TC 35/SC 9; ISO: Geneva, Switzerland, 2006.
45. *ISO 1518-1*; Paints and Varnishes—Determination of Scratch Resistance. Technical Committee: ISO/TC 35/SC 9; ISO: Geneva, Switzerland, 2023.
46. *EN 828*; Adhesives—Wettability—Determination by Measurement of Contact Angle and Surface Free Energy of Solid Surface. European Committee for Standardization: Brussels, Belgium, 2014.
47. *ISO 834-2*; Fire-Resistance Tests—Elements of Building Construction—Part 2: Requirements and Recommendations for Measuring Furnace Exposure on Test Samples. Technical Committee: ISO/TC 92/SC 2; ISO: Geneva, Switzerland, 2019.
48. *ISO 17872*; Paints and Varnishes—Guidelines for the Introduction of Scribe Marks through Coatings on Metallic Panels for Corrosion Testing. Technical Committee: ISO/TC 35/SC 9; ISO: Geneva, Switzerland, 2019.
49. *ISO 2812-1*; Paints and Varnishes—Determination of Resistance to Liquids—Part 1: Immersion in Liquids Other than Water. Technical Committee: ISO/TC 35/SC 9; ISO: Geneva, Switzerland, 2017.
50. *ISO 4628-8*; Paints and Varnishes—Evaluation of Degradation of Coatings—Designation of Quantity and Size of Defects, and of Intensity of Uniform Changes in Appearance—Part 8: Assessment of Degree of Delamination and Corrosion around a Scribe or Other Artificial Defect. Technical Committee: ISO/TC 35/SC 9; ISO: Geneva, Switzerland, 2012.
51. Czachor-Jadacka, D.; Pilch-Pitera, B.; Byczyński, Ł.; Kisiel, M.; Zioło, A. Hydrophobic polyurethane powder clear coatings with lower curing temperature: Study on the synthesis of new blocked polyisocyanates. *Prog. Org. Coat.* **2021**, *159*, 106402. [CrossRef]
52. Byczyński, Ł.; Dutkiewicz, M.; Januszewski, R. Thermal behaviour and flame retardancy of polyurethane high-solid coatings modified with hexakis(2,3-epoxypropyl)cyclotriphosphazene. *Prog. Org. Coat.* **2017**, *108*, 51–58. [CrossRef]
53. *ISO 4628 1-10*; Paints and Varnishes—Evaluation of Degradation of Coatings—Designation of Quantity and Size of Defects, and of Intensity of Uniform Changes in Appearance. Technical Committee: ISO/TC 35/SC 9; ISO: Geneva, Switzerland, 2016.
54. Tansuğ, G.; Tüken, T.; Özyılmaz, A.T.; Erbil, M.; Yazıcı, B. Mild steel protection with epoxy top coated polypyrrole and polyaniline in 3.5% NaCl. *Curr. Appl. Phys.* **2007**, *7*, 440–445. [CrossRef]

materials

MDPI

Article

Preparation and Characterization of Multilayer NiTi Coatings by a Thermal Plasma Process

Sneha Samal *, Jakub Zeman, Stanislav Habr, Oliva Pacherová, Jaromír Kopeček and Petr Šittner

FZU-Institute of Physics of Czech Academy of Science, Na Slovance 1999/2, 18200 Prague, Czech Republic; zemanja@fzu.cz (J.Z.); habr@fzu.cz (S.H.); pacher@fzu.cz (O.P.); kopecek@fzu.cz (J.K.); sittner@fzu.cz (P.Š.)

* Correspondence: samal@fzu.cz; Tel.: +420-266-05-2361

Abstract: The deposition of multilayer coating of NiTi is carried out by a thermal plasma spraying process on a stainless steel substrate. The deposition of melted NiTi particles creates an adhesion layer on the substrate with the subsequent formation of multilayer coating with a certain thickness. Six layers of coating are created to achieve a certain thickness in terms of the sprayed sample. This paper aims to investigate multilayer NiTi coatings created through a thermal plasma process. The key variable feed rate was considered, as well as its effect on the microstructure characteristics. The shape memory effect associated with the coating properties was analyzed in detail. The variable feed rate was considered one of the most important parameters in the thermal plasma spraying process due to its ability to control the quality and compactness of the coating structure. The coatings were characterized by examining their microstructure, thermal, chemical, and microhardness. The indent marks were made/realized along the cross-section surface for the analysis of crack propagation resistance and wear properties. The coating's surface did not display segmentation crack lines. Nevertheless, the cross-sectional surfaces showed evidence of crack lines. There were eutectic zones of the interlamellar structure observed in the structure of the coating. The plasma-sprayed samples from thermo-mechanical analysis of the hysteresis curve provide strong confirmation of the shape memory effect.

Keywords: thick NiTi coating; thermal plasma process; deposition; characterization

1. Introduction

To obtain high thrust and efficient performance in aerospace alloys, NiTi coatings have been implemented on surface alloys and metal surfaces; for example, [1–4] studied the application of ceramic barrier coatings for metals and alloys, which can be used in aggressive environments under temperatures close to their melting point. Metals have excellent mechanical properties and corrosion resistance, and are, therefore, useful in industrial applications [5,6]. However, despite technological advancements in various sectors, it is very challenging to meet the demand with stainless steel performance. NiTi alloy has gained popularity due to its superelasticity, shape memory, and wear resistance; it can be considered to have good potential as a coating for effectively improving surface properties [7]. The implementation of NiTi powder as a coating layer can effectively improve a material's performance, thus extending its durability over its lifetime as service part in the field of wear behavior [8–10]. NiTi coatings are implemented as functional protective layers in cavitation and to prevent erosion behavior by improving the surface properties [11–13]. However, current research mostly focuses on the application of NiTi as bulk material in various aspects of biomedical industries, such as stents and guide wire [14]. Researchers are mostly focused on fatigue behavior, corrosion resistance, and improving the lifetime of NiTi as the primary material to understand superelasticity and martensitic phase transformation processes [15,16]. However, in the field of additive manufacturing, a limited number of researchers [17–22] have studied alternative routes, such as the powder metallurgy route of additive manufacturing, laser melting, and direct energy deposition processes for bulk NiTi materials. The process of powder metallurgy results in the disintegration of the composition from the stoichiometric composition of the 50Ni:50Ti alloy, which could lead to deviations in the behavior of shape memory alloys [23,24]. Proper material design plays an important role in the advancement of the engineering field. Thermal plasma spray technology opens up the possibility of spraying multiple layers in various patterns to enhance a material's surface properties in various potential application areas [25–27].

The purpose of coating stainless steel is that it provides insights into multilayer coatings designed and fabricated using thermal plasma technology. The powder feeding rate varied for each multilayer coating structure on the substrate of stainless steel with fixed experimental parameters using the plasma spraying process. Recently, a thermal plasma process has emerged in the processing of powder coatings on the surfaces of materials that could enhance the performance of the surface properties. A lack of understanding about the compactness of the coating structure arises from the effect of thermal plasma spraying parameters, as identified in our previous studies [28–30]. In this work, two samples were chosen with variable feed rates to create multilayer coating with varying quality and compactness. The multilayer coating was fabricated on the surface of a stainless steel substrate. The lamellar interface between various layers of the coating was examined in terms of dense porosity, and segmentation crack lines. The difference in the coefficient of thermal expansion (CTE) between the substrate and coating layer may influence interfacial cracks during the spraying process [31,32]. The quality of the coating layers, which determines the compactness and porosity generated during the thermal plasma process, may have a strong influence on the performance of the material [33]. Therefore, in this study, multilayer coating was prepared by varying the powder feed rates on the stainless steel substrate. The adhesion between the substrate and the coating and that between the segmented layers was studied and analyzed in consideration of the overall performance of the multilayer coating structure.

This work provides good insight into the preparation of NiTi layers through the use of plasma spraying, both in terms of preparation methods and characterization. This paper aims to investigate multilayer NiTi coatings fabricated through the use of the thermal plasma process. The primary variable feed rate was taken into consideration, along with its impact on the microstructure characteristics. The shape memory effect associated with the coating properties was analyzed in detail. The NiTi coating could be used as a protective layer for the material, enhancing SE and SMA behavior against wear, erosion, and corrosion.

2. Materials and Methods

2.1. Materials

The functional coating was deposited by using NiTi powder as feedstock raw material and stainless steel as the substrate. NiTi powders were purchased, rather by choosing the selective composition of each particle with a major composition of 50 at. % of Ni and Ti with a purity of 99.5% and particle sizes ranging from 20 to 63 μm, from American Elements, MERELEX Corporation, Los Angeles, CA, USA. There were other elements, such as Fe, Al, Cu, Si, Ca, C, O, and N, present in minor amounts (less than $\leq$ 0.01–0.005) [31]. Figure 1a,b displays the particle size and distribution of NiTi powder in a larger view, and one enlarged particle shows the surface features with a red cross mark, where Energy-Dispersive Analysis (EDX) was performed. Figure 1c shows the EDX peaks with the composition of Ni and Ti within the inset table. The dimension of $60 \times 20 \times 3$ mm were considered for the stainless steel.

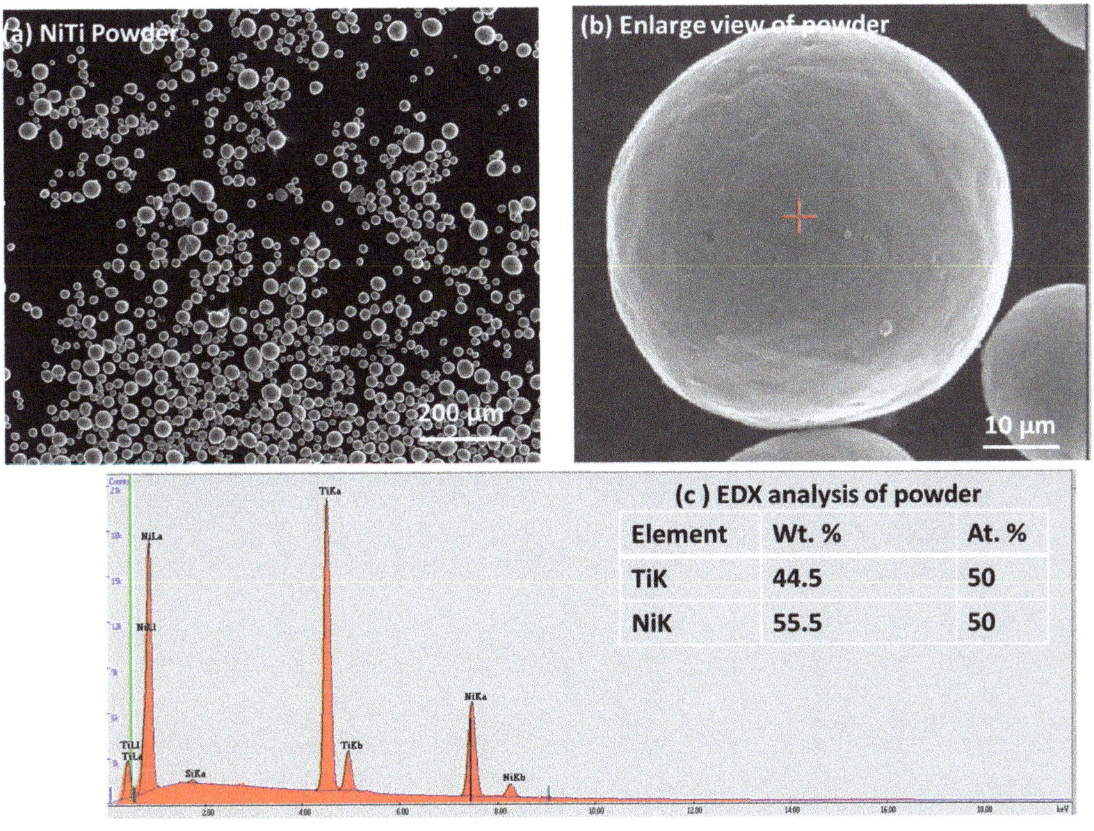

Figure 1. (**a**) Particles of NiTi considered for thermal plasma coating; (**b**) one-particle view with red cross showing EDX location; (**c**) EDX analysis of elements [7] (open access from MDPI).

2.2. Experimental Methods

The samples were fabricated by the facility of thermal plasma spraying (RF-ICP) from IPP, Czech Republic. The coatings were deposited by using natural argon gas with a mixture of hydrogen. Argon was the chosen carrier gas, which was used as a powder feeder to the plasma system. The spraying was performed at a plasma power of 9 kW for both samples with lamellar layers of deposition that developed along the thickness at two different feed rates. A spraying speed of 1 mm/s was chosen for both samples, with preheating for 60 s for each spraying pass. Both samples show six lamellar layers in coating thickness, with the same amount of powder spray and constant power. The variable parameter was the feed rate of powder used in the thermal plasma spraying method. This difference affects the effect, porosity, and interlamellar cracks within the segmented layers, determining the structure of the coating. Figure 2 shows the thermal plasma deposition process. The particles melt along the path of the thermal plasma source, leading to particles impacting the surface. During impact, multiple splats are formed, forming layers of coating. Figure 2 shows the RF-ICP plasma process for coating deposition with particles and the splat, showing the formation of coating layers. Table 1 represents the spraying parameters for the preparation of two selected samples. Table 1 represents the experimental parameters for the preparation of samples 1 and 2 with different feed rates of powder spraying. Argon was used as the plasma gas to create the plasma arc that induces the melting of particles and deposition of the coating layers. The substrate underwent preheating before the application of powders in a plasma arc. The spraying was performed for 60 s six times to maintain thick multilayered coating layers. A 15 s break was considered for the cooling process in between each spraying.

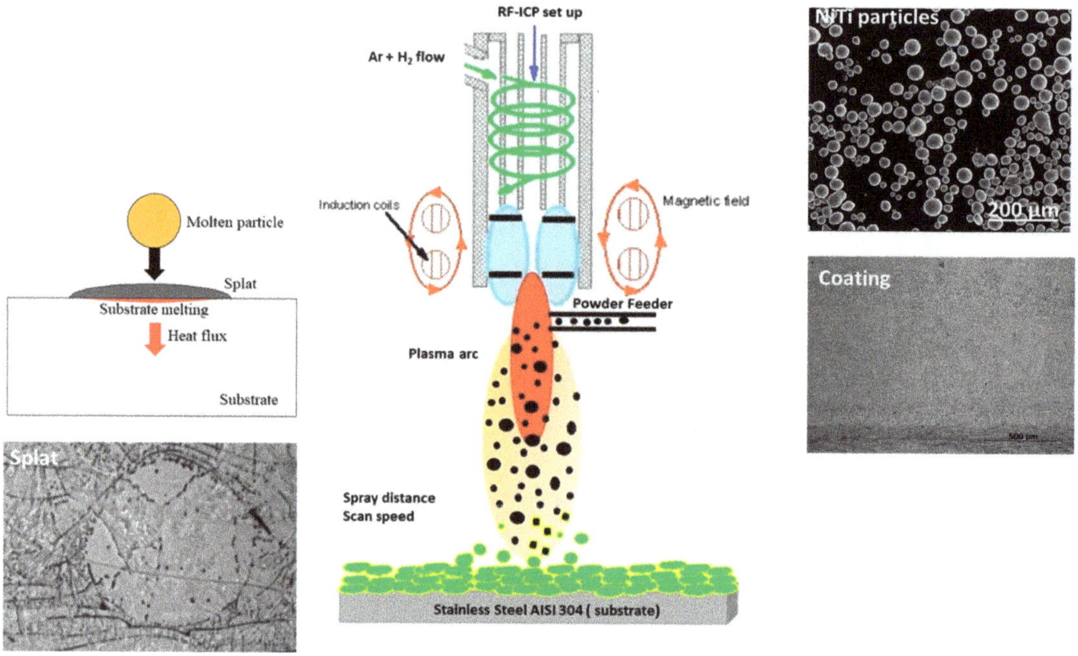

Figure 2. Schematic representation of the spraying process showing powder deposition on the stainless steel substrate [34] (open access from MDPI).

Table 1. Spraying parameter for the multilayered coating for both samples (1 and 2).

Sample	Substrate	Plasma Power (kW)	Powder Feed Rate (g/min)	Moving Speed (mm/s)	Microhardness on Cross-Section (HV)
1	Stainless steel	9	2.1	1	257.4
2			4.2		254.8

Ar (sheath gas, 35, Central, 10) + H_2 (1.0); carrier gas Ar, 8; preheat: 60 s; spraying plan: 60 × 6; net powder spray time: 360 s.

2.3. Plasma-Sprayed Sample Characterization

Various techniques have been used to investigate the physical, microstructural, thermal, and mechanical behavior of the coatings. The surface image and quality of the coating samples were investigated by examining the cross-sectional images of the coatings. Initially, the coatings were separated by cutting the coating layers without and with coating by using an electric discharge machine (FI240 SLP). The surface image was investigated using Tescan FERA 3 (Tescan, Brno, Czech Republic) scanning electron microscopes (SEM), the latter equipped with a field emission cathode. The analyses were performed using either secondary electron imaging for topography or backscattered electrons for chemical Z contrast. Energy-dispersive X-ray spectroscopy (EDS) and line analysis were carried out using the EDAX system (EDAX, Ametek Inc., Mahwah, NJ, USA) with an Octane Super 60 mm^2 detector to determine the homogeneity of the chemical composition. The voltage used for EDS analysis was 15 kV, which was used to ensure the visibility of the K line for Ni. The phases in the plasma-sprayed samples were revealed through the use of an X-ray diffractometer using the PANalytical X'Pert Pro diffractometer (Malvern, UK) with Kα Co radiation. The hardness of the samples was determined by indenting the cross-section surfaces. A load of 1.961 N was used for 10 s for each indent point. Five indents were used for the calculation of the average hardness of the samples. Differential scanning calorimetry (DSC) using the DSC 25 (TA Instruments, New Castle, DE, USA) instrument at a scanning rate of 5 K/min in a temperature range from −100 to +100 °C (powders) and from 0 to +150 °C (samples) was performed. For these tests, small samples were placed and measured in an Al crucible (approximately 13.3 mg). The samples were tested in a three-point bending test at a fixed load of 100 mN for the shape memory effect. The experiment was performed according to the ASTM E831 standard [35] by using a thermo-mechanical analyzer (TMA, Linseis, Germany). The samples were tested within the temperature range of −25 to +130 °C.

3. Results

The microstructure determines the quality of the coating that is evaluated from the cross-section of the image. The porosity and interlamellar cracks along the various coating layers on the surface of the substrate are determined by the quality of the coating. The elemental composition and line analysis along the coating layers have been analyzed. Various phases of NiTi that formed in the coating layers are derived from the phase diagram of Ni and Ti as a function of the temperature. The transformation temperature of the phases in relation to heating and cooling is investigated in the coating samples. The transformation temperature frame was chosen for the sample bending behavior at a constant load to induce shape memory behavior.

3.1. *Microstructural Observation in Coating Layers*

Figure 3a,b presents the cross-section images in the SE mode in both samples. Sample 1 shows minor porosity along the cross-section, which may have developed due to a lower feed rate. The low feeding rate may create a void within the coating layers due to entrapped air or oxygen within the layers. The porosity is prominently visible in sample 1 along the coating layers in the cross-section. When analyzing the coating's cross-section, the segmentation crack lines are visible instead of on the coating's surface. A segmentation crack line is observed in the middle of sample 1 cross-section surface. The crack line appears due to the weak interconnection of particles along the lamellar layer. However, this crack disappears with the better connection of particles at a higher feed rate. Sample 2 shows a crack line at the edge of the top surface that may have been generated during cooling. The various locations of the coating layer are revealed by etching the surface before using an optical microscope. Figure 4a,d displays the overall surface and the lower, middle, and top surface layers. The etching reveals the magnified surface image of the coating layers in various zones. The bottom surface shows larger splat droplets, which are arranged linearly along the thickness of the coating layer. However, the crack line is observed in the overall sample, which corresponds to the location within the upper part of the sample. This may arise due to the inter-surface separation between two adjacent layers due to the lower feed rate. The middle layers show dense, compact layers with minor porosity, which arises from the interconnected neighborhoods of the splat particles. The middle layer shows the columnar lines along the deposition height of the coating layers. Some interfacial cracks are found in the coatings, which may attributed to stress mismatch between the layers. The interfacial cracks arise from oxidized impurities within the layers that arise from the porosity regions. Figure 5a,d displays the overall surface and the bottom, middle, and top surfaces of the coating layers. There are compact layers of coating observed in the bottom layer, although the homogenized, compact microstructure also shows a continuation in the middle layer. There is a very thin crack line observed in the middle layers; however, the overall sample shows good integrity in terms of the coating layers. The segmented crack line shifted to the edge of the upper surface, which may arise from the cooling difference in sample 2. However, sample 2 is more compact and denser within the microstructure, with good layers in the coating surface that may have been achieved from the 4.2 g/min feed rate with very minor porosity.

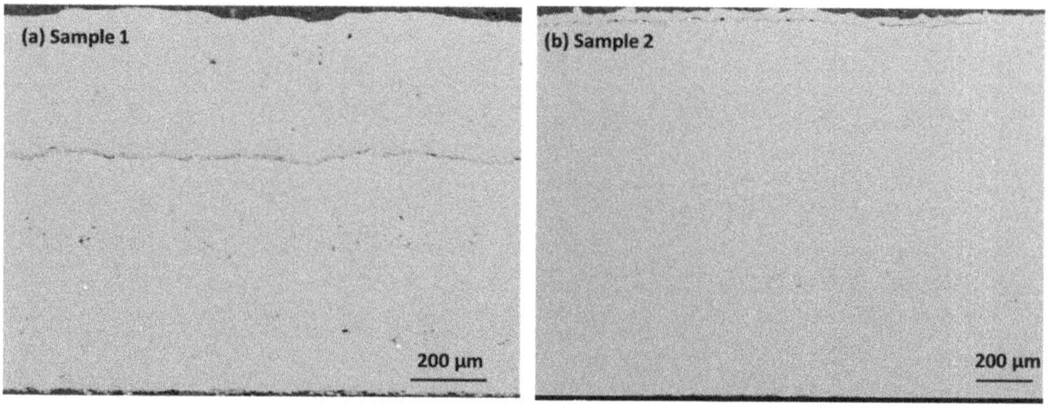

Figure 3. (**a**,**b**) Cross-section images along the cross-section for samples 1 and 2.

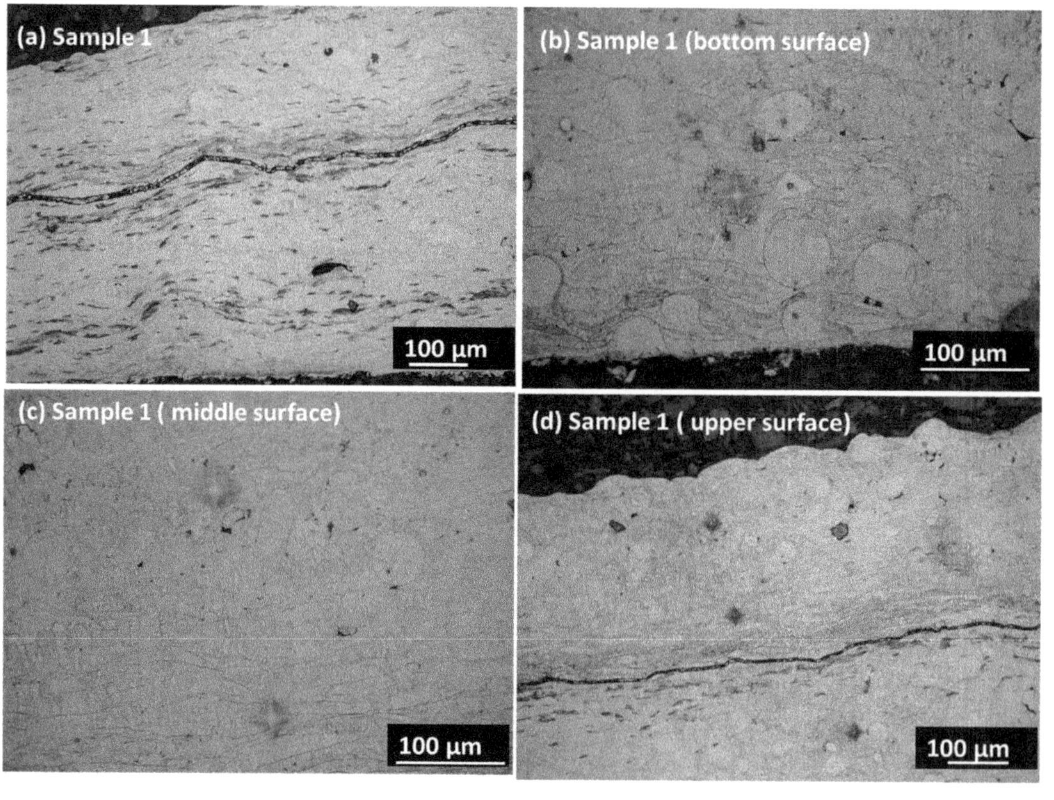

Figure 4. (**a**–**d**) Optical microscope image of sample 1; (**a**) overall surface; (**b**) bottom surface; (**c**) middle area; (**d**) upper surface area.

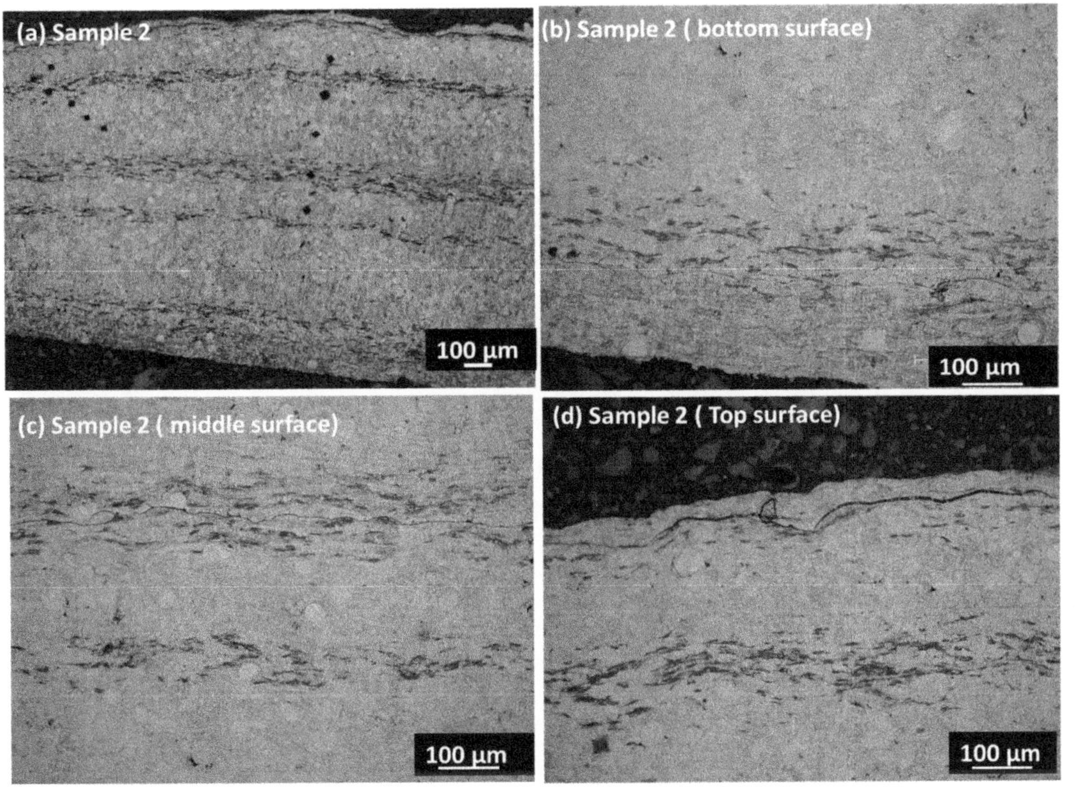

Figure 5. (**a**–**d**) Optical microscope image of sample 2; (**a**) overall surface; (**b**) bottom surface; (**c**) middle area; (**d**) upper surface area.

3.2. Elemental Analysis within Coating Layers Using Image and Line Analysis

The presence of various elements within the coating layers is investigated by using image and line analysis using the SEM technique (Figure 6). The elements of Ni, Ti, and others are present in samples 1 and 2 when examining the EDX peaks. There are 5% and 2% C K in samples 1 and 2, respectively. Sample 1 displays 58% TiK, while sample 2 displays 61% TiK. In both samples 1 and 2, NiK is nearly equal. Sample 1 has more contaminants than sample 2, such as higher C concentration. Other elements, such as O, Al, Fe, and Si, are very minor. The line analyses for major elements such as Ni and Ti and minor elements such as C and O were performed for samples 1 and 2 (Figure 7). It has been observed that the Ni content is lower than the Ti content in sample 1, and it remains constant along the coating layers, with the minimum content of C and O. Nonetheless, sample 2 shows that, with a minimum content of C and O, Ni and Ti contents overlap in the same quantity over the coating layers, as shown in Figure 7a,b. Because both samples have different scales, the signal displays different ratios.

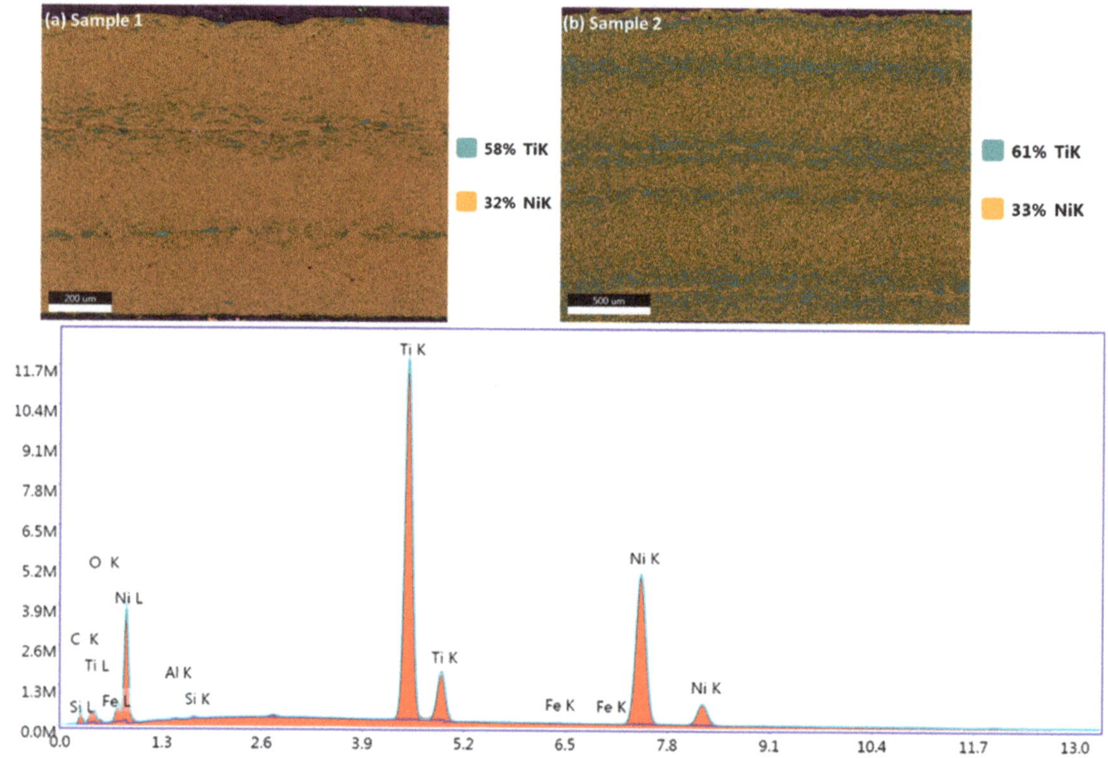

Figure 6. Elemental mapping of sample 1 and below shows the compositional analysis in Wt. and at.%.

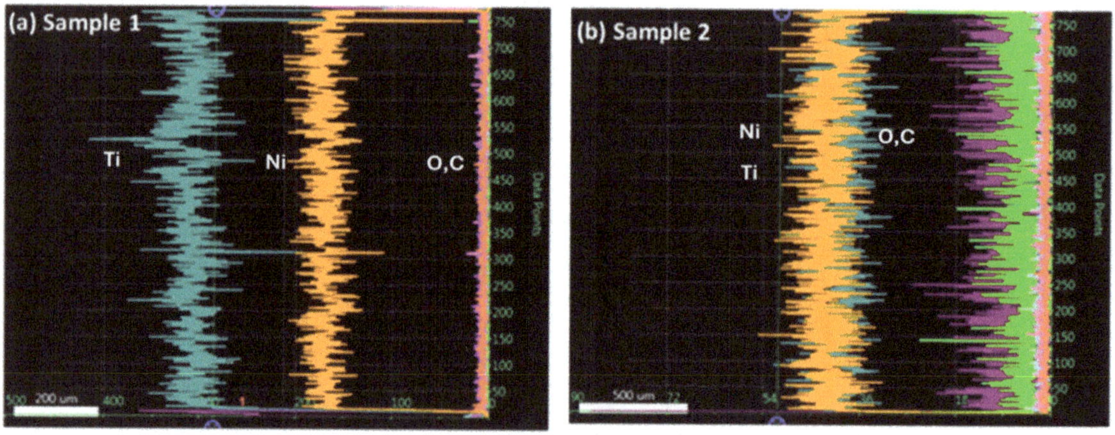

Figure 7. Line analysis in samples 1 and 2 shows the presence of a shift in the composition of Ni, Ti, C, and O content.

3.3. Phase Identification in the Coating Structure

Figure 8 displays the peaks of various phases of austenite and martensite for NiTi powder and both samples 1–2. NiTi powder shows the presence of austenite (64 Wt. %) and martensite (36 Wt. %) [19]. However, there is a deviation in the phases out of austenite and martensite from equilibrium towards intermetallic phases in both plasma-sprayed

NiTi coating layers. According to the Ni-Ti phase diagram, the melting point of NiTi is 1310 °C, at which 50% of Ni or Ti is present in the NiTi phase [12]. When the temperature shifts below 984 °C, NiTi is combined with other phases to create $NiTi_2$. The temperature causes NiTi particles to melt, which, in turn, causes reactions and interactions that lead to the creation of different phases. This corresponds to the intermetallic phases present in both coating layers in samples 1 and 2, with some porosity. The intermetallic phases are found mostly in the region of intersplat areas on the connection boundary between the layers. However, the variable parameter of different feed rates controls the compactness and generates porosity along the coating layers. The influence of intermetallic phases on the major phases of austenite and martensite is determined by the thermal response of the samples. This finding leads to determining the transformation temperature of phases during cooling and heating cycles.

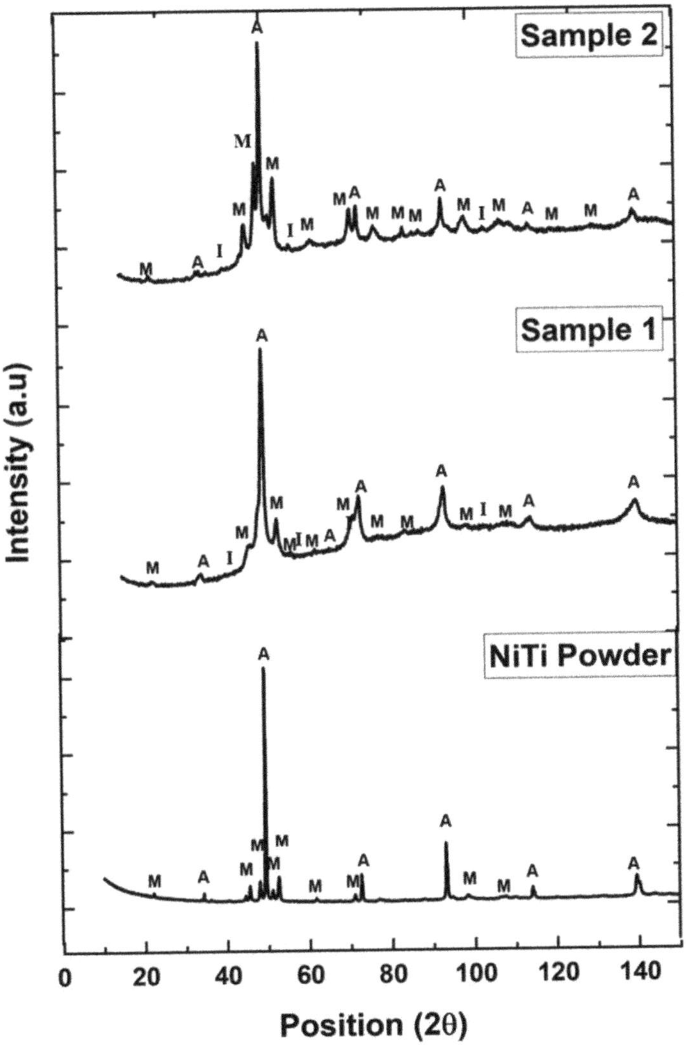

Figure 8. Phases of A: austenite (B2); M: martensite (B19′); and I: intermetallic ($NiTi_2$) present in NiTi powder and plasma-sprayed samples 1 and 2.

3.4. Thermal Characterization of the Samples

Figure 9a–c displays the transformation temperatures of the powder and coating samples. The powder shows the presence of austenite, R-phase, and martensite during cooling and heating cycles. The formation of the R-phase initiates at 39 °C and ends at 3.3 °C, which leads to the formation of the martensite phase at −12 °C and ends at −44 °C during the cooling cycle in NiTi powder. The powder exhibits the martensite phase below −44 °C. However, upon heating, the R-phase starts at a temperature of −4.4 °C, which proceeds toward the final temperature of 27 °C. Further, the austenite phase starts at 34 °C and ends at 62 °C. Figure 9a shows the NiTi powder in the austenite phase after 62 °C during the heating cycle (Figure 9a). However, in the case of the plasma-sprayed samples, the austenite start and finish temperatures shift towards higher temperatures. Sample 1 (Figure 9b) demonstrated the fact that the austenite phase begins at 57 °C and ends at 68 °C, without clear R-phase and martensite peaks during the cooling cycle. The narrow range of transformation temperatures may lead to narrow hysteresis in the shape memory effect. However, sample 2 (Figure 9c) shows that the austenite start temperature is 40 °C with a final temperature of 70 °C in the heating cycle, indicating that the R-phase start temperature is 92 °C and ends at 40 °C. These findings are considered to show the temperature window for a thermo-mechanical cycle under a constant load in the cooling and heating cycle.

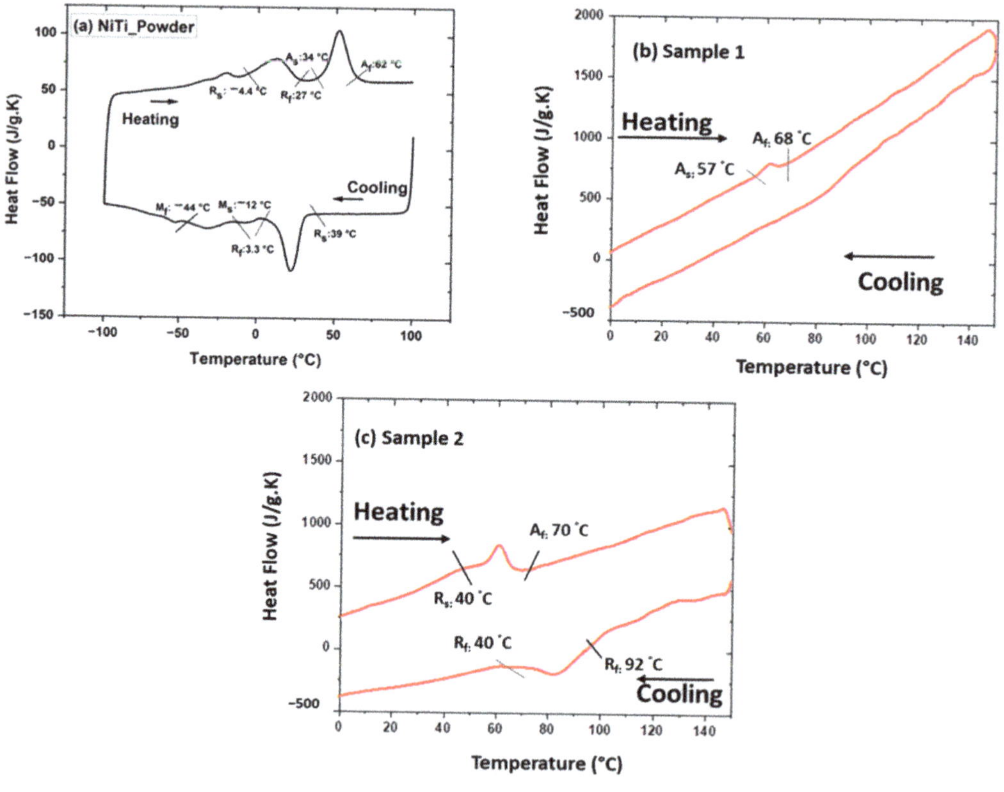

Figure 9. DSC graph of (**a**) NiTi powder; (**b**) sample 1; and (**c**) sample 2 showing transformation temperature during the cooling and heating cycle. NiTi powder shows austenite, R-Phase, and martensite phases; however, the martensite phase is missing in plasma-sprayed samples 1 and 2.

3.5. Thermo-Mechanical Characterization of Samples

Figure 10 presents the thermo-mechanical characterization of both samples 1 and 2. Sample 1 exhibits a narrow hysteresis region that arises from the narrow transformation temperature. The very irregular region of hysteresis could be caused by the presence of intermetallic phases, such as pores and segmented crack lines, and by the non-uniform composition of the coating layers. However, sample 2 exhibits a broad hysteresis curve, which could be caused by the phases' wide range of transformation temperatures. Both samples show recovery with displacement on the heating cycle, which signifies the shape memory behavior. The return of the sample may not coincide well with the same position, indicating the accumulation of residual stress during the thermal cycles. The transformation temperature in samples 1 and 2 is consistent with the phase's start and finish temperatures, as determined by using the coefficient of thermal expansion (CTE). The first peak corresponds to the start R_f temperature during the cooling cycle (R_f: 40 °C), and the second one corresponds to the final austenite transformation temperature (Af: 70 °C) during the heating cycle for sample 2.

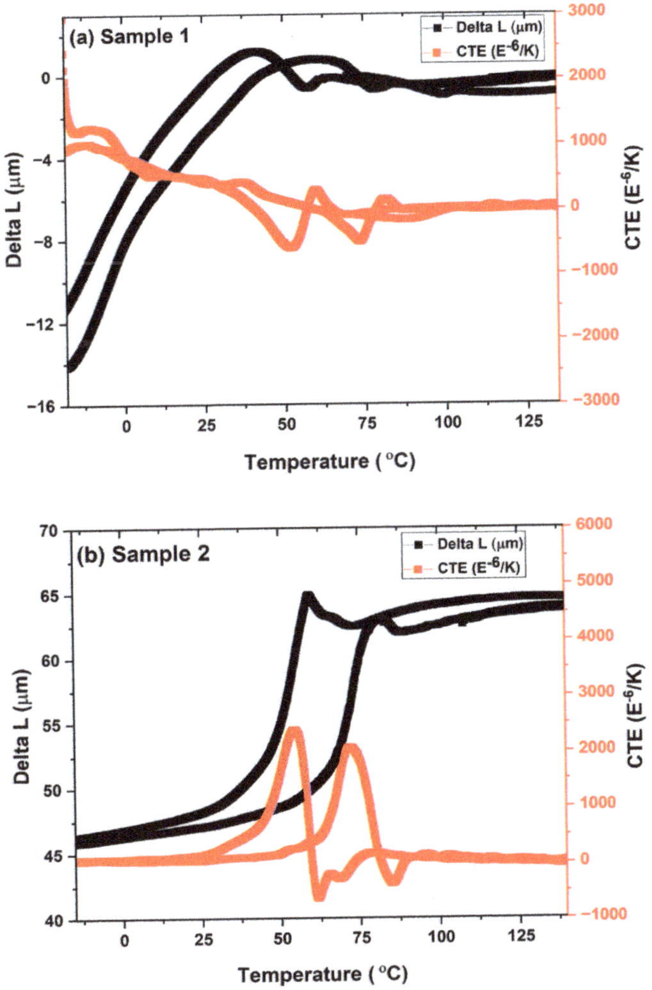

Figure 10. Thermo-mechanical analysis of sample 1 and 2. Delta L and coefficient of thermal expansion (CTE) as a function of temperature.

4. Discussion

Temperature has a significant role in the melting of the particles and the cooling of the substrate in the spraying process. Molten particles begin to spread on the substrate's surface. Preheating promotes the spreading of particles on the surface of the substrate in more significant ways. Figure 11a–c presents the spreading of molten particles on the substrate surface from initial contact to the final position following the second stage of spreading and then, finally, the splashing of the droplet within the central depression region. Upon impact of a droplet on the substrate surface, a localized mechanical deformation is produced that promotes interlocking adherence of splats to the surface. The surface viscosity of the molten particles promotes spreading on the surface (Figure 11b). There is a little depression in the impact area of the substrate surface (Figure 11c). Nevertheless, in this experiment, spreading along the substrate's surface can result in undetectable micro-deformation in the nearby area. The preheating of the substrate helps ensure less difference in temperature from the melting of the droplet toward the deposition temperature. As a result, there is less chance of phase deviation as a function of temperature, which was predicted from the phase diagram of Ni-Ti. The deposition process is composed of several stages, from initial melting to final deposition. Thermal plasma arc initiates the melting of particles that reach the arc zone, which begins the first stage of the reaction process. In the second stage, melted particles accelerate and impact the surface of the substrate. The third stage displays the spreading of splats on the surface.

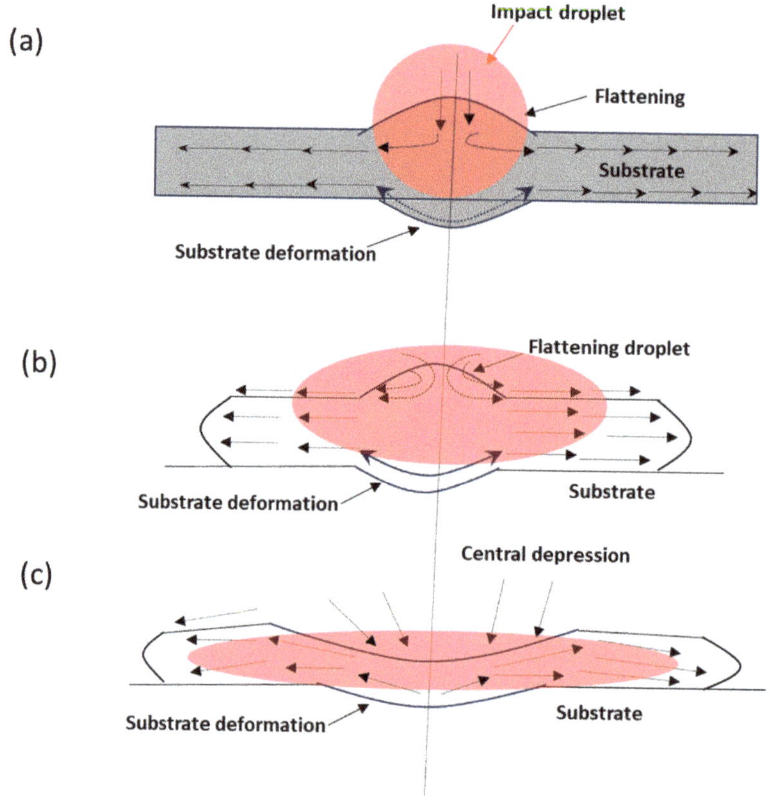

Figure 11. Schematic diagram of molten NiTi particles impacting the stainless steel substrate, showing flattening and spreading behavior. (**a**) Initial stage of contact of a molten particle on the surface of the substrate; (**b**) spreading of the splat; (**c**) droplet splashing with central depression.

Figure 12 shows the various stages of the deposition process during the melting of the particles in the plasma arc. When plasma gases such as Ar + H_2 create an arc at a power of 9 kW, the charge explodes at high temperatures. The powder feeder injects the particles at a feed rate of 2.1 g/min or 4.2 g/min into the plasma arc. When the powder enters a high-temperature plasma arc zone, it begins to melt quickly, converting to droplets. The droplets travel with high velocity from the plasma zone toward the substrate. The molten droplets splash upon the substrate with impact force. The impact force causes the particles to disperse laterally, and the substrate temperature causes them to solidify quickly, forming a single splat. Various splats deposited on the substrate form a coating layer (Figure 12a). However, the degree of the melting of the particles is different under different plasma powers, which results in a significant difference in the microstructure and mechanical properties of the coating layers. According to the experiment results, it can be considered that the deposition behavior is based on three categories. At a lower power of 9 kW, the particles are close to the semi-molten mode. The particles that pass through the outer zone of the plasma arc form a partially molten splat in the coating layers. The partially molten splat creates voids within the coating layers that generate porosity. When the feed rate increased from 2.1 to 4.2 g/min at a power of 9 kW, the porosity was reduced. The compact structure of the coating could be formed at a higher feeding rate with higher input power (Figure 12c).

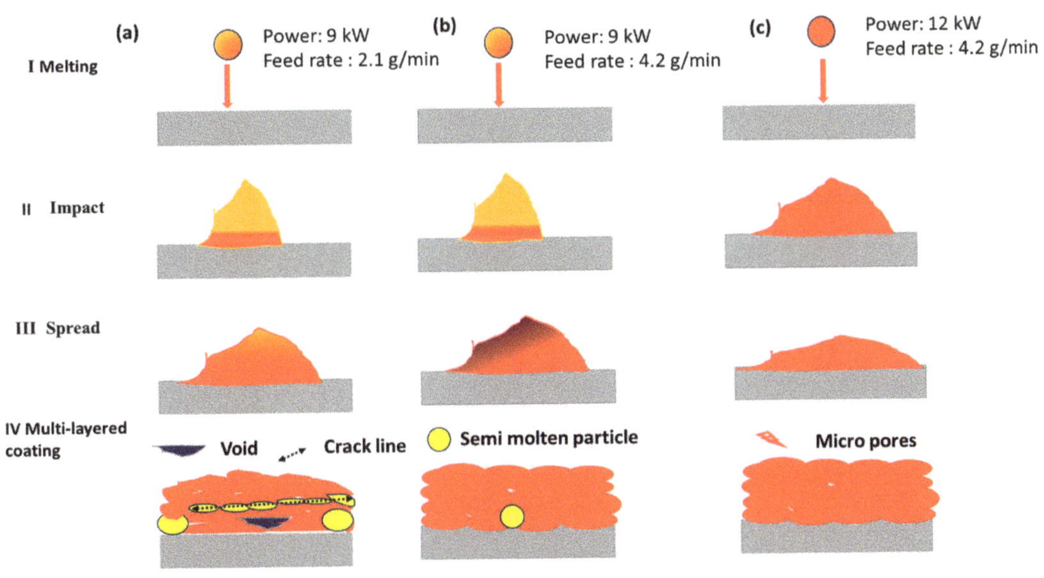

Figure 12. Various stages (**a**–**c**) of particles from melting to impact, with spreading and final formation of the coating layers.

When the particle melts in the plasma arc during spraying, it creates partially molten splats with circular and elliptical shapes in the coating layers. Figure 13 shows the various forms of splat forms produced in the plasma spraying process during the deposition of the coating layers. A specific number of splats are considered in the coating layers, with partially and completely splashed molten splats produced during the deposition process. The melting parameters depend upon the particle size and position in the plasma arc during flow and deposition on the substrate. The eccentric shape of the melted particle varies from minimum to optimum forms of circular shape. On the other hand, the circular shape shifts away from being perfectly round to being just slightly less round than normal. In the case of solidity, the case shifted from circular to eccentric.

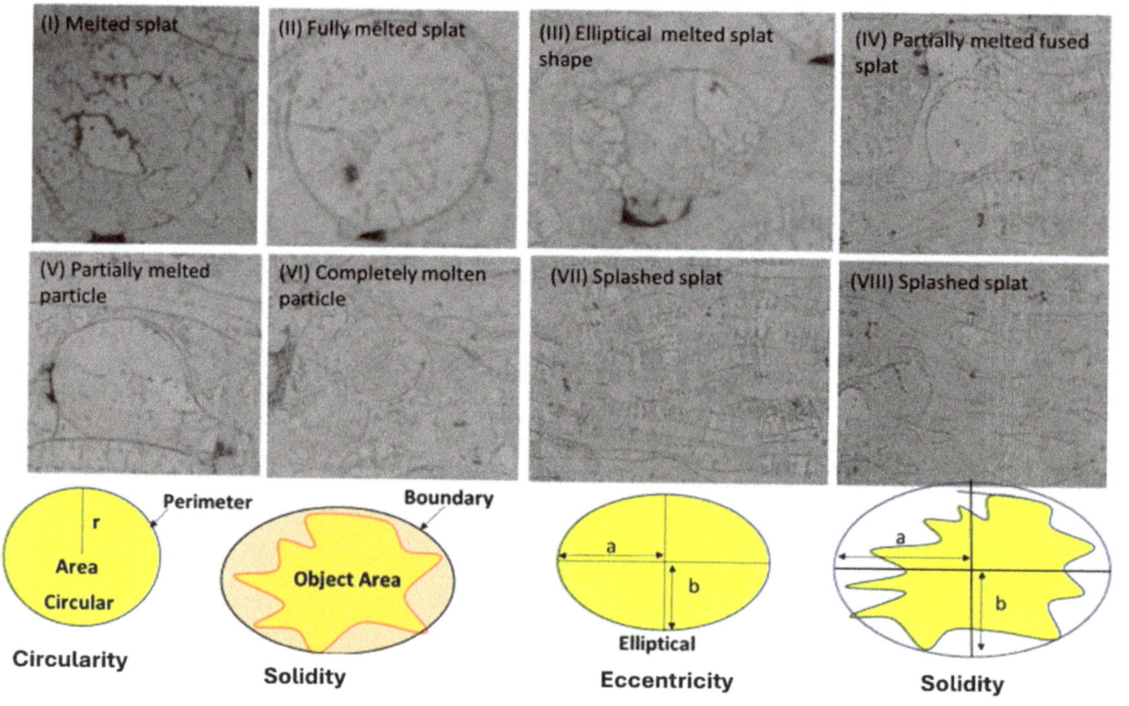

Figure 13. Various morphologies of melted droplets of spraying and evolution of the corresponding melted splat of various areas, where b and a represent the axis in elliptical and partially molten splats.

As a result, the circularity, eccentricity, and solidity of the shape are determined by using the following equations [32]:

$$\text{Circularity} = \frac{4\pi\text{Area}}{\text{Perimeter}^2} \tag{1}$$

$$\text{Eccentricity} = \sqrt{1 - \frac{b^2}{a^2}} \tag{2}$$

$$\text{Solidity} = \frac{\text{object area}}{\text{Bounding area}} \tag{3}$$

Utilizing Equations (1)–(3), the various splats and their correlations with the circular, eccentric, and solid areas are calculated and plotted in Figure 14. The uncertainties vary in terms of the number of splats taken into consideration.

The number of splats has an inverse relationship with the solidity and circularity of the splat shape. When the splat shape is close to circularity, the uncertainty error value is lower; when the splat shape deviates from circularity, the uncertainty error value is larger. Solidity is calculated using the object area to the bounding area ratio, which produces a uniform uncertainty error number for all splats that are considered. The most significant divergence in uncertainty error could result from the different elongated forms of the sprayed splat that developed during cooling.

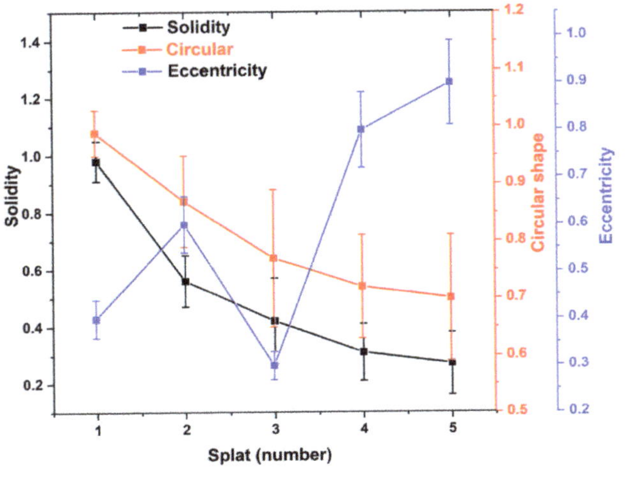

Figure 14. Splats with various areas from circular, elliptical to solid areas in the spraying process.

5. Conclusions

Multilayer NiTi coatings were prepared using the thermal plasma spraying process using different feed rates. The plasma spraying method was implemented with two variable feed rates that influenced the compactness of the coating layer and determined the pores in the structure. Feedstock NiTi powders with a feed rate of 4.2 g/min create a dense coating layer without any crack line. Both plasma-sprayed samples show higher transformation temperatures for phases during the cooling and heating thermal cycle. This shifting toward higher temperature may be attributed to the intermetallic compounds such as $NiTi_2$, which was confirmed in the phase analysis. The shape memory effect of the plasma-sprayed samples exhibits a narrow region of response that may arise from the coating layers with porosity and cracks. However, sample 2 shows a better hysteresis region with shape memory effect with corresponding transformation temperature. The dense layer of the coating prevents any diffusion of foreign elements or impurities into the system. This coating without porosity shows a shape memory effect with a similar response to NiTi shape memory alloys. This process could open the possibilities for using material surface protection using a functional coating of NiTi. Future studies will focus on the impact resistance, scratch, and wear behavior of the coating to emphasize the shape memory effect of NiTi alloy.

Author Contributions: Conceptualization, S.S. and P.Š.; methodology, S.S., J.Z., J.K. and S.H.; formal analysis, S.S. and O.P.; investigation, writing—original draft preparation, writing—review and editing, S.S.; funding acquisition, P.Š. All authors have read and agreed to the published version of the manuscript.

Funding: This research was funded by the Czech Science Foundation (CSF) projects 22-20181 S (P.Š.) and Project Solid-21 with support from the Institute of Physics, Prague, Czech Republic (SOLID21: CZ.02.1.01/0.0/0.0/16_019/0000760, SOLID21-Fyzika Pevných Látek Pro 21. Století, Fyzikální Ústav AV ČR, v. v. i. (2018–2023). CzechNanoLab project LM2023051 is funded by MEYS CR.

Institutional Review Board Statement: Not applicable.

Informed Consent Statement: Not applicable.

Data Availability Statement: The data is confidential due to the project policy. However it will be available on reader request.

Acknowledgments: The first author would like to acknowledge the RF-ICP facility of the Institute of Plasma Physics of the Czech Academy of Science, Prague, for the preparation of the samples.

Conflicts of Interest: The authors declare no conflicts of interest.

References

1. Grilli, M.L.; Valerini, D.; Slobozeanu, A.E.; Postolnyi, B.O.; Balos, S.; Rizzo, A.; Piticescu, R.R. Critical Raw Materials Saving by Protective Coatings under Extreme Conditions: A Review of Last Trends in Alloys and Coatings for Aerospace Engine Applications. *Materials* **2021**, *14*, 1656. [CrossRef]
2. Barwinska, I.; Kopec, M.; Kukla, D.; Senderowski, C.; Kowalewski, Z.L. Thermal Barrier Coatings for High-Temperature Performance of Nickel-Based Superalloys: A Synthetic Review. *Coatings* **2023**, *13*, 769. [CrossRef]
3. Bakan, E.; Vaßen, R. Ceramic Top Coats of Plasma-Sprayed Thermal Barrier Coatings: Materials, Processes, and Properties. *J. Therm. Spray Technol.* **2017**, *26*, 992–1010. [CrossRef]
4. Falcón, J.C.P.; Echeverría, A.; Afonso, C.R.M.; Carrullo, J.C.Z.; Borrás, V.A. Microstructure assessment at high temperature in NiCoCrAlY overlay coating obtained by laser metal deposition. *J. Mater. Res. Technol.* **2019**, *8*, 1761–1772. [CrossRef]
5. Armstrong, M.; Mehrabi, H.; Naveed, N. An overview of modern metal additive manufacturing technology. *J. Manuf. Process.* **2022**, *84*, 1001–1029. [CrossRef]
6. Kohyama, A.; Grossbeck, M.L.; Piatti, G. The application of austenitic stainless steels in advanced fusion systems: Current limitations and future prospects. *J. Nucl. Mater.* **1992**, *191–194 Pt A*, 37–44. [CrossRef]
7. Frenzel, J.; George, E.P.; Dlouhy, A.; Somsen, C.; Wagner, M.X.; Eggeler, G. Influence of Ni on martensitic phase transformations in NiTi shape memory alloys. *Acta Mater.* **2010**, *58*, 3444–3458. [CrossRef]
8. Bitzer, M.; Rauhut, N.; Mauer, G.; Bram, M.; Vaßen, R.; Buchkremer, H.P.; Stöver, D.; Pohl, M. Cavitation-resistant NiTi coatings produced by low-pressure plasma spraying (LPPS). *Wear* **2015**, *328–329*, 369–377. [CrossRef]
9. Espitia, L.; Dong, H.; Li, X.-Y.; Pinedo, C.; Tschiptschin, A. Cavitation erosion resistance and wear mechanisms of active screen low temperature plasma nitrided AISI 410 martensitic stainless steel. *Wear* **2015**, *332–333*, 1070–1079. [CrossRef]
10. Nie, M.H.; Jiang, P.F.; Zhou, Y.X.; Li, Y.L.; Zhang, Z.H. Studies on the 316/NiTi functionally gradient ultra-thick coatings fabricated with directed energy deposition: Microstructure, crystallography and wear mechanism. *Appl. Surf. Sci.* **2023**, *630*, 157497. [CrossRef]
11. Farvizi, M.; Ebadzadeh, T.; Vaezi, M.; Yoon, E.; Kim, Y.-J.; Kang, J.; Kim, H.; Simchi, A. Effect of starting materials on the wear performance of NiTi-based composites. *Wear* **2015**, *334–335*, 35–43. [CrossRef]
12. Chang, F.T.; Lo, K.H.; Man, H.C. NiTi cladding on stainless steel by TIG surfacing process: Part I. Cavitation erosion behavior. *Surf. Coat. Technol.* **2003**, *172*, 308–315. [CrossRef]
13. Oliveira, J.P.; Miranda, R.M.; Fernandes, F.M. Welding and Joining of NiTi Shape Memory Alloys: A Review. *Prog. Mater. Sci.* **2017**, *88*, 412–466. [CrossRef]
14. Verschuur, E.M.L.; Homs, M.Y.V.; Steyerberg, E.W.; Haringsma, J.; Wahab, P.J.; Kuipers, E.J.; Siersema, P.D. A new esophageal stent design (Niti-S stent) for the prevention of migration: A prospective study in 42 patients. *Gastrointest. Endosc.* **2006**, *63*, 134–140. [CrossRef] [PubMed]
15. Alarcon, E.; Heller, L.; Chirani, S.A.; Šittner, P.; Kopeček, J.; Saint-Sulpice, L.; Calloch, S. Fatigue performance of superelastic NiTi near stress-induced martensitic transformation. *Int. J. Fatigue* **2017**, *95*, 76–89. [CrossRef]
16. Frenzel, J. On the Importance of Structural and Functional Fatigue in Shape Memory Technology. *Shap. Mem. Superelasticity* **2020**, *6*, 213–222. [CrossRef]
17. Elahinia, M.; Moghaddam, N.S.; Andani, M.T.; Amerinatanzi, A.; Beth, A.B.; Reginald, F.H. Fabrication of NiTi through additive manufacturing: A review. *Prog. Mater. Sci.* **2016**, *83*, 630–663. [CrossRef]
18. Samal, S.; Zeman, J.; Habr, S.; Pacherová, O.; Chandra, M.; Kopeček, J.; Šittner, P. Evaluation of Microstructure–Porosity–Hardness of Thermal Plasma-Sprayed NiTi Coating Layers. *J. Manuf. Mater. Process.* **2023**, *7*, 198. [CrossRef]
19. Bolelli, G.; Cannillo, V.; Lugli, C.; Lusvarghi, L.; Manfredini, T. Plasma-sprayed graded ceramic coatings on refractory materials for improved chemical resistance. *J. Eur. Ceram. Soc.* **2006**, *26*, 2561–2579. [CrossRef]
20. Bonizzoni, G.; Vassallo, E. Plasma physics and technology; industrial applications. *Vaccum* **2002**, *64*, 327–336. [CrossRef]
21. Singh, S.; Jinoop, A.N.; Palani, I.A.; Paul, C.P.; Tomar, K.P.; Prashanth, K.G. Microstructure and mechanical properties of NiTi-SS bimetallic structures built using Wire Arc Additive Manufacturing. *Mater. Lett.* **2021**, *303*, 130499. [CrossRef]
22. Hu, L.; Xue, Y.; Shi, F. Intermetallic formation and mechanical properties of Ni-Ti diffusion couples. *Mater. Des.* **2017**, *130*, 175–182. [CrossRef]
23. Zhu, J.; Ding, Z.; Borisov, E.; Yao, X.; Brouwer, J.C.; Popovich, A.; Hermans, M.; Popovich, V. Healing cracks in additively manufactured NiTi shape memory alloys. *Virtual Phys. Prototyp.* **2023**, *18*, e2246437. [CrossRef]
24. Zou, Z.; Jia, L.; Yang, L.; Shan, X.; Luo, L.; Guo, F.; Zhao, X.; Xiao, P. Role of internal oxidation on the failure of air plasma sprayed thermal barrier coatings with a double-layered bond coat. *Surf. Coat. Technol.* **2017**, *319*, 370–377. [CrossRef]
25. Samal, S. Study of porosity on titania slag obtained by conventional sintering and thermal plasma process. *JOM* **2016**, *68*, 3000–3005. [CrossRef]

26. Dhiman, R.; McDonald, A.G.; Chandra, S. Predicting splat morphology in a thermal spray process. *Surf. Coat. Technol.* **2007**, *201*, 7789–7801. [CrossRef]
27. Mulero, M.A.; Zapata, J.; Vilar, R.; Martınez, V.; Gadow, R. Automated image inspection system to quantify thermal spray splat morphology. *Surf. Coat. Technol.* **2015**, *278*, 1–11. [CrossRef]
28. Samal, S.; Tyc, O.; Cizek, J.; Klecka, J.; Lukáč, F.; Molnárová, O.; de Prado, E.; Weiss, Z.; Kopeček, J.; Heller, L.; et al. Fabrication of Thermal Plasma Sprayed NiTi Coatings Possessing Functional Properties. *Coatings* **2021**, *11*, 610. [CrossRef]
29. Velmurugan, C.; Senthilkumar, V.; Biswas, K.; Yadav, S. Densification and microstructural evolution of spark plasma sintered NiTi shape memory alloy. *Adv. Powder Technol.* **2018**, *29*, 2456–2462. [CrossRef]
30. Samal, S.; Sulovský, M.; Kopeček, J.; Šittner, P. Deposition and characterization of plasma sprayed NiTi powder on stainless steel and graphite substrate. *Mater. Chem. Phys.* **2023**, *307*, 128146. [CrossRef]
31. Zhe, X.Y.; Li, X.H.; Wang, Q.; Zhang, Y.; Song, X.D. Substrate temperature dependence of splat morphology for plasma-sprayed cast iron on aluminum surface. *Surf. Coat. Technol.* **2015**, *283*, 234–240. [CrossRef]
32. Yu, Z.X.; Huang, J.B.; Wang, W.Z.; Yu, J.Y.; Wu, L.M. Deposition and properties of a multilayered thermal barrier coating. *Surf. Coat. Technol.* **2016**, *288*, 126–134. [CrossRef]
33. Zhai, H.; Ou, M.; Cui, S.; Li, W.; Zhang, X.; Cheng, B.; He, D.; Li, X.; Cai, A. Characterizations the deposition behavior and mechanical properties of detonation sprayed Fe-based amorphous coatings. *J. Mater. Res. Technol.* **2022**, *18*, 2506–2518. [CrossRef]
34. Samal, S.; Kopeček, J.; Šittner, P. Interfacial Adhesion of Thick NiTi Coating on Substrate Stainless Steel. *Materials* **2022**, *15*, 8598. [CrossRef]
35. *ASTM E831-19*; Standard Test Method for Linear Thermal Expansion of Solid Materials by Thermomechanical Analysis. ASTM: West Conshohocken, PA, USA, 2019.

materials

MDPI

Article

Self-Healing Thermal-Reversible Low-Temperature Polyurethane Powder Coating Based on Diels–Alder Reaction

Katarzyna Pojnar [1,*], Barbara Pilch-Pitera [2,*], Shahla Ataei [3], Patrycja Gazdowicz [2], Beata Mossety-Leszczak [2], Beata Grabowska [4] and Artur Bobrowski [4]

1 Doctoral School of Engineering and Technical Sciences, Rzeszow University of Technology, ul. Powstańców Warszawy 12, 35-959 Rzeszów, Poland
2 Faculty of Chemistry, Rzeszow University of Technology, ul. Powstańców Warszawy 6, 35-959 Rzeszów, Poland; 163711@stud.prz.edu.pl (P.G.); mossety@prz.edu.pl (B.M.-L.)
3 Harsin Chemical Homa Company, Garmsar 3593147487, Iran; sh.ataei.k@gmail.com
4 Faculty of Foundry Engineering, AGH University of Science and Technology, ul. Reymonta 23, 30-059 Kraków, Poland; beata.grabowska@agh.edu.pl (B.G.); arturb@agh.edu.pl (A.B.)
* Correspondence: d521@stud.prz.edu.pl (K.P.); barbpi@prz.edu.pl (B.P.-P.)

Abstract: This work focused on obtaining a low-temperature powder coating characterized by self-healing properties. To achieve this, acrylic resin, blocked polyisocyanates (bPICs) with 1,2,4-triazole, and unsaturated commercial resin were used. The synthesis of bPICs with triazole enabled the low-temperature curing and reversible Diels–Alder (DA) reaction at 160 °C. The chemical structure of bPICs was confirmed using ^{1}H-NMR. The occurrence of the DA and retro-DA (rDA) reactions in the crosslinked polymer, at temperatures of 60–85 °C and 90–130 °C, respectively, was confirmed using Differential Scanning Calorimetry (DSC), Thermogravimetric Analysis (TGA), and FT-IR spectroscopy. The self-healing properties of the powder coating were examined using polarized optical microscopy. Additionally, the occurrence of the DA and rDA reactions between triazole and unsaturated polyester resin was investigated through repeated self-healing tests.

Keywords: low-temperature-curing powder coatings; reversible Diels–Alder reaction; self-healing

1. Introduction

The growing market demand for increasingly eco-friendly products and coating processes necessitates the exploration of innovative solutions. Currently, the most environmentally sustainable options available are powder coatings and paints, which do not emit volatile organic compounds (VOCs) like solvent-based products and do not contain biocides to protect them from microorganism growth in packaging, as is the case with water-based products [1,2]. For these reasons, powder coatings play a crucial role in the chemical industry. The "5E" standards, such as efficiency, economy, energy savings, environmental compliance, and excellence of finish, of powder coatings are fulfilled [3]. A significant drawback of standard high-temperature powder coating systems is the high energy consumption during the curing process, which typically occurs at temperatures ranging from 180 to 220 °C [4]. To address this issue and reduce energy usage, ongoing research aims to lower the curing temperature. Innovations in this area include the development of special crosslinking agents, resins, and additives that enable curing at temperatures below 160 °C or through UV radiation. [5,6].

Acrylic resins allow for the development of low-temperature powder coating systems and UV-cured coatings [7,8]. Additionally, they possess excellent properties, including superior weatherability and high chemical, thermal, and mechanical resistance. This is demonstrated by their hydrophobic properties, color durability, and high resistance to scratches, sand abrasion, and damage caused by stones [9]. Due to the presence of functional groups within acrylic resin, powder coatings can be divided into three categories containing carboxyl, epoxy, and hydroxyl groups [10]. Acrylic resin containing carboxyl groups, due to issues with flexibility, is most commonly found in a hybrid system. Okada et al. described an acrylic/polyester hybrid powder coating [11]. Polyester powder coatings, featuring both -COOH and -OH functionality, offer advantages such as good appearance and favorable mechanical properties like toughness and flexibility. However, their drawback lies in insufficient weather durability [11]. Acrylic resins containing epoxy groups are characterized by good mechanical properties such as rigidity, toughness, and adhesion [12]. Additionally, they enable UV curing through cationic polymerization [8]. Acrylic resins containing hydroxyl groups, in reaction with blocked polyisocyanates (bPICs), can enable curing temperatures below 160 °C (low-temperature systems). Additionally, compared to acrylic resins containing carboxyl or epoxy groups, they exhibit the best flexibility properties [13]. This is one of the reasons for choosing them for use in this research.

In the literature, the known blocking agents are imidazoles, amides, oximes, phenols, triazoles, imides, pyrazoles, hydroxyamic acid esters, and active methylene compounds [14,15]. Czachor-Jadacka et al. described crosslinking agents based on blocked polyisocyanates that induce curing at reduced temperatures. An example is methyl ethyl ketoxime (MEKO)-blocked PICs or butanone oxime, which, compared to ε-caprolactam-blocked PICs, undergo deblocking at temperatures below 160 °C rather than 180 °C [16].

The Diels–Alder (DA) reaction, as a non-autonomic process, is one of the most well-known and common reactions used in the structure design of self-healing polymers. Due to the thermal reversibility of the DA bonds, the microcracks are healed as many times as needed by undergoing a cooling/heating cycle, followed by the reforming of the DA bonds [17]. The reaction involves a [4 + 2] cycloaddition between an electron-rich conjugated diene and a substituted alkene (a poor electron dienophile). This forms strong dynamic covalent bonds, resulting in the formation of a substituted cyclohexane as the DA adduct [18]. Due to the low energy consumption for the formation of this cyclohexane ring, it also provides the possibility of forming and functionalizing different molecules.

In summary, the healing process through crosslink formation includes two steps:

(1) Upon the cracks being formed on a microscopic scale, healing may be achieved by forming a network through the discharge of crosslinks formed between the diene and dienophile in that place. As the temperature increases (around 120–160 °C), the equilibrium reaction shifts from the side of DA bonds formed toward the breaking of these bonds, which leads to an increase in molecular mobility and a higher concentration of active diene and dienophile groups. At a certain temperature, the contact between the cracked surfaces is facilitated by decreasing the crosslinking density, which in turn helps to close the microscopic cavities.

(2) The sample is annealed to the temperature required for the DA reaction and retrieval of bonds to form a DA moiety. Since the crosslinks are formed through an equilibrium reaction between a diene and dienophile and the formation of a DA linkage is exothermic, upon the temperature decrease (50–70 °C), the equilibrium shifts to the reversed direction, i.e., the more bonded state, and as a result, dynamic covalent crosslinks are formed at the crack interface during the [4 + 2] cycloaddition DA reaction. Thus, this process can be repeated at a suitable and effective temperature until the crack is filled [19,20].

The DA reaction has been used in many works to obtain the self-healing effect of coatings, including polyurethane coatings in the work of Farshchi N. et al. [21]. In order to achieve the self-healing effect, an additional self-healing agent was introduced containing furfuryl alcohol and maleic anhydride adduct. The innovation of this study is the use of a blocking agent to create a self-healing adduct, which will additionally eliminate its emission into the environment.

The aim of this study was to examine the effect of the addition of unsaturated polyester resin to the low-temperature-curing polyurethane powder coating formulation on the self-healing properties. The key feature of this coating was the ability to cure at a low temperature. This was achieved by incorporating a triazole blocked polyisocyanates and an acrylic resin containing hydroxyl groups. To start the curing process, the unblocking reaction was performed at 110–130 °C. During the curing process, the unblocked -NCO groups reacted with the -OH groups of the acrylic resin, resulting in the formation of low-temperature-curing powder coatings. In addition, at the same time, the triazole released from the unblocking reaction was able to react with the commercial unsaturated polyester resin, which allowed the DA reaction to occur. Various analytical techniques such as DSC, FTIR, polarized optical microscopy, repeated self-healing test, and TGA were used to evaluate the effectiveness of low-temperature curing and the DA reaction. The obtained results confirmed that the new powder coatings exhibit a low VOC emission, a self-healing effect, and better physical–mechanical properties.

2. Experimental Section

2.1. Reagents

Raw materials employed in the synthesis of the acrylic resin: 2-hydroxyethyl methacrylate (HEMA) (Merck, Darmstadt, Germany), methyl methacrylate (MMA) (Sigma Aldrich, Darmstadt, Germany), *n*-butyl acrylate (BA) (Sigma Aldrich, Darmstadt, Germany), and azobisisobutyronitrile (AIBN; Sigma Aldrich, Darmstadt, Germany).

Raw materials utilized in the synthesis of blocked polyisocyanate (bPIC): isophorone diisocyanate (IPDI; Evonik Industries, Essen, Germany), dibutyltin dilaurate (Sigma Aldrich, Darmstadt, Germany), glycerin (Chempur, Piekary Śląskie, Poland), α,ω bis (hydroxyethyleneoxypropylene)polydimethylsiloxane (KF6000) with LOH = 120 mg KOH/g (Shin-Etsu, Tokio, Japan), and 1,2,4-triazole as a blocking agent (Sigma Aldrich, Darmstadt, Germany).

Commercial unsaturated resins: UVECOAT 3003 (Meth)acrylated epoxy/polyester resin (Allnex, Frankfurt, Germany).

2.2. Synthesis of Acrylic Resin

The acrylic resin was synthesized via bulk free radical polymerization, employing the following monomers: 2-hydroxyethyl methacrylate (HEMA), methyl methacrylate (MMA), *n*-butyl acrylate (BA), and 1.7% azobisisobutyronitrile (AIBN). Therefore, the acrylic resin was named according to the names of the monomers used, HEMA, MMA, and BA, in a molar ratio of 1:5:2, e.g., HEMA/5MMA/2BA [22].

2.3. Synthesis of Blocked Polyisocyanates (bPICs)

IPDI and dibutyltin dilaurate catalyst (0.1 wt% based on diisocyanate) were introduced into a three-necked flask. Additionally, a reflux condenser, thermometer, glass stirrer, nitrogen inlet, and dropping funnel were used. Simultaneously, glycerin (Chempur, Piekary Śląskie, Poland) and KF-6000 (Shin-Etsu, Tokio, Japan) were mixed in a beaker before being slowly added to the diisocyanate in the flask. The reaction mixture was then maintained at 90 °C, stirred, and refluxed for a duration of 1.5 h. Triazole, serving as a blocking agent, was added to the obtained polyisocyanate at a 1:1 ratio of -NCO to -NH, for 1 h at 115 °C under stirring. The names of the samples were designated using the first letters of the compound, for example, IGKF/T indicates a blocked polyisocyanate synthesized from IPDI ("I"), glycerin ("G"), and KF-6000 ("KF") and blocked with triazole ("T").

2.4. Preparing Self-Healing Powder Coating Composition and Coatings

The self-healing powder coating composition consisted of an acrylic resin HEMA/5MMA/BA, bPIC, and an appropriate selected commercial unsaturated resin, UVECOAT 3003. The hydroxyl group contained in acrylic resin in the reaction with the deblocked polyisocyanate was used in the curing process of powder coatings. A co-rotating twin-screw mini extruder EHP 2 × 12 Sline from Zamak (Cracow, Poland) was used to homogenize the mixture. The temperature settings of the extruder were as follows: Zone I—75 °C;

Zone II—90 °C; Zone III—100 °C; and adapter—110 °C. In the next steps, the composition was cooled, pulverized, and passed through a 100 μm sieve. The resulting powder coatings were applied to the Q-panels using the CORONA electrostatic method with a WAGNER PEM X1 gun (Wagner, Alstatten, Switzerland). Subsequently, the powder coatings were cured at 160 °C for 15 min. After the curing process, the powder coatings were named, for example, L_HEMA/5MMA/2BA/IGKF/T/UVECOAT 3003 indicates a coating made from the resin HEMA/5MMA/2BA (qualitative composition), IGKF/T (bPIC), and UVECOAT 3003 (chosen commercial resin) (Table 1).

Table 1. Composition of the powder coatings.

Symbol of Powder Coating	HEMA/5MMA/2BA [%]	IGKF/T [%]	UVECOAT 3003 [%]
L_HEMA/5MMA/2BA/IGKF/T/	84.4	15.6	-
L_HEMA/5MMA/2BA/IGKF/T/UVECOAT 3003	69.6	12.8	17.6

3. Characteristics of the Methods Used

3.1. Nuclear Magnetic Resonance Spectroscopy (NMR)

The Bruker Avance II 500 MHz spectrometer (Bruker BioSpin, Rheinstetten, Germany), equipped with a 5 mm nitrogen-cooled dual (BB-1H) cryoprobe, was used for the experiments. Chemical shifts are reported in parts per million (ppm) relative to tetramethylsilane (TMS) as the internal standard, with deuterated chloroform (CDCl3) used as the solvent. Data analysis was conducted using NMR Topspin 2.1 pl 8 software (Bruker BioSpin, Rheinstetten, Germany).

3.2. Differential Scanning Calorimetry (DSC)

The thermal properties of the powder compositions were analyzed using a Mettler Toledo type 822e differential scanning calorimeter (DSC) (Mettler Toledo, Columbus, OH, USA), operated with Stare System software 16.20. Heating was conducted at a rate of 10 °C/min. Aluminum crucibles containing samples weighing 0.015 g were placed into the measurement chamber. Measurements were performed in a nitrogen atmosphere at a flow rate of 60 cm^3/min, covering a temperature range from 0 to 160 °C.

3.3. FT-IR Measurements

The IR spectra were acquired using a Thermo Scientific Nicolet 6700 FT-IR spectrophotometer (ThermoFisher Scientific, Waltham, MA, USA) equipped with a helium–neon (HeNe) laser. Spectra were recorded over the range of 700–4000 cm^{-1} with a resolution of 4 cm^{-1}. The data are presented as transmittance (%) versus wavenumber ν (cm^{-1}).

3.4. Polarized Optical Microscopy

A VHX-7000 polarized optical microscope with an EA-300 laser by Keyence (Osaka, Japan) was employed for this study. The sample of cured and cracked powder coatings was positioned on a heating microscope table. The heating was conducted at a rate of around 20 °C/min, and the powder coating self-healing process was evaluated.

3.5. Thermogravimetric Analysis (TGA)

The TGA and DTG were performed using a Mettler Toledo TGA/DSC instrument (Greifensee, Switzerland) equipped with Stare System software. TGA experiments were conducted under a nitrogen atmosphere, ranging from 25 to 600 °C at a heating rate of 10 °C/min. The experimental parameters were as follows: sample weight of approximately 5 mg, gas flow rate of 50 cm^3/min, and an open alumina pan with a volume of 150 μL.

3.6. Polymerization Test

The test involved rubbing the powder coatings with a swab soaked in MEK (methyl ethyl ketone) lightly back and forth 30 times in each direction. The evaluation was conducted according to the guidelines outlined in the "Technical requirements of the QUALICOAT quality label" [23]. Assessment took place 30 min after rubbing, and coatings were categorized based on the following criteria:

Coating appears matte and soft;
Coating appears matte and can be scratched with a nail;
Slight gloss reduction;
No noticeable changes.

3.7. Flow Test

In order to conduct the flow test, the PN-EN ISO 8130-11 standard was used [24]. The test plates were 6.6 mm deep, and 0.4 g of powder coatings was added. The test plates with the powder coating samples were then placed in an oven at a temperature of 160 °C for 20 min, positioned at a 60° angle from the horizontal. Finally, the distance between the bottom edge of the recess and the furthest point reached by the molten powder coating was evaluated.

3.8. Roughness

Roughness values were evaluated using a Mar Surf PSI profilometer (Göttingen, Germany) in accordance with PN-EN ISO 12085 [25]. The R_a parameter, which represents the arithmetic mean of the roughness profile deviations from the baseline, and the R_z parameter, which represents the arithmetic mean of the 5 highest profile peaks minus the arithmetic mean of the 5 lowest profile valleys, were measured [25]. The measurement was conducted automatically by moving the needle along the surface of the coating.

3.9. Thickness and Gloss

A micro-TRI-gloss tester from BYK-Gardner GmbH (Geretsried, Germany) was used in accordance with PN-EN ISO 2813 for gloss measurement and PN-EN ISO 2808 for thickness measurement [26,27]. Gloss measurements were taken at angles of 20°, 60°, and 85°. The same device was used for measuring thickness. The results for these parameters were obtained by averaging the measurements from ten trials for each sample.

3.10. Adhesion to Steel

According to PN-EN ISO 2409, adhesion to steel was evaluated by using a multi-cut tool manufactured by Byk Gardner (Geretsried, Germany) with six cutters spaced 2 mm apart [28]. In order to indicate parameters, a scale from 0 to 5 was used. A score of 0 indicated the best surface adhesion, with completely smooth incision edges, while a score of 5 indicated the worst adhesion, with more than 65% of the incision network damaged.

3.11. Hardness

The Konig Pendulum tester from BYK-Gardner GmbH (Geretsried, Germany), in accordance with PN-EN ISO 1522, was used to determine the relative hardness of the powder coatings [29]. The relative hardness was calculated by dividing the arithmetic mean of the number of pendulum oscillations for the tested sample by the glass constant, which is 171 pendulum oscillations. Three measurements were performed for each coating.

3.12. Scratch Resistance

The Clemen Tester from Elcometer (Manchester, UK) was used to check the scratch resistance of the powder coatings, according to PN-EN ISO 1518 [30]. The sample was positioned facing upwards. The device, without any load applied, was placed on the coating, and the test panel was moved outward at a speed of 30 mm/s. The scratch resistance was determined as the lowest load applied to the tool at which a scratch appeared on the cured coating.

3.13. Cupping

The cupping properties were determined according to the PN-EN ISO 1520 standard [31]. The results were obtained using a manual SP4300 tester by TQC (Capelle aan den Ijssel Miasto, Holandia, The Netherlands). The spherical drawing punch was used to indent a clamped sheet until the coating cracked. The point of crack initiation was then recorded. To ensure reliability, three measurements were performed on the same cured coating.

3.14. Water Contact Angle (WCA)

The water contact angle (WCA) was measured in accordance with PN-EN ISO 19403-6:2020-08 using an optical goniometer OCA15 EC from DataPhysics (Filderstadt, Germany), which was equipped with a digital camera [32]. The final result was determined by averaging multiple measurements, ensuring a comprehensive representation of the coating's water contact angle.

4. Results and Discussion

In order to develop low-temperature and self-healing powder coatings, acrylic resin containing a hydroxyl group (-OH), blocked polyisocyanates (bPICs), commercial unsaturated resin (UVECOAT 3003), and additives were used. The first step of this research consisted of synthesizing and characterizing the acrylic resin and bPIC. Then, the isocyanate groups were unblocked at a temperature of 160 °C for 15 min. Therefore, the obtained powder coatings are classified as low-temperature systems. Subsequently, the focus shifted to examining the reversibility of the DA reaction, followed by the evaluation of the performance properties of the powder coatings.

4.1. Characterization of Acrylic Resin and bPICs

4.1.1. Acrylic Resin

2-hydroxyethyl methacrylate (HEMA) was used as the main component in the production of acrylic resin due to its hydroxyl group, which facilitates the crosslinking process with polyisocyanates. The addition of *n*-butyl acrylate (BA) in the mixture enhances the flexibility characteristics of powder coatings, while methyl methacrylate (MMA) exhibits contrasting properties to BA. The stiffness and thermal stability of the acrylic resin are enhanced by the use of MMA. The hydroxyl number value of acrylic resin was of LOH = 40 mg KOH/g. The number-average molecular mass (M_n) was 7870 Da, and the dispersity index was 1.95. The viscosity (30.2 Pa*s, cone 6 at a temperature of 140 °C) and the temperature glass transition (T_g = 40.4 °C) were also determined. The acrylic resin used in this study was thoroughly characterized in a previous study [22].

4.1.2. Blocked Polyisocyanates (bPICs)

The blocking of polyisocyanates were performed using 1,2,4-triazole, a compound capable of undergoing a Diels–Alder (DA) reaction with unsaturated polyester resin. The blocked polyisocyanates included cycloaliphatic isophorone diisocyanate (IPDI), glycerol, polysiloxane KF-6000, and 1,2,4-triazole (Figure 1). Powder coatings based on IPDI exhibit a lower tendency to yellow under light exposure compared to those made with aromatic diisocyanates, making them suitable for outdoor applications with direct exposure to weather conditions. Glycerol was used to enhance the functionality of polyisocyanates, while polysiloxane KF-6000 acted as a modifier to improve the physicochemical properties of the powder coatings [33].

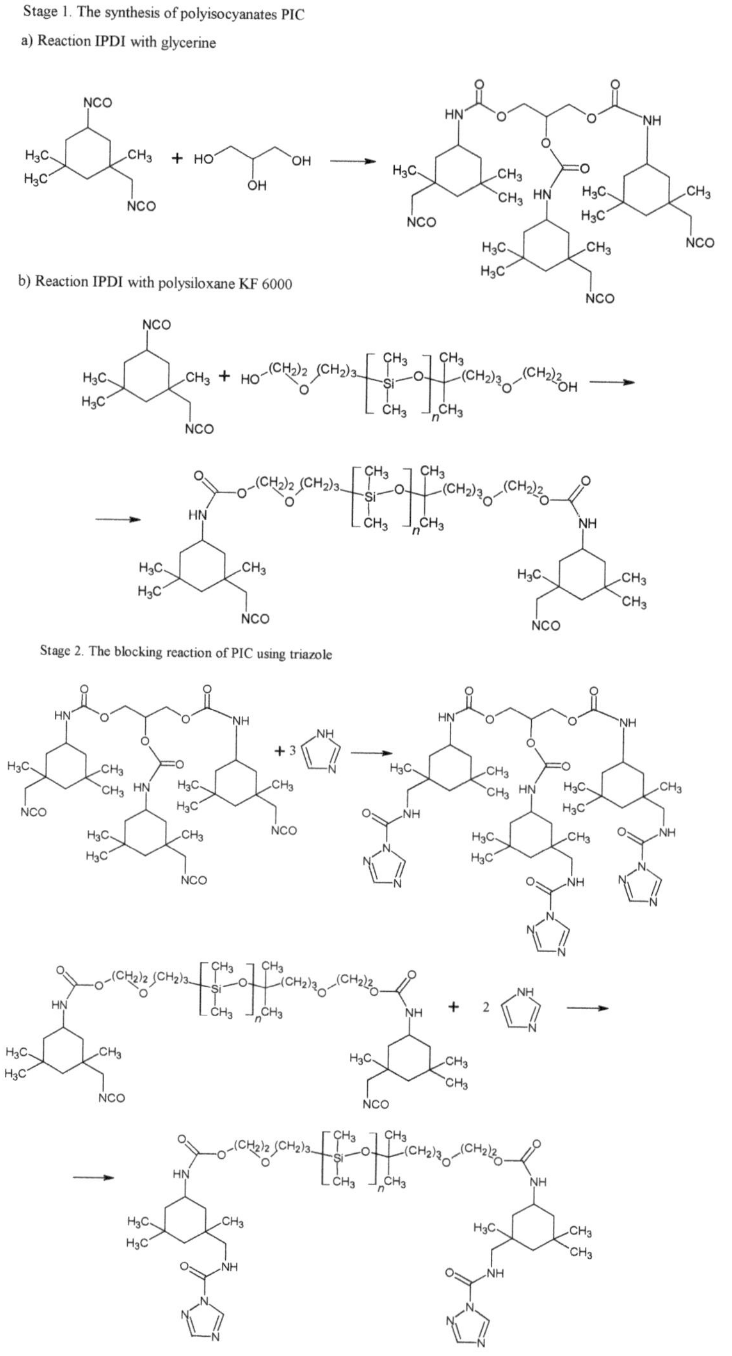

Figure 1. Scheme of reaction of bPIC's.

By using ^{1}H-NMR, the chemical structure of bPICs was confirmed (Figure 2). Peaks observed at 6.78 and at 7.26 ppm (labeled as "A" and "B") indicated the presence of protons from the urethane groups, formed through the interaction between polysiloxane KF-6000 or glycerin and the -NCO groups of IPDI [34]. The double signals arising from the urethane groups reflect the ability of both the primary (at 6.78 ppm) and secondary (at 7.26 ppm) isocyanate groups in IPDI to participate in their formation. The hydrogen atoms of the methylene group adjacent to the urethane bond formed by the primary -NCO group of IPDI appear at 3.16–3.23 ppm (designated as "C"). The signal at 3.4–3.5 ppm (assigned as "D") originated from the proton on a cycloaliphatic carbon adjacent to a urethane bond formed by the secondary -NCO group of IPDI. In this study, based on the ^{1}H-NMR spectrum of blocked PICs, we were unable to determine which -NCO group of IPDI reacted first. The primary isocyanate group, due to its proximity to the methyl group, the cyclohexane ring, and the β-methyl substituent, is effectively shielded. At 4.15–4.19 ppm, signals (assigned as "E") originating from the CH_2 groups of polysiloxane or glycerin, adjacent to the urethane bonds, were detected. Signals from the methyl and methylene groups of IPDI were observed within the range of 0.9–1.2 ppm (assigned as "F"). The signal from the CH group of the 1,2,4-triazole moiety was assigned as "G" and observed at 8.21 ppm.

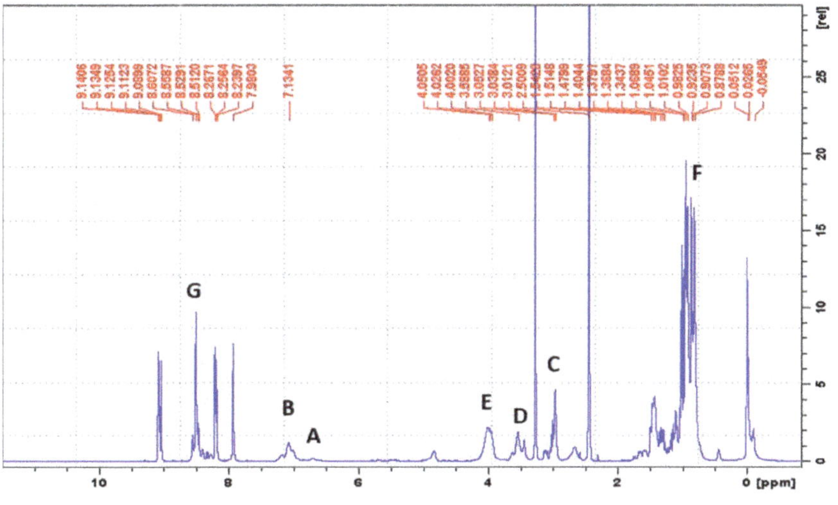

Figure 2. ^{1}H-NMR spectrum of IGKF/T.

Additionally, the curing agent was evaluated for thermal stability (determining T_g) and viscosity. The glass transition temperature (T_g) of IGKF/T was found to be 39.41 °C, while the viscosity was determined to be 10.95 Pa*s (cone 4, at a temperature of 140 °C, speed 10 rpm). These parameters, similar to those of the resin, are crucial in the subsequent stages of manufacturing and storage of powder coatings.

4.2. Curing Process and Self-Healing Properties of Powder Coatings

The curing process and self-healing properties of powder coatings were also examined.

As shown in Figure 3, the deblocking reaction takes place at 160 °C for 15 min (Figure 3a, stage I). Following the deblocking of bPICs, the crosslinking reaction is initiated by the interaction between the free isocyanate groups in PICs and the hydroxyl groups of acrylic resin, leading to the formation of powder coatings at a reduced temperature (Figure 3b, stage II). The self-healing mechanism is then activated through the DA reaction between the triazole released from bPICs and unsaturated resin, leading to the formation of a DA moiety (Figure 4).

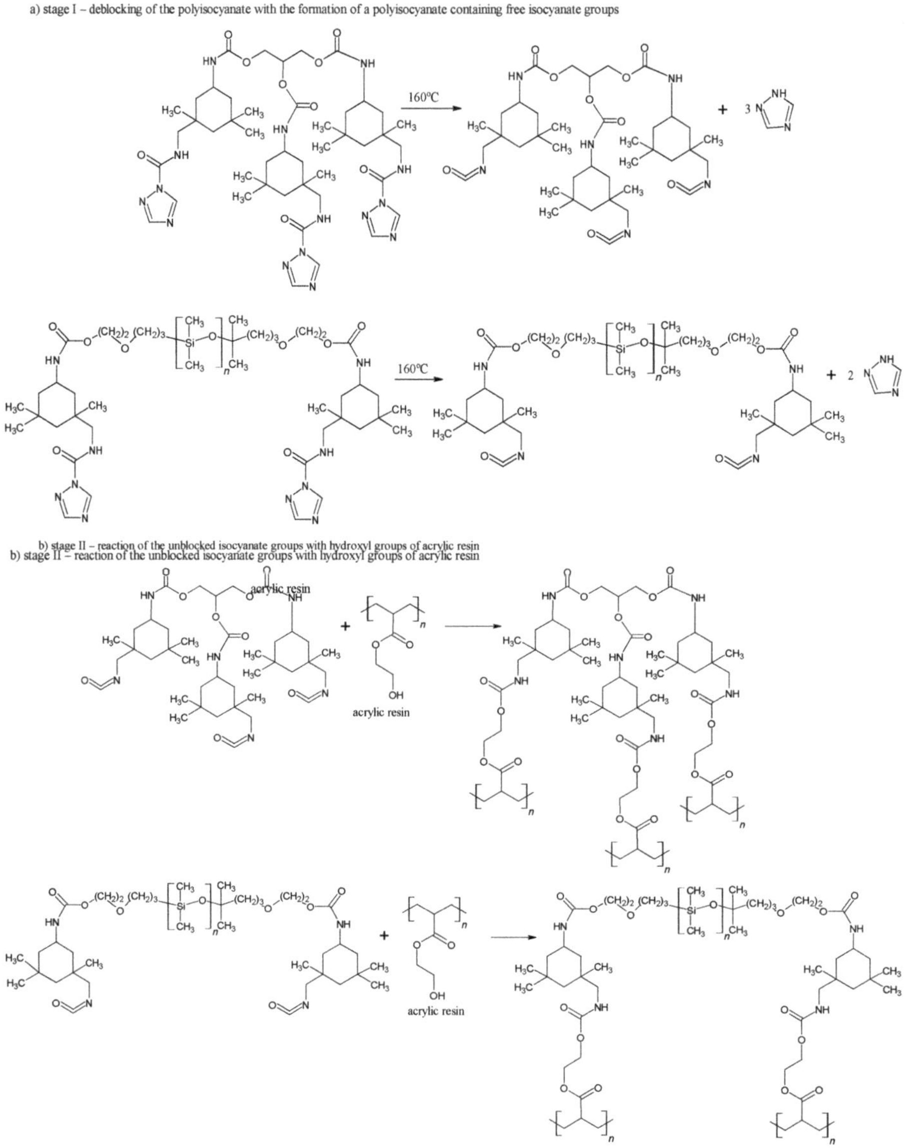

Figure 3. Schematic of the deblocking process for the bPICs (IGKF/T/) at 160 °C, along with the crosslinking reaction between –OH groups (originating from acrylic resin) and –NCO caused by PICs.

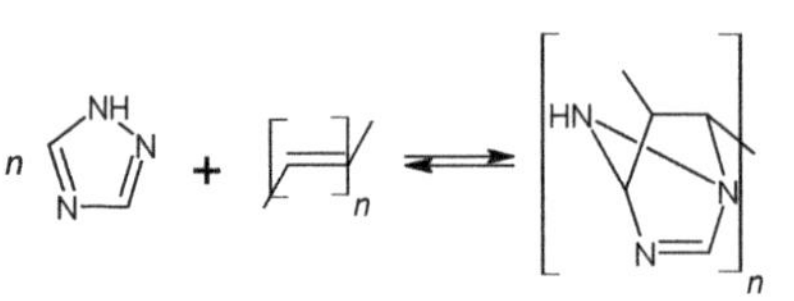

Figure 4. The self-healing mechanism of powder coating through a reversible DA reaction between triazole and unsaturated resin.

The course of the curing process and the reversible DA reaction of powder coatings were monitored using DSC and FT-IR techniques.

Figure 5 shows the DSC thermograms for unsaturated polyester resin UVECOAT3003, UVECOAT 3003 resin mixed with triazole T/UVECOAT3003, UVECOAT 3003 resin mixed with bPIC IGKF/T/UVECOAT 3003, and powder coating L_HEMA/5MMA/2BA/IGKF/T/UVECOAT 3003.

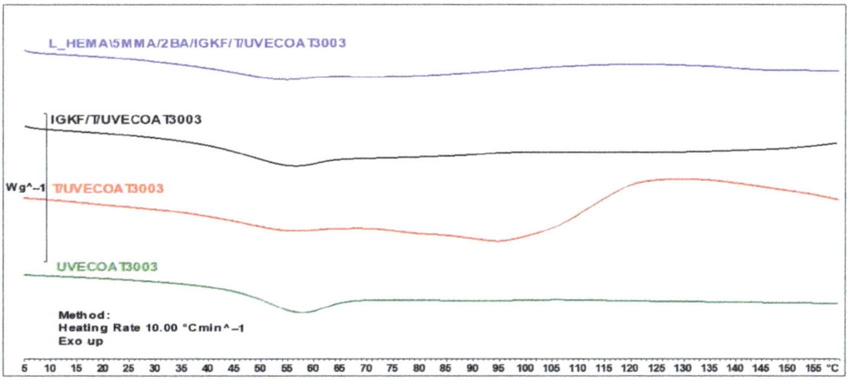

Figure 5. DSC curves of UVECOAT 3003, T/UVECOAT 3003, IGKF/T/UVECOAT 3003, and L_HEMA/5MMA/2BA/IGKF/T/UVECOAT 3003.

At the beginning of all thermograms in the range between 40 °C and 70 °C, an endothermic peak appears, indicating the transition of the resins and powder coating components to a flexible state, accompanied by enthalpy relaxation. In the thermogram of the UVECOAT 3003 resin, apart from the glass transition, no further changes are observed under the influence of controlled heating.

In the case of the sample containing UVECOAT 3003 resin and 1,2,4-triazole (T/UVECOAT3003), in addition to the glass transition, a broad exothermic peak in the range of 60–85 °C is visible, confirming the course of the DA reaction between triazole and unsaturated double bonds of UVECOAT 3003 resin. Then, an endothermic peak appears on this thermogram in the range of 90–130 °C, confirming the course of the r/DA reaction. According to the literature, the DA reaction between furfuryl alcohol and maleic anhydride adduct took place in a similar temperature range: DA at 50–60 °C (exothermic peak) and rDA at a temperature of 120–150 °C (endothermic peak) [21,35].

In the IGKF/T-UVECOAT3003 sample, in addition to the DA and rDA reactions, an endothermic deblocking process of polyisocyanate takes place in the range of 110–130 °C. The thermal effects of these processes overlap, and for this reason, the thermogram of this sample is flatter.

Under the influence of heating, in the powder coating sample L_HEMA/5MMA/2BA/IGKF/T/UVECOAT 3003, in addition to the DA, rDa, and deblocking reactions, a crosslinking process takes place between the deblocked isocyanate groups and the hydroxyl groups of the acrylic resin. This process is visible on the thermogram in the range of 120–150 °C as a broad exothermic peak.

The FT-IR analysis was used to confirm the blocking reaction of PICs and curing reaction of powder coatings (Figure 6). The FT-IR spectrum of bPICs showed no absorption in the range of 2250–2270 cm^{-1}, which corresponds to the asymmetric C–N stretching vibration in the -NCO groups of diisocyanate. Moreover, no absorption at 3126 cm^{-1} originating from the N-H stretching vibration of triazole was observed. The absence of these bands indicates that the -NCO groups from the PICs were completely blocked by the triazole. The strong twin absorption bands in the range of 1500–1540 cm^{-1} in the spectrum of IGKF/T correspond to the C=N- aromatic stretching vibration of triazole [35].

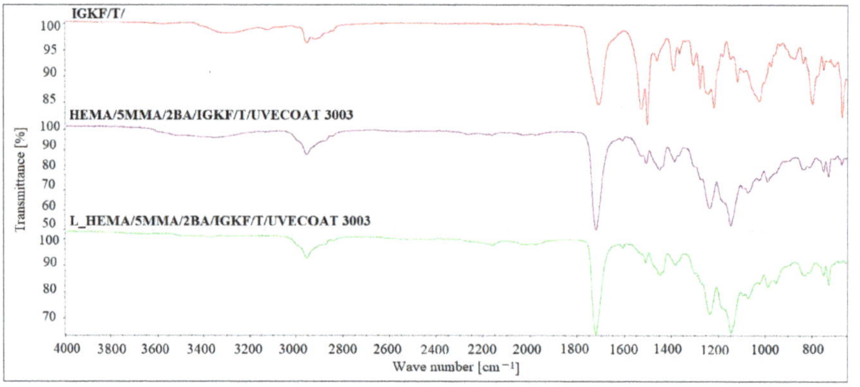

Figure 6. The FTIR spectra of IGKF/T/, L_HEMA/5MMA/2BA/IGKF/T/UVECOAT 3003 non-crosslinked and crosslinked powder coating.

The spectra of both non-crosslinked and crosslinked L_HEMA/5MMA/2BA/IGKF/T/UVECOAT 3003 are nearly identical, showing a urethane -NH stretching absorption at 3330 cm^{-1}, a urethane -NH bending absorption at 1521 cm^{-1}, and stretching vibrations of carbonyl groups (C=O) at 1700 cm^{-1}. The urethane -NH bending absorption at 1521 cm^{-1} overlaps with the twin C=N- aromatic stretching vibration peaks derived from triazole. These twin peaks in powder coating spectra are weak because the IGKF/T content is low (12.8%). The intensity of these peaks in the spectrum of the crosslinked coating is slightly lower than in the non-crosslinked one, and this may indicate the partial evaporation of triazole from the coating during heating, which was confirmed by TGA analysis. This slight change in intensity indicates that an r/DA reaction has occurred. Otherwise, the intensity of this signal should decrease by at least half because there is only one C=N- bond in the DA adduct (Figure 4). In the range of 1600 cm^{-1}, a weak signal is visible coming from the C=C stretching vibrations of the unsaturated UVECOAT 3003 resin. This signal is present in the spectrum of the crosslinked and non-crosslinked coating, which proves that the r/DA reaction has occurred and the C=C groups have been recreated and are capable of the DA reaction again.

The presence of polysiloxane is confirmed by Si-O-Si absorption in the 1020–1100 cm^{-1} range, while Si-CH_3 shows absorption at 1220 cm^{-1} and 800 cm^{-1} [36]. The cured coatings confirm the progression of the reaction between the hydroxyl groups (-OH) of the acrylic resin and the isocyanate groups (-NCO) derived from the crosslinking agent within this range (3400–3600 cm^{-1}). The characteristic C-O stretching vibrations are observed in the range of 1100–1200 cm^{-1}. Absorption bands in the range of 2800–3000 cm^{-1} indicate the presence of aliphatic groups (-CH_2, -CH_3), while signals at 1411 cm^{-1} are characteristic of the C-H vibrations of methyl groups. Additionally, absorbance at 1230 cm^{-1} corresponds to the asymmetric stretching vibrations of C=O and O-CH_2 bonds originating from the ester groups of the polyester resin [16].

In order to confirm the self-healing properties of the powder coating, polarized optical microscopy equipped with a heating table was employed. Before investigating the temperature required for healing the powder coating, a scratch was created under a load of 300 g. Subsequently, the temperature and time required for the self-healing of the coating were evaluated. It was observed that at 160 °C for 15 min, the crack was successfully stuck together. The results are presented in Table 2.

Table 2. Examination and comparison of the self-healing ability of the cracks created on the powder coatings at room temperature and 160 °C.

Name of Sample	Room Temperature [23 °C]	Self-Healing Temperature [160 °C]
L_HEMA/5MMA/2BA/IGKF/T/	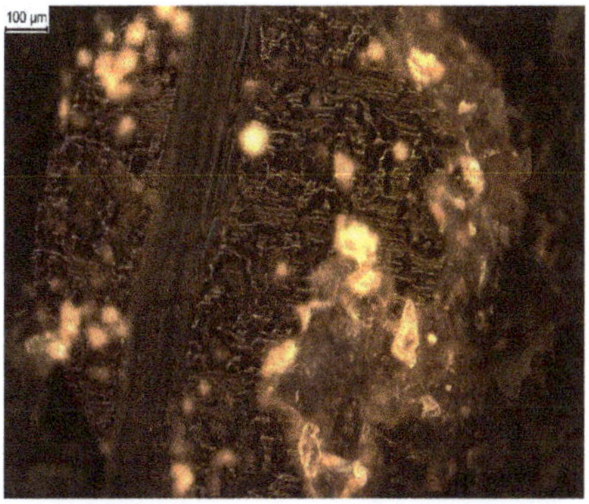	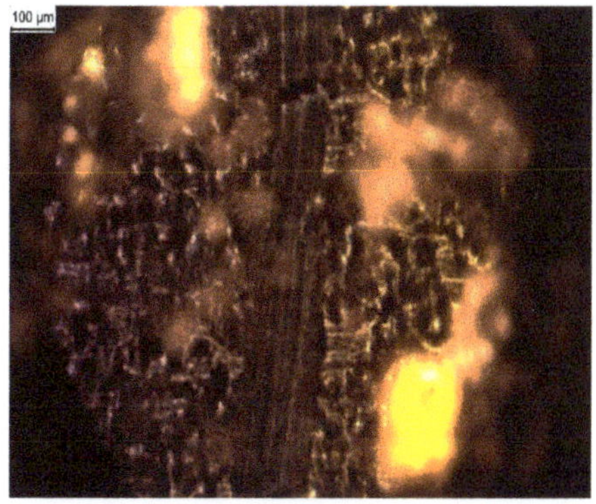
L_HEMA/5MMA/2BA/IGKF/T/ UVECOAT 3003	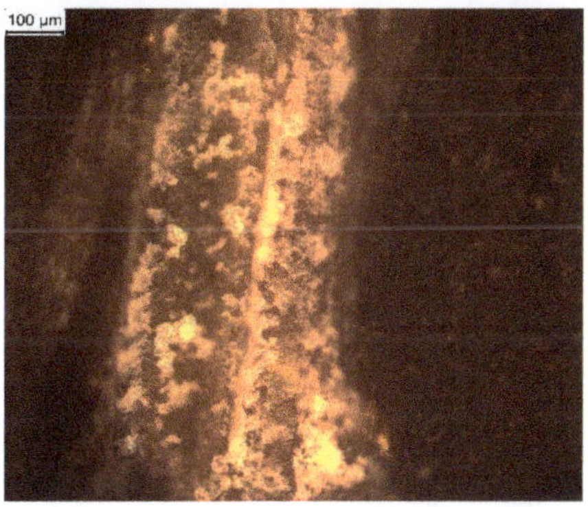	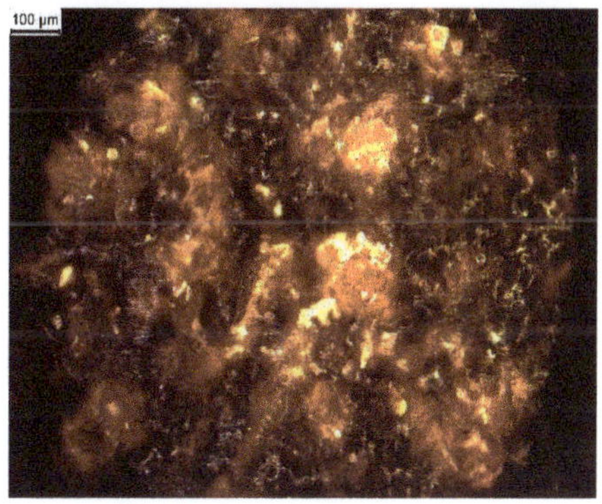

The powder coatings containing unsaturated polyester resin were characterized by self-healing properties. This means that under the influence of temperature, triazole with unsaturated resin created a DA adduct, allowing for the restoration of the powder coating.

In order to test whether the coating is capable of repeated self-healing, the tests were performed by using polarized optical microscopy. The L_HEMA/5MMA/2BA/IGKF/T/ UVECOAT 3003 coating was sprayed onto a Teflon tray and cured. Then, the created cured powder coating was cooled, cut into small pieces (Figure 7a), and heated again to 160 °C for 15 min. The powder coating stuck together (Figure 7b). When the operation was repeated a second time, the coating did not solidify completely (Figure 7c), which indicates a partial loss of self-healing ability.

The observed partial loss of self-healing ability results from the partial release of 1,2,4-triazole (diene) from the coating, which was proven based on the results of the TGA analysis (Figures 8 and 9). The weight loss at a temperature above 160 °C in the sample containing unsaturated polyester resin (dienophile) (L_HEMA/5MMA/2BA/IGKF/T/UVECOAT 3003) is lower (7.63%) than in the sample not containing the dienophile (10.54%), which indicates that some of the 1,2,4-triazole remains in the coating and takes part in the Diels–Alder reaction and some of it evaporates. This remaining amount of triazole in the coating is sufficient to ensure the complete self-healing of the coating during the first test, but during the next test, there may not be enough 1,2,4-triazole to ensure the complete self-healing of the coating.

Figure 7. Image of sample of L_HEMA/5MMA/2BA/IGKF/T/UVECOAT 3003 (**a**) before healing, (**b**) after "healing", and (**c**) after "healing" for the second time.

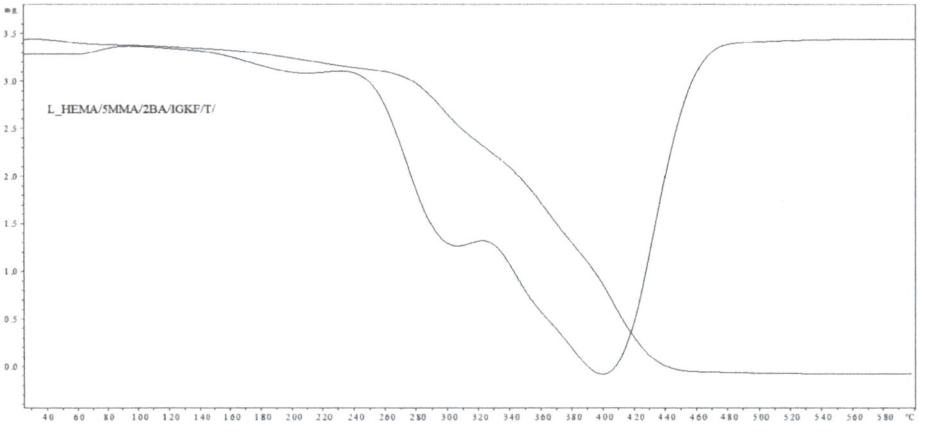

Figure 8. TGA and DTG curves of L_HEMA/5MMA/2BA/IGKF/T/.

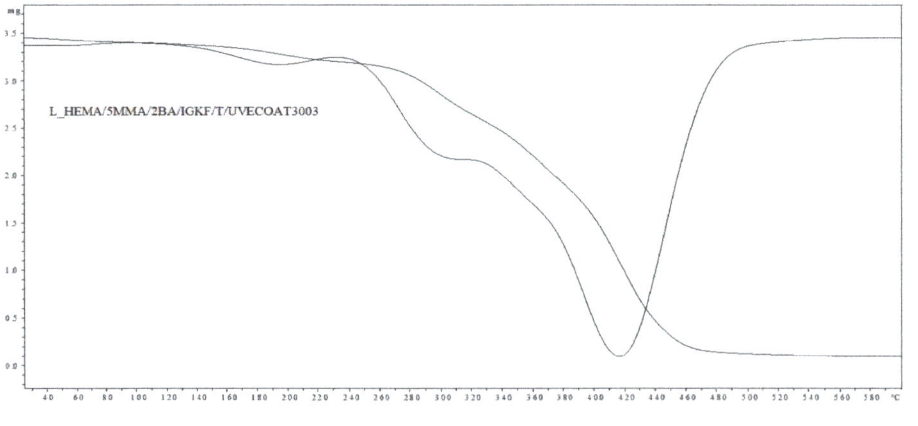

Figure 9. TGA and DTG curves of L_HEMA/5MMA/2BA/IGKF/T/UVECOAT 3003.

The thermal stability of L_HEMA/5MMA/2BA/IGKF/T/ and L_HEMA/5MMA/2BA/IGKF/T/UVECOAT 3003 powder coatings was investigated with TGA and DTG analysis (Figures 8 and 9).

In both cases, the coating compositions began to lose mass at temperatures above 160 °C. The mass loss in this range is 10.54% for the sample L_HEMA/5MMA/2BA/IGKF/T/ and 7.63% for L_HEMA/5MMA/2BA/IGKF/T/UVECOAT 3003. The maximum mass loss rate occurs at the temperature of Tmax1 = 195 °C for sample L_HEMA/5MMA/2BA/IGKF/T and at Tmax1 = 185 °C for sample L_HEMA/5MMA/2BA/IGKF/T/UVECOAT 3003, appropriately. This mass loss is related to the deblocking of triazole and the formation of polyisocyanate. The mass loss in this temperature range was slightly lower in the sample L_HEMA/5MMA/2BA/IGKF/T/UVECOAT 3003, which could be due to the reaction of triazoles with unsaturated resin UVECOAT 3003 and the formation of the DA adduct.

Furthermore, in the next step, the T_{max2} for L_HEMA/5MMA/2BA/IGKF/T/UVECOAT 3003 reaches 295 °C (with a weight loss of 34.72%). Meanwhile, this temperature for L_HEMA/5MMA/2BA/IGKF/T is 300 °C with a mass loss percentage of 63.92%. This step is related to the degradation of urethane bonds and triazole [36]. Slower weight loss in the range 240–380 °C for the L_HEMA/5MMA/2BA/IGKF/T/UVECOAT3003 coating suggests its higher thermal stability, which may be related to its higher density of crosslinking by DA adducts.

The third step (T_{max3} = 404 °C and T_{max3} = 418 °C for resin without and with UVECOAT 3003, respectively) is associated with the degradation of the acrylic and polyester resin as well as polysiloxane KF-6000 segments.

4.3. Characterization of Powder Coatings

The obtained powder coatings were also tested in terms of physical–mechanical properties. Table 3 lists the parameters for the investigated powder coatings.

Table 3. Summary of physical–mechanical parameters of the powder coatings.

Symbol of Coatings	**L_HEMA/5MMA/2BA/IGKF/T/**	**L_HEMA/5MMA/2BA/IGKF/T/ UVECOAT 3003 (After Curing)**	**L_HEMA/5MMA/2BA/IGKF/T/ UVECOAT 3003 (after Crack and Self-Healing Process)**
Flowability [cm]	1.10	1.25	Not measured
Roughness: Ra/Rz	4.9/12.8	1.5/8.0	4.3/14.35
Gloss 60° [GU]	30.36	37.86	27.26
Thickness [um]	56.85	68.74	65.86
Relative hardness [-]	0.54	0.73	0.47
Adhesion to steel [0—good; 5—bad]	2	1	3
Scratch resistance [g]	300	800	250
Cupping [mm]	4.6	4.5	1.2
Water contact angle [deg]	107.23	84.13	82.19

A polymerization test was conducted to confirm the degree of crosslinking in the coating according to the technical requirements of Qualicoat [23].

No noticeable changes were observed after wiping the samples back and forth 30 times in each direction with a cotton swab soaked in methyl ethyl ketone (MEK), indicating the complete crosslinking of all samples. Comparing the reference powder coatings (L_HEMA/5MMA/2BA/IGKF/T/) with L_HEMA/5MMA/2BA/IGKF/T/UVECOAT3003 powder coatings, the sample containing unsaturated resin showed better hardness, scratch resistance, and adhesion.

Comparing the reference sample (L_HEMA/5MMA/2BA/IGKF/T/) with L_HEMA/5MMA/2BA/IGKF/T/UVECOAT3003 powder coatings, the coating containing unsaturated resin showed higher hardness, scratch resistance, and adhesion. Moreover, the L_HEMA/5MMA/2BA/IGKF/T/UVECOAT3003 coating showed much lower roughness and higher gloss than L_HEMA/5MMA/2BA/IGKF/T, which proves that the components have been correctly selected in terms of chemical structure because the improvement in these parameters shows their good compatibility. The water contact angle of the coating with the addition of unsaturated polyester resin is lower than that of the L_HEMA/5MMA/2BA/IGKF/T sample, which is a consequence of the higher hydrophilicity of the polyester resin compared to the acrylic resin. Column 3 of Table 3 contains the properties of the L_HEMA/5MMA/2BA/IGKF/T/UVECOAT3003 coating after the self-healing process, measured in the place where the scratch was made. However, the physico-mechanical properties at the self-healed site decreased even though visually the coating showed no significant differences. The gloss decreased slightly, and the roughness increased, but the mechanical properties deteriorated, in particular the adhesion to the substrate, cupping, and scratch resistance.

5. Conclusions

As a result of this research, a low-temperature, self-healing polyurethane powder coating was developed, which uses a releasing blocking agent in the self-healing process. In the case of classic polyurethane powder coatings, the blocking agent completely evaporates from the coating, but in our case, it was used in the self-healing process. The key is to select the blocking agent so that it has a diene moiety and at the same time is deblocked at a low temperature. In our research, 1,2,4-triazole used as a blocking agent worked well, and after deblocking, it underwent a Diels–Alder reaction with an unsaturated polyester resin and then, as a result of further heating, a retro-Diels–Alder reaction. However, a multiple self-healing test is not recommended because the coating did not heal completely after the second cooling cycle. The reason for this was the partial evaporation of triazole from the coating, which was confirmed by thermal analysis tests. The consequences of this were worse physical and mechanical parameters of the coating at the point of cracking after the self-healing process. However, these properties were better compared to the reference sample, which indicates that performing self-regeneration of the coating once is beneficial, but this process cannot be repeated many times due to the deterioration of the coating properties. The concept of this research opens up new opportunities to develop advanced polyurethane powder coatings with self-healing properties and limited blocking agent emissions. However, the blocking agent should be selected appropriately so that it is blocked at low temperatures and is difficult to evaporate from the coating; then, the self-healing process can be repeated more times.

Author Contributions: Conceptualization, K.P., B.P.-P. and S.A.; Methodology, K.P., B.P.-P. and S.A.; Investigation, K.P. and B.P.-P., P.G., B.M.-L., B.G. and A.B.; Data curation, S.A.; Writing—original draft, K.P.; Writing—review & editing, K.P., B.P.-P. and S.A. All authors have read and agreed to the published version of the manuscript.

Funding: This research was funded by the Minister of Science and Higher Education Republic of Poland within the program "Regional Excellence Initiative", grant number RCD.RB.24.002.

Institutional Review Board Statement: Not applicable.

Informed Consent Statement: Not applicable.

Data Availability Statement: All data is contained within the article.

Acknowledgments: The authors would like to thank Allnex, Evonic Degussa, Shin-Etsu, and BYK-Chemie for sending free samples of raw materials.

Conflicts of Interest: Author Shahla Ataei was employed by the company Harsin Chemical Homa Company. The remaining authors declare that the research was conducted in the absence of any commercial or financial relationships that could be construed as a potential conflict of interest.

References

1. *Directive 2004/42/EC of the European Parliament and of the Council of April 21, 2004 on the Limitation of Emissions of Volatile Organic Compounds Due to the Use of Organic Solvents in Certain Paints and Varnishes and Vehicle Refinishing Products, and Amending Directive 1999/13/EC*; European Parliament: Strasbourg, France, 2004.
2. Directive 2000/60/EC of the European Parliament and of the Council of 23 October 2000 Establishing a Framework for Community Action in the Field of Water Policy, Document 32000L0060. Available online: https://eur-lex.europa.eu/eli/dir/2000/60/oj (accessed on 1 July 2024).
3. Jones, F.N.; Nichols, M.E.; Pappas, S.P. *Organic Coatings: Science and Technology*; Wiley: New York, NY, USA, 2017.
4. Tassel, X.; Barbry, D.; Tighzert, L. A New Blocking Agent of Isocyanates. *Eur. Polym. J.* **2000**, *36*, 1745–1751. [CrossRef]
5. Lv, X.; Zhang, H.; Zhang, H.; Shao, Y.; Zhu, J. Nanosized curing catalyst via nano-templated plasma PVD for attaining superior low-cure powder coating films. *Prog. Org. Coat.* **2024**, *192*, 108499. [CrossRef]
6. Czachor-Jadacka, D.; Pilch-Pitera, B. Progress in Development of UV Curable Powder Coatings. *Prog. Org. Coat.* **2021**, *158*, 106355. [CrossRef]
7. Pojnar, K.; Pilch-Pitera, B.; Roś, N.; Florczak, Ł. Low-temperature powder paint modified with graphene oxide. *Ochr. Przed Koroz.* **2024**, *67*, 30–39. [CrossRef]
8. Pojnar, K.; Pilch-Pitera, B.; Kisiel, M.; Zioło, A.; Kędzierski, M. UV-cured powder transparent coatings based on oligo(meth)acrylic resins. *Polimery* **2024**, *69*, 11–24. [CrossRef]
9. Pojnar, K.; Pilch-Pitera, B.; Patil, R. Progress in the development of acrylic resin-based powder coatings—An overview. *Polimery* **2024**, *3*, 143–158. [CrossRef]
10. Path, U.; Schwalm, R.; Schwalm, M. *Acrylic Resin*; Vincentz Network GmbH & Co. KG: Hanover, Germany, 2011.
11. Zhou, Z.; Xu, W.; Fan, J.; Ren, F.; Xu, C. Synthesis and characterization of carboxyl group-containing acrylic resin for powder coatings. *Prog. Org. Coat.* **2008**, *62*, 179–182. [CrossRef]
12. Guo, X.; Ge, S.; Wang, J.; Zhang, X.; Zhang, T.; Lin, J.; Zhao, C.X.; Wang, B.; Zhu, G.; Guo, Z. Waterborne acrylic resin modified with glycidyl methacrylate (GMA): Formula optimization and property analysis. *Polymer* **2018**, *143*, 155. [CrossRef]
13. Pojnar, K.; Pilch-Pitera, B. Correlation between the Chemical Structure of (Meth)Acrylic Monomers and the Properties of Powder Clear Coatings Based on the Polyacrylate Resins. *Materials* **2024**, *17*, 1655. [CrossRef] [PubMed]
14. Gedan-Smolka, M.; HäEuûler, L.; Fischer, D. Thermal deblocking of masked low molecular isocyanates I. Aliphatic isocyanates. *Thermochimica Acta* **2000**, *351*, 95–105. [CrossRef]
15. Sonnenschein, M. Introduction to polyurethane chemistry. In *Polyurethanes: Science, Technology, Markets, and Trends*; ACS Publications: Washington, DC, USA, 2014.
16. Pilch-Pitera, B.; Kisiel, M.; Gumieniak, J. Hydrophobic UV-Curable Powder Clear Coatings: Study on the Synthesis of New Crosslinking Agents Based on Raw Materials Derived from Renewable Sources. *Materials* **2021**, *14*, 4710. [CrossRef] [PubMed]
17. Gandini, A. The furan/maleimide Diels–Alder reaction: A versatile click–unclick tool in macromolecular synthesis. *Prog. Polym. Sci.* **2013**, *38*, 1–29. [CrossRef]
18. Karami, Z.; Zolghadr, M.; Zohuriaan-Mehr, M.J. Self-healing Diels–Alder engineered thermosets. In *Self-Healing Polymer-Based Systems*; Elsevier: Amsterdam, The Netherlands, 2020.
19. Ren, J.; Dong, X.; Duan, Y.; Lin, L.; Xu, X.; Shi, J.; Jia, R.; Wu, D.; He, X. Synthesis and self-healing investigation of waterborne polyurethane based on reversible covalent bond. *J. Appl. Polym. Sci.* **2022**, *139*, 52144. [CrossRef]
20. Motora, K.G.; Wu, C.M.; Chang, C.C.; Liao, J.H. NIR Light Stimulated Self-Healing Reduced Tungsten Oxide/Polyurethane Nanocomposite Based on the Diels−Alder Reaction. *Macromolar Mater. Eng.* **2021**, *306*, 2100438. [CrossRef]
21. Farshchi, N.; Gedan-Smolka, M.; Stommel, M. Preparation and Characterization of Self-Healing Polyurethane Powder Coating Using Diels–Alder Reaction. *Polymers* **2021**, *13*, 3803. [CrossRef] [PubMed]
22. Pojnar, K.; Pilch-Pitera, B.; Byczyński, Ł.; Zając, W.; Walczak, M.; Kramek, A. Polyacrylate resins containing fluoroalkyl groups for powder clear coatings. *Prog. Org. Coat.* **2022**, *172*, 107116. [CrossRef]
23. Qualicoat. *Specifications for a Quality Label for Liquid and Powder Organic Coatings on Aluminium for Architectural Applications*, 16th ed.; Qualicoat: Zurich, Switzerland, 2019.
24. *PN-EN ISO 8130-11*; Coating Powders—Part 11: Inclined-Plane Flow Test. ISO: Geneva, Switzerland, 2019.
25. *PN-EN ISO 12085*; Geometrical Product Specifications (GPS)—Surface Texture: Profile Method—Part 2: Terms, Definitions and Surface Texture Parameters. ISO: Geneva, Switzerland, 2018.
26. *PN-EN ISO 2813*; Paints and Varnishes—Determination of Specular Gloss of Non-Metallic Paint Films at 20 Degrees, 60 Degrees and 85 Degrees. ISO: Geneva, Switzerland, 2014.

27. *PN-EN ISO 2808*; Paints and Varnishes—Determination of Film Thickness. ISO: Geneva, Switzerland, 2019.
28. *PN-EN ISO 2409*; Paints and Varnishes—Cross-Cut Test. ISO: Geneva, Switzerland, 2013.
29. *PN-EN ISO 1522*; Paints and Varnishes—Pendulum Damping Test. ISO: Geneva, Switzerland, 2011.
30. *PN-EN ISO 1518*; Paints and Varnishes—Scratch Hardness Tests—Part 1: Constant-Loading Method. ISO: Geneva, Switzerland, 2016.
31. *PN-EN ISO 1520*; Paints and Varnishes—Cupping Test. ISO: Geneva, Switzerland, 2019.
32. *PN-EN ISO 19403-6:2020-08*; Paints and Varnishes—Wetting—Part 3: Determination of Surface Tension of Liquids by the Hanging Drop Method. ISO: Geneva, Switzerland, 2020.
33. Czachor-Jadacka, D.; Pilch-Pitera, B.; Byczyński, Ł.; Kisiel, M.; Zioło, A. Hydrophobic polyurethane powder clear coatings with lower curing temperature: Study on the synthesis of new blocked polyisocyanates. *Prog. Org. Coat.* **2021**, *159*, 106402. [CrossRef]
34. Prabhakar, A.; Chattopadhyay, D.K.; Jagadeesh, B.; Raju, K.V.S.N. Structural investigations of polypropylene glycol (PPG) and isophorone diisocyanate (IPDI)-based polyurethane prepolymer by 1D and 2D NMR spectroscopy. *J. Polym. Sci. Part A Polym. Chem.* **2005**, *43*, 1196–1209. [CrossRef]
35. Trivedi, M.K.; Tallapragada, R.M.; Branton, A.; Trivedi, D.; Nayak, G.; Mishra, R.K.; Jana, S. Characterization of Physical, Spectral and Thermal Properties of Biofield Treated 1,2,4-Triazole. *Mol. Pharm. Org. Process Res.* **2015**, *3*, 1000128. [CrossRef]
36. Czachor-Jadacka, D.; Pilch-Pitera, B.; Kisiel, M.; Thomas, J. Polyurethane powder coatings with low curing temperature: Research on the effect of chemical structure of crosslinking agent on the properties of coatings. *Prog. Org. Coat.* **2023**, *182*, 107662. [CrossRef]

materials

MDPI

Article

Reduction in Powder Wall Friction by an a-C:H:Si Film

Christof Lanzerstorfer [1,*], Christian Forsich [1], Francisco Delfin [1,2], Manuel C. J. Schachinger [1] and Daniel Heim [1]

[1] School of Engineering, University of Applied Sciences Upper Austria, Stelzhamerstraße 23, 4600 Wels, Austria; christian.forsich@fh-wels.at (C.F.); francisco.delfin@fh-wels.at (F.D.); manuel.schachinger@fh-wels.at (M.C.J.S.); daniel.heim@fh-wels.at (D.H.)

[2] Regional Faculty of Concepción del Uruguay (UTN-FRCU), National University of Technology, Ing. Pereyra 676, Concepción del Uruguay 3260, Argentina

* Correspondence: c.lanzerstorfer@fh-wels.at; Tel.: +43-50804-4332

Abstract: The wall friction angle is an important parameter in powder flow. In a recent study for various powders, a reduction in the wall friction angle for steel was demonstrated by the application of an a-C:H:Si film on the steel surface. This work presents the results of a study of this effect in more detail regarding the influence of the powder material, the wall normal stress and the particle size of the powder for mass median diameters from 4 μm to approximately 150 μm. The wall friction angles were measured using a Schulze ring shear tester for three different powder materials: aluminum oxide, calcium carbonate and silicon carbide. The results showed little difference with respect to powder chemistry. For the coarser powders, the reduction in the wall friction angle due to the a-C:H:Si coating was highest (10° to 12°) and rather stress-independent, while for the fine and medium-size powders the reduction was lower and stress-dependent. With increasing wall normal stress, the reduction in the wall friction angle increased. These results can be explained by the friction reduction mechanism of a-C:H:Si, which requires a certain contact pressure for superficial graphitization.

Keywords: powder; wall friction; diamond-like carbon

1. Introduction

In the design of particulate material handling and conveying equipment, friction between the particulate material and the wall material plays a crucial role. The friction between the powder and the wall material can be described using the wall friction angle φ_W or the wall friction coefficient μ [1]. The wall friction coefficient is the ratio of the wall shear stress τ_W to the wall normal stress σ_W and the wall friction angle is the arctangent of the wall friction coefficient. Powder wall friction depends on both the powder properties and the wall material. The particle size of the powder is an important parameter with respect to wall friction. Lower values of the wall friction angle for granular material with larger particle size were reported for fine granular materials, like wet gypsum [2], and for fly ash from biomass combustion plants [3] as well as for steel mill dedusting residues [4]. In contrast, for large particles with a diameter of more than several hundred micrometers an increase in wall friction with the particles size was reported [5]. Another important parameter for wall friction is the moisture content of the powder. For various food powders, the wall friction angle increased with the moisture content [6]. However, it is not always certain whether an increased moisture content of the powder will increase or decrease the wall friction angle [2,5]. Further powder properties determined to influence wall friction are the particle size distribution, the surface structure of the particles, the particle hardness and the chemical composition [1].

The most important property of the wall material with respect to wall friction is the surface roughness [7]. With increasing roughness of the wall material, increasing wall friction angles were reported for salt [8], for limestone powder [9] and for polyethylene pellets [10]. Other parameters of the wall material that influence wall friction are its hardness [11] and its chemical composition [1].

The outstanding properties of diamond-like carbon (DLC), such as a high degree of hardness and chemical inertness, are two of the reasons for the wide use of DLC-based films in different applications. DLC-based films possess very good tribological properties such as low friction coefficients and low wear rates [12]. Additionally, the material is very corrosion-resistant and biologically compatible [13]. Therefore, DLC-based coatings find widespread use for friction and wear reduction on sliding mechanical parts [14,15]. To further improve the already beneficial properties of DLC, silicon doping may be applied as silicon only forms tetrahedral (sp^3-type) bonding and thus increases the hardness of the amorphous carbon [16,17].

The low friction coefficient of DLC-based coatings is explained by graphitization of the DLC, where the formed debris is acting as a third body, and by the formation of a transfer film onto the counterpart. These processes need the transfer of energy into the contact zone and depend on the sliding speed and sliding distance and also on the contact pressure [18–21].

While the benefit of DLC-based coatings in friction reduction for sliding and rotating parts has been investigated widely [22–26], the use of DLC-based coatings for powder wall friction reduction has been suggested only recently [27]. However, in the literature search, no further publications were found on the topic of powder wall friction reduction through DLC-based coatings. In [27], a reduction in the wall friction angle by the a-C:H:Si film of 2°–14° was reported in comparison with a stainless steel material of similar surface roughness. The level of reduction was not constant but depended on the type of powder material, as well as on the wall normal stress.

The aim of the present study was to investigate the reduction in wall friction for various powders by employing an a-C:H:Si film and determine the influencing parameters. The wall friction angle of three different materials, aluminum oxide, calcium carbonate and silicon carbide, with three different mass median diameters in the range of 4.0 μm to approximately 150 μm for each material was measured for a-C:H:Si-coated stainless steel, uncoated stainless steel with a similar roughness and an ultra-high molecular weight polyethylene (UHMWPE) material with a somewhat higher wall roughness.

2. Materials and Methods

The a-C:H:Si film was deposited on a stainless-steel sample (EN 1.4301) in a commercial hot wall plasma-assisted chemical vapor deposition (PACVD) reactor manufactured by Rübig GmbH & Co KG (Wels, Austria). The a-C:H:Si film produced consisted of an amorphous hydrogenated carbon film modified by embedded Si in the carbon matrix. Silicon was chosen as a dopant to increase the hardness and thus the wear resistance of the film. The reactor vessel of the PACVD reactor has an inner diameter of 400 mm and a height of 600 mm, which was evacuated to a base pressure of 1 Pa using a screw pump and a roots blower controlled by a throttle valve. A pulsed DC voltage of 350 V, with a frequency of 1.5 kHz and a duty cycle of 25% was applied to the reactor in such way that the substrate plate acted as cathode and the wall of the chamber acted as anode. The gas mixture consisted of 15% acetylene and 1% hexamethyldisiloxane (HMDSO), as carbon and silicon precursors, respectively, with argon completing the gas mixture, under an operating pressure of 200 Pa. The process temperature was set to 450 °C, which was measured by thermocouples inserted in the core of dummy samples placed on the deposition plate. Prior to the main process, a thin silicon-based interlayer was deposited to improve adhesion and performance of the coating. Deposition time was 35 h, with an estimated deposition rate of about 1 μm/h. Further information concerning the PACVD system has been published previously [28].

Powder samples of aluminum oxide, calcium carbonate and silicon carbide were used in the wall friction tests. The three powder materials differ significantly in terms of hardness. While the Mohs hardness of calcium carbonate is 3 the Mohs hardness of aluminum oxide and silicon carbide is 9 and 9.5, respectively. From each powder material, aluminum oxide, calcium carbonate and silicon carbide, three samples with different particle sizes were used: (1) fine powders with a mass median diameter d_{50} of approximately 4 μm; (2) medium-size powders with a d_{50} in the range of 14 to 32 μm; and (3) coarse powders with a d_{50} of approximately 150 μm. The fine silicon carbide is F 1200-D and the medium-size silicon carbide is P 600, both from ESK-SiC GmbH (Frechen, Germany). The coarse silicon carbide is 100 mesh silicon carbide from abcr GmbH (Karlsruhe, Germany). The fine calcium carbonate is the certified reference material BCR-116 purchased from the Institute for Reference Materials and Measurements of the European Commission for shear tests. The medium-size calcium carbonate is calcium carbonate Normapur from VWR International (Radnor, PA, USA) and the coarse calcium carbonate was made by sieving crushed limestone using 160 μm and 63 μm sieves. The size fraction 63–160 μm was used as the coarse calcium carbonate. The fine aluminum oxide and the medium-size aluminum oxide were obtained from KUVAG GmbH (Neumarkt, Austria) without a specification and the coarse aluminum oxide is fused brown aluminum oxide for sandblasting from Washington Mills.

For the measurement of the wall friction angle of the particulate materials, an RST-XS ring shear tester from Schulze was used. In this test, a sample of the wall material forms the bottom ring of the wall friction shear cell. The measuring method is described in detail in the standard ASTM D6773 [29]. In addition to the wall material with the a-C:H:Si film, a wall sample of uncoated stainless steel EN 1.4301 and a sample of UHMWPE were used in the tests.

The wall yield locus results from corresponding values of the normal stress and the shear stress. The kinematic angle of wall friction is the slope of a straight line running through the origin and a point of the wall yield locus [1]. In the tests, each measurement was repeated five times. In the results, the average wall friction angle and its standard deviation are shown. The measurements were performed using the whole measuring range of the shear tester. The used values of wall normal stress (240 Pa, 600 Pa, 2000 Pa, 6000 Pa and 20,000 Pa) are approximately equidistant on a logarithmic scale.

The particle size distribution of the powder samples was measured by laser diffraction using a HELOS/RODOS instrument from Sympatec GmbH (Clausthal-Zellerfeld, Germany) with dry sample dispersion. The calibration of the instrument was checked with a Sympatec SiC-P600′06 reference material. Additionally, microscopic images of the various powder samples were taken with a TESCAN (Brno, Czech Republic) scanning electron microscope (SEM), type MIRA3.

The moisture content of the powders was determined by drying at 105 °C before and after the measurement of the wall friction angle. The surface roughness of the wall samples was measured along two perpendicular axes with a surface roughness tester from Mitutoyo (Kawasaki, Japan), type SJ-201p. The instrument scans the unevenness of the surface with a measuring feeler. The vertical displacement of the feeler pin is recorded and converted to the standardized output values R_a (arithmetical mean roughness value) and R_z (mean roughness depth). The Nano Indenter XP measurement system from MTS was used to measure the hardness of the a-C:H:Si film and the stainless steel sample. The measurements were carried out at a maximum indentation load of 30 mN using a Berkovich-type diamond indenter. The hardness of the UHMWPE sample was measured with a Shore durometer from Zwick Roell (Ulm, Germany).

Further characterization of the a-C:H:Si coating was carried out using SEM (TESCAN MIRA 3) coupled with EDX analysis (Oxford AZtecEnergy XT, Oxford Instruments, Abingdon, UK) to examine the surface morphology as well as the elemental composition of the DLC. In addition, Raman spectroscopy (Horiba XploRA Plus, Kyoto, Japan), equipped with a 532 nm laser, was used to analyze the structure and bonding of the carbon material. The obtained Raman spectra, showing the characteristic D and G bands, were subsequently fitted by Gaussian–Lorentzian functions to obtain relevant Raman parameters such as the position and FWHM of the G band, the I(D)/I(G) ratio and the baseline slope of the spectrum.

3. Results

3.1. Powder Samples

The measured data of the powder samples used in the tests are summarized in Table 1. The humidity of all samples was very low and did not change during the wall friction measurements.

Table 1. Powder samples.

Particulate Material	Mass Median Diameter d_{50}	Moisture
Al_2O_3 fine	3.95 μm	0.26%
Al_2O_3 medium	32.3 μm	<0.05%
Al_2O_3 coarse	140 μm	<0.05%
$CaCO_3$ fine	3.87 μm	0.19%
$CaCO_3$ medium	14.0 μm	0.15%
$CaCO_3$ coarse	147 μm	<0.05%
SiC fine	3.96 μm	<0.05%
SiC medium	25.4 μm	<0.05%
SiC coarse	173 μm	<0.05%

Figure 1 shows the particle size distribution of the aluminum oxide, calcium carbonate and silicon carbide powders. The size distributions of the three different samples of the fine powder and the three coarse powders were very similar, while there were some differences in the size distributions of the medium-size powder samples.

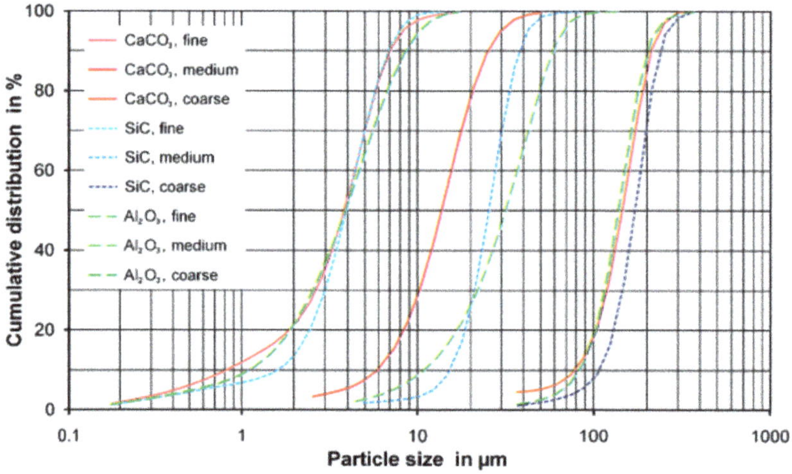

Figure 1. Cumulative size distribution of the aluminum oxide, calcium carbonate and silicon carbide powder samples.

Microscopic images of the various aluminum oxide, calcium carbonate and silicon carbide powder samples are shown in Figure 2. All samples consisted mainly of angular and relatively compact particles.

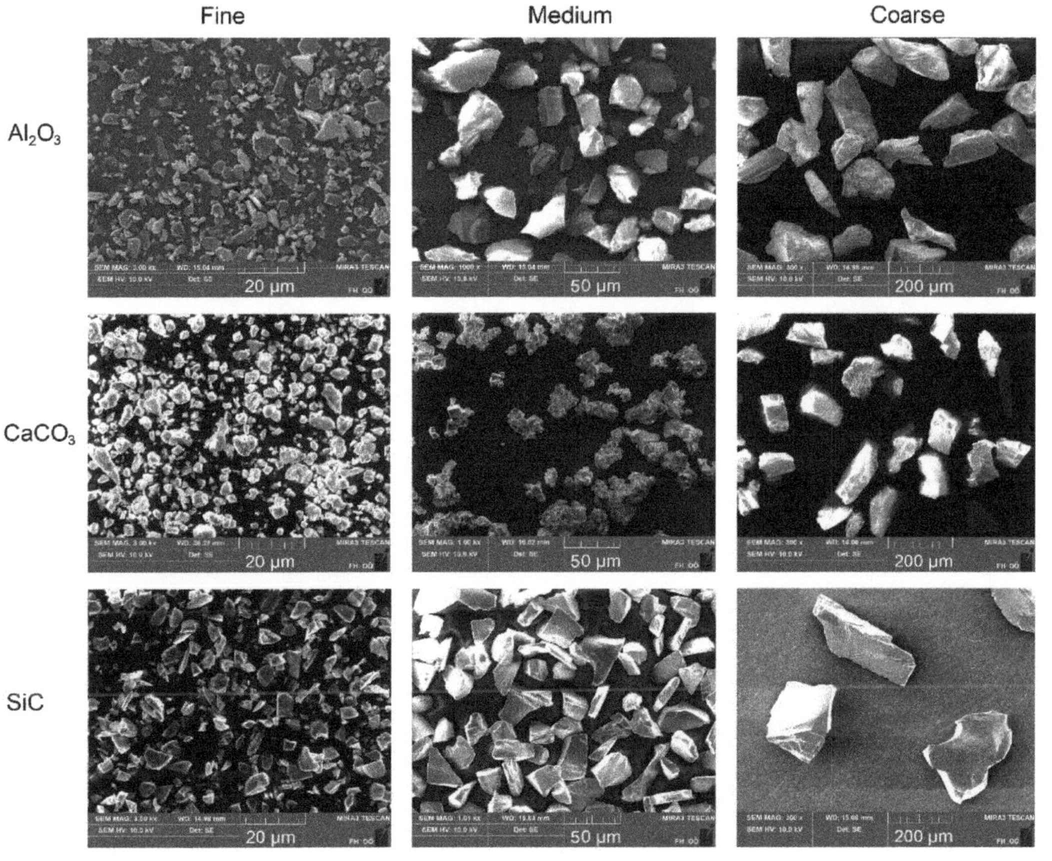

Figure 2. Microscopic images of the aluminum oxide, calcium carbonate and silicon carbide powder samples.

3.2. Wall Samples

The values of the mean roughness depth R_z and the arithmetical mean roughness value R_a were quite similar for the wall sample with the a-C:H:Si film and the uncoated 1.4301 sample, whereas the values for the UHMWPE material were approximately twice as high. The hardness of the a-C:H:Si film was approximately seven times higher compared to the hardness of the uncoated 1.4301 material. The measured results of the surface roughness and the hardness are summarized in Table 2.

Table 2. Wall materials used in the wall friction test.

Wall Material	R_a [2]	R_z [3]	Hardness
Stainless steel 1.4301	0.46 µm	3.3 µm	1.5 GPa
a-C:H:Si film	0.51 µm	3.1 µm	10.2 GPa
UHMWPE	1.04 µm	6.5 µm	65.2 Shore D (0.07 GPa) [1]

[1] For better comparison, the hardness of the UHMWPE was also measured with the nanoindentation instrument; [2] arithmetical mean roughness value; [3] mean roughness depth.

Surface characterization via SEM, showing the characteristic globular morphology, and an EDX line-scan of the cross-section of the a-C:H:Si film with the chemical composition as a function of the distance are shown in Figure 3. The different areas of the cross-section correspond to the substrate material, the silicon-based interlayer as well as the a-C:H:Si film. The line-scan further demonstrates the homogenous distribution of silicon inside the a-C:H:Si along the cross-section, with an average silicon content of 15.7 at.-% according to EDX.

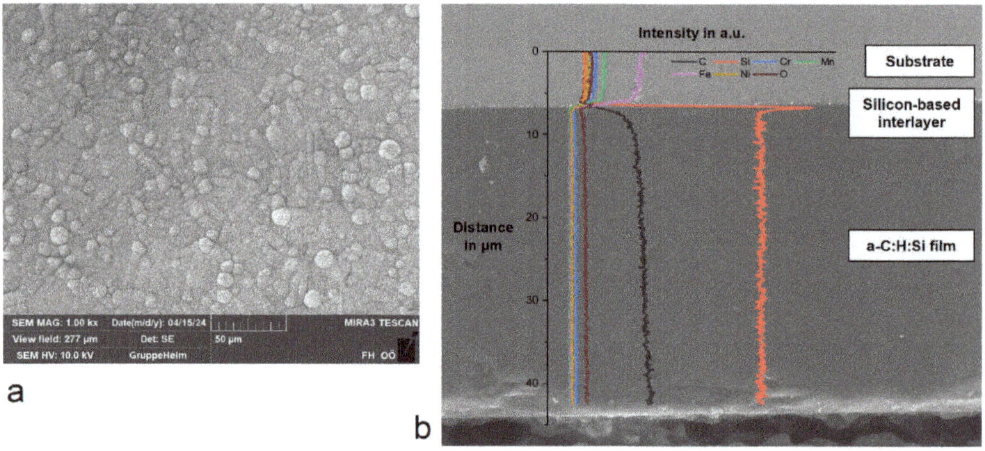

Figure 3. (**a**) Scanning electron microscope image showing the morphology of the a-C:H:Si film; (**b**) EDX line-scan of the cross-section of the a-C:H:Si film.

Typically, carbon Raman bands manifest at 1580 cm^{-1} and 1350 cm^{-1} corresponding to the G and D bands, respectively. However, in the acquired spectrum of the a-C:H:Si shown in Figure 4, these peaks exhibit a notable shift towards the center of the spectrum, with the G band positioned at 1555 cm^{-1} and the D band at 1389 cm^{-1}. The lower position of the G band, along with its broadening (FWHM(G) = 120 cm^{-1}) indicate high topological as well as structural disorder, the latter resembling the distortion of bonding lengths and angles of the carbon network [30]. The increase in structural and in topological disorder may both be partly attributed to the incorporation of Si atoms into the amorphous carbon as silicon exhibits a larger atomic radius than carbon and only forms tetrahedral (Si-C sp^3) bonding, ultimately distorting the amorphous carbon network [17].

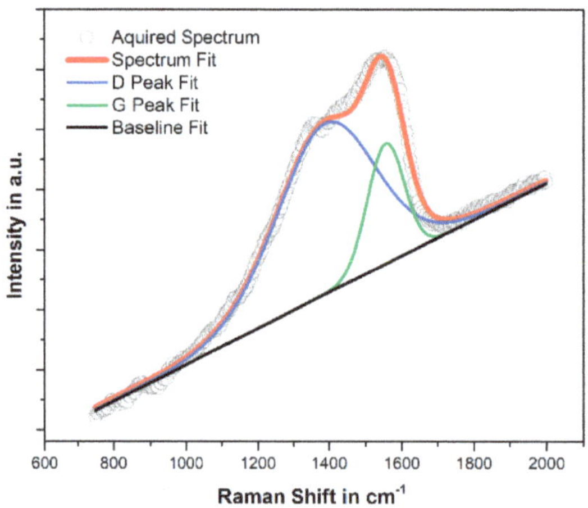

Figure 4. Deconvoluted Raman spectrum of the a-C:H:Si coating, obtained using a 532 nm excitation wavelength.

Even further, the hydrogen content estimated with the baseline slope was about 41%, according to the method described in [31]. The downward shift of the G band, alongside with the significant degree of structural disorder present in the material, suggests a high fraction of total sp^3 hybridization. However, considering the substantial presence of hydrogen due to the HMDSO doping, which terminates the C-C sp^3-bonded carbon backbone as well as the Si-C sp^3 bonding, lead to the conclusion that the number of diamond-like C-C sp^3 bonds and Si-C sp^3 bonds relative to the total fraction of sp^3 bonding is relatively low. This is consistent with the reduced hardness values of the obtained a-C:H:Si in comparison to other DLC films, as hardness is mainly a function of tetrahedral carbon bonding [32].

3.3. Particle Size versus Roughness of Wall Material

According to the literature [33], the ratio of particle size to wall material roughness can be used to characterize the wall friction because particles smaller than the roughness depth in contact with the wall can become trapped in grooves. With a substantial fraction of particles smaller than the mean roughness depth of the wall material, a significant proportion of the movement of the bulk of the powder relative to the wall takes place by means of flow within the powder [34]. This results in an apparent increase in wall friction, because the angle of internal friction of a powder is usually much higher than the wall friction angle between the particles and the wall material [7].

For the fine aluminum oxide, calcium carbonate and silicon carbide powders the fraction of the volume of the powder which is smaller to the mean roughness depth of the wall material was in the range of approximately 30% for the a-C:H:Si film and the 1.4301 material to approximately 75% for the UHMWPE wall material. For the medium-size powders, it was less than 5% for all wall materials and powders except when pairing calcium carbonate with UHMWPE wall material, where the fraction was approximately 10%. For the coarse powder samples, the fractions smaller than the mean roughness depth of the wall material were even smaller.

3.4. Particle Hardness versus Hardness of Wall Material

The Mohs hardness of calcium carbonate of 3 corresponds to approximately 1.3 GPa, the Mohs hardness of aluminum oxide and silicon carbide of 9 and 9.5 corresponds to 20 GPa and 27 GPa, respectively [35]. Consequently, the hardness of aluminum oxide and silicon carbide was much greater than the hardness of each of the wall materials. By contrast, the hardness of calcium carbonate was similar to that of the stainless steel 1.4301 wall material and, therefore, markedly lower than the hardness of the a-C:H:Si film.

3.5. Wall Friction Angles

Figure 5 shows the wall friction angle of the various aluminum oxide, calcium carbonate and silicon carbide powder samples with the a-C:H:Si film and the two other wall materials in dependence of the wall normal stress. Generally, the reproducibility of the measurements was very good. The relative standard deviation of the measured wall friction angles was in the range of 1% to 4%.

The wall friction angles decrease with increasing wall normal stress, which is in accordance with various published data [36–39]. However, this effect was not equally pronounced for all wall materials and powder particle sizes.

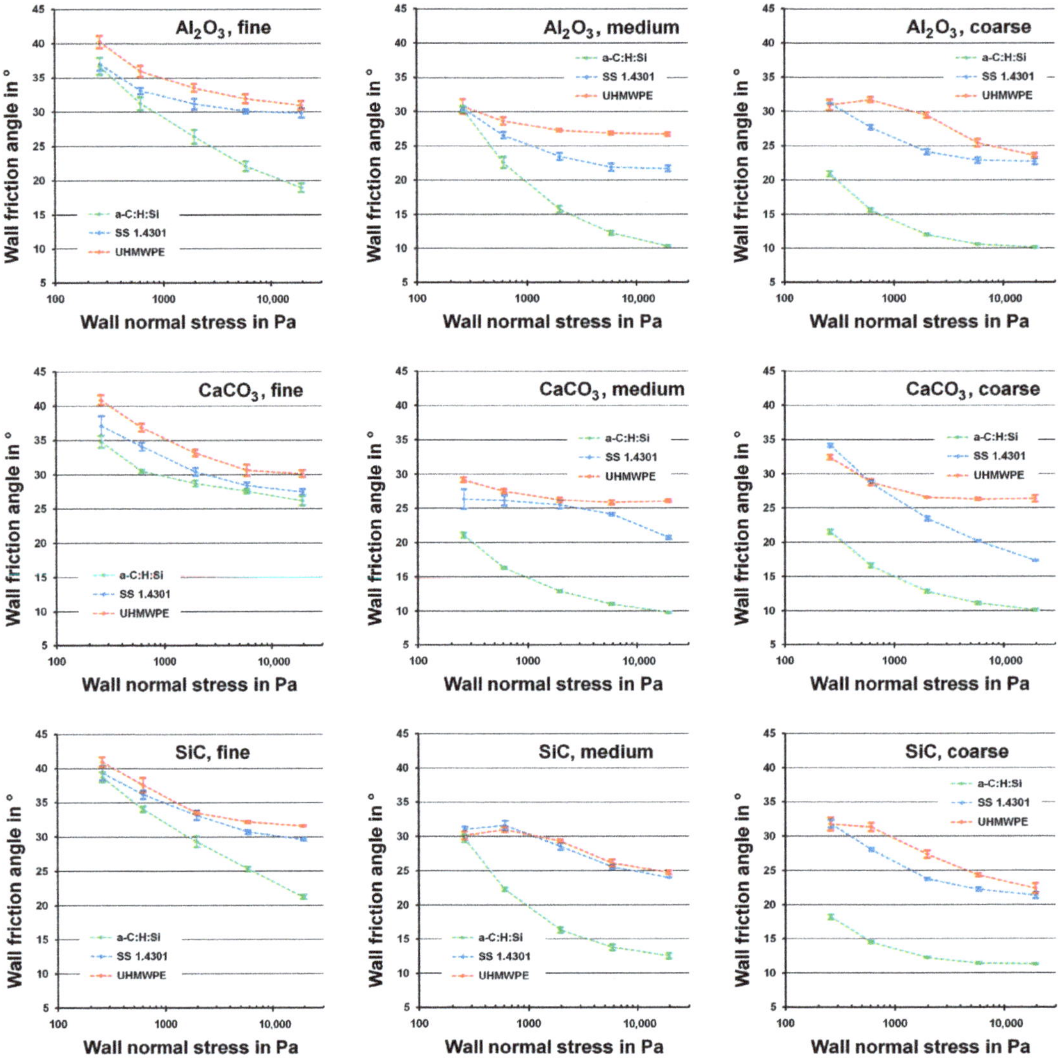

Figure 5. Wall friction angle as a function of the wall normal stress for the various aluminum oxide, calcium carbonate and silicon carbide powder samples.

3.5.1. Influence of the Wall Material on the Wall Friction Angle

The wall material significantly influenced the wall friction angle. In all measurements, the wall friction angles with the a-C:H:Si-coated material were smaller than the wall friction angles with stainless steel 1.4301. The wall friction angles with UHMWPE were a few degrees larger than with stainless steel 1.4301 in all cases, except for the coarse powders at the lowest value of the wall normal stress.

The observed reduction in wall friction by the a-C:H:Si film was not constant. It was usually lower for smaller values of the wall normal stress and for smaller particles.

3.5.2. Influence of the Particle Size on the Wall Friction Angle

Figure 6 shows the dependence of the wall friction angle on the particle size. The wall friction angles with the stainless steel 1.4301 material were higher for the fine powders and lower for the medium-size and the coarse powders. Higher values of the wall friction angle for very fine powders are in accordance with the findings from other studies [2–4]. Since the influence of the material of the powder was not very high (Figure 6a) average values were calculated for the different powder materials for the various levels of wall normal stress (Figure 6b). Independent of the wall normal stress, the wall friction angle was approximately 5° smaller for the medium-size powders compared to the fine powders. For the coarse powders, the wall friction angle was even slightly lower at high values of the wall normal stress, had the same value for medium values of the wall normal stress and was somewhat higher for low values of the wall normal stress. For the UHMWPE wall material, similar trends were found with slightly higher values of the wall friction angle (Figure 6c).

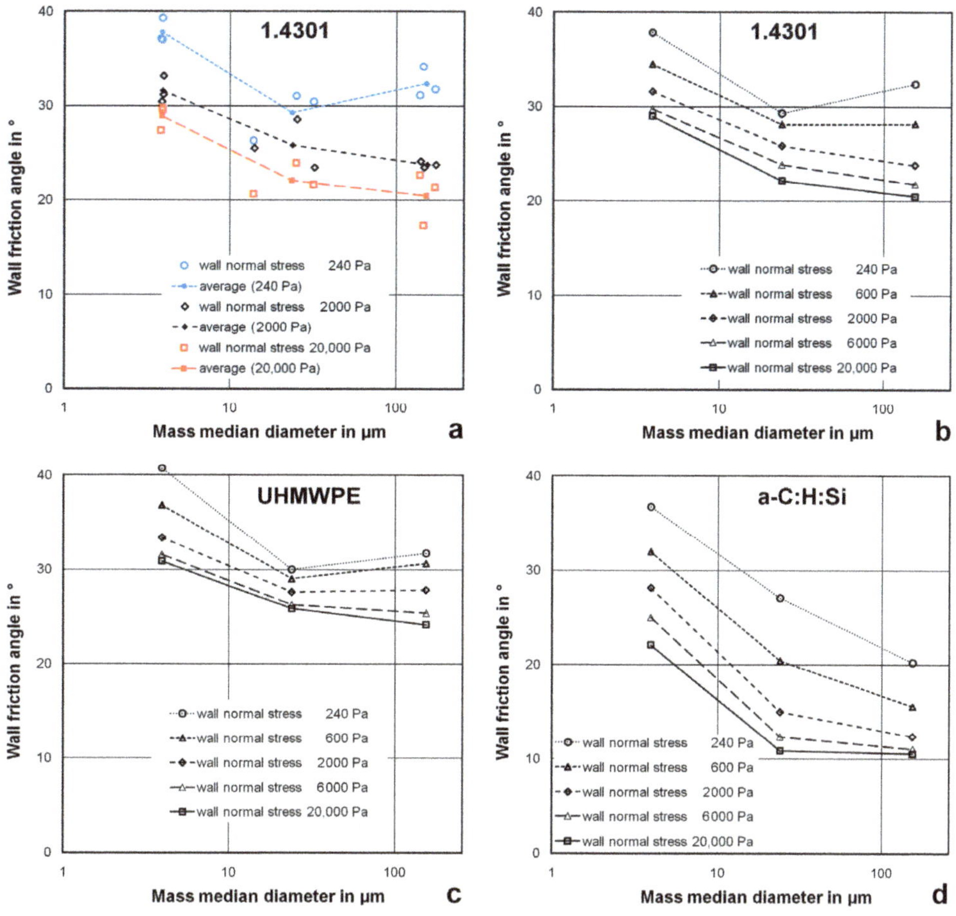

Figure 6. Dependence of the wall friction angle on the particle size; (**a**) wall material 1.4301, data points for different powder materials; (**b**–**d**) average values for fine, medium-size and coarse powder materials.

The dependence between particle size and wall friction angle for the a-C:H:Si film was significantly different (Figure 6d). Firstly, the wall friction angle's dependence on the wall normal stress was more pronounced for the fine and medium-size powders. Secondly, the reduction in wall friction for the fine particle size and the medium particle size was higher, and, thirdly, there was a further reduction in the wall friction for the medium particle size and the large particle size. In contrast to stainless steel 1.4301 and UHMWPE, this further reduction was especially high at lower values of the wall friction angle.

3.5.3. Influence of the Powder Material on the Wall Friction Angle

The measured wall friction angles for the very hard powder materials of aluminum oxide and silicon carbide were very similar, while for the softer calcium carbonate, slightly different results were obtained (Figure 5). However, the influence of the powder material on the wall friction angle was rather small compared to the influence of the wall material and the particle size of the powder.

4. Discussion

The biggest influence on the wall friction angle was found to be the wall material. A significant reduction in wall friction for the a-C:H:Si film was observed in comparison to stainless steel 1.4301, while in most cases the wall friction angle for UHMWPE was somewhat higher than with stainless steel 1.4301. Because of the small influence of the powder material, average wall friction angles were calculated for the three materials. Figure 7 shows the differences between the average wall friction angle for the a-C:H:Si film and with the UHMWPE material to the average wall friction angle for the stainless steel 1.4301 as a function of the wall normal stress. The average wall friction angles with UHMWPE were, with one exception, 1° to 4° degrees larger than with 1.4301. This difference was almost independent of the wall normal stress. The higher wall friction angles for UHMWPE might be partly explained by the higher surface roughness values of this material.

In all measurements, the wall friction angle for the a-C:H:Si-coated material was lower than the wall friction angle for stainless steel 1.4301. For the coarse powders, the reduction in the wall friction angle was 12° at a wall normal stress of 240 Pa, the lowest value of the wall normal stress measured. At higher values of the wall normal stress, the reduction decreases steadily to 10°. This is a result of the fact, that the wall friction angle for the a-C:H:Si-coated material gradually approaches a value of approximately 10°, while the wall friction angle for the stainless steel 1.4301 decreases further.

For the fine powders, at low values of the wall normal stress the reduction in wall friction was very small. However, the reduction increased steadily with increasing wall normal stress and reached approximately 6° at a wall normal stress of 20,000 Pa.

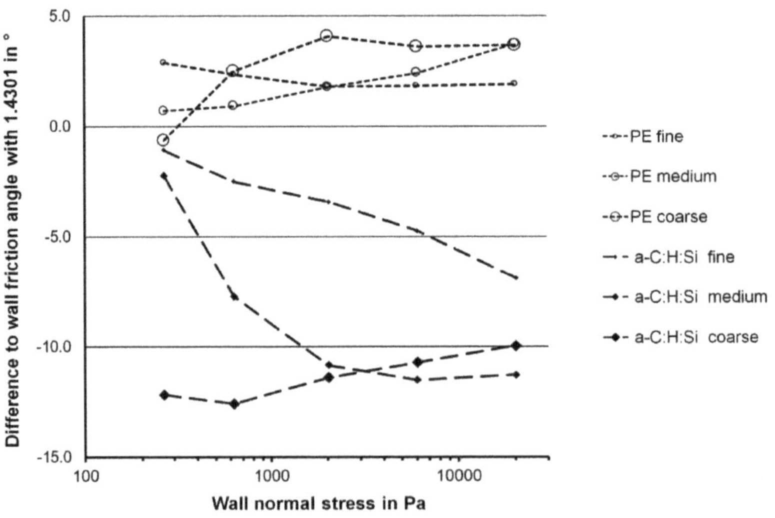

Figure 7. *Cont.*

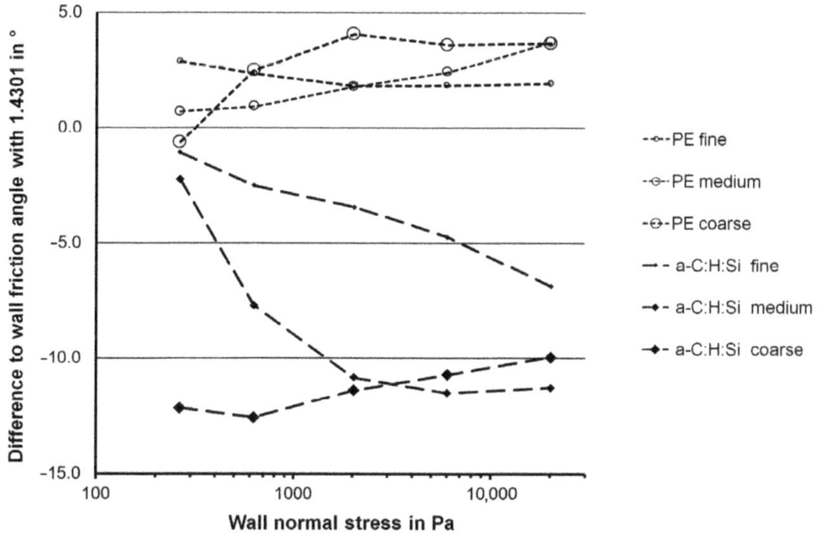

Figure 7. Differences between the average wall friction angles for the specified wall material to the average wall friction angle for 1.4301 for the three powder sizes (fine, medium and coarse) as a function of the wall normal stress.

For the medium-size powders, the wall friction reduction at low wall normal stress was very small, too, but reached 11° at a wall normal stress of approximately 2000 Pa. At higher values of the wall normal stress the reduction stayed nearly constant at the same level as it was found for the coarse powders.

An explanation of the difference in wall friction reduction might be the differences in the contact stress between the particles and the wall material. The number of contact points per unit area between the wall material and the powders increases from the coarse powders (average d_{50} of approximately 150 μm) to the medium-size powders (average d_{50} of approximately 25 μm) and the fine powders (average d_{50} of approximately 4.0 μm). Parallel to the increase in contact points, the average stress per contact point at a given value of the wall normal stress decreases.

A higher contact pressure produces an enhancement of the graphitization process of the a-C:H:Si coating, which is ultimately responsible for the low friction coefficient. The Raman spectrum in Figure 8 highlights alterations on the surface of the a-C:H:Si coating compared to its original state as depicted in Figure 4. Notably, the characteristic bands are separated due to a shift of the G band towards higher wavenumbers, specifically from 1555 cm^{-1} to 1581 cm^{-1}, which corresponds to a decrease in topological disorder in the film. Moreover, a significant reduction in the FWHM of the G band, from 120 cm^{-1} to 101 cm^{-1}, indicates a pronounced decrease in structural disorder [30]. Additionally, a reduction in the estimated hydrogen content from 41% to 34% was observed. All these mechanisms are attributed to the conversion of sp^3 sites into more ordered graphite-like sp^2 C-C bonds, i.e., the onset of graphitization, providing an explanation for the phenomenon of the reduction in the wall friction angle during the testing. Furthermore, the steep decrease in topological and structural disorder in the material may also be partly ascribed to the removal of silicon in the a-C:H:Si network due to tribological loading [17].

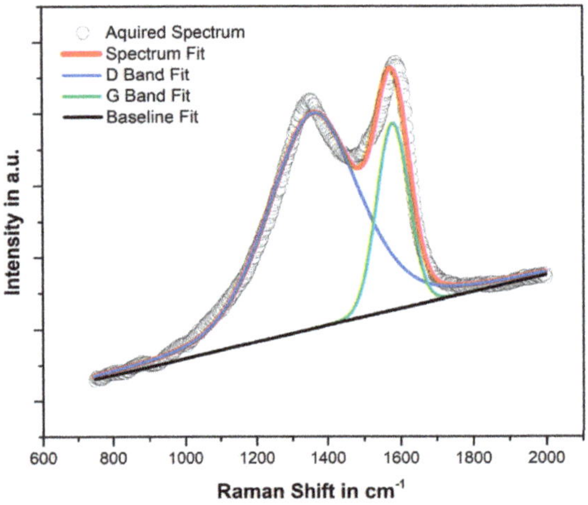

Figure 8. Deconvoluted Raman spectrum of the a-C:H:Si coating showing graphitization, obtained using a 532 nm excitation wavelength.

5. Conclusions

This study shows that a-C:H:Si coating can reduce wall friction of powders significantly. The effect of the a-C:H:Si film in wall friction reduction was dependent on wall normal stress and particle size, but nearly independent of the powder material. Therefore, the reduction in wall friction angle depends on the contact stress of the particles. For coarser powders, a significant wall friction reduction was found even at low values of the wall normal stress. For finer particles, markedly reduced wall friction angles were found only at somewhat higher values of the wall normal stress.

Author Contributions: Conceptualization, C.L.; methodology, C.L. and C.F.; validation, C.L., C.F. and D.H.; investigation, C.L., C.F., F.D. and M.C.J.S.; resources, C.L. and D.H.; data curation, C.L., C.F., F.D. and M.C.J.S.; writing—original draft preparation, C.L.; writing—review and editing, C.L., C.F., F.D., M.C.J.S. and D.H.; visualization, C.L., F.D. and M.C.J.S.; supervision, C.L.; project administration, C.L. All authors have read and agreed to the published version of the manuscript.

Funding: This research received no external funding.

Data Availability Statement: The raw data supporting the conclusions of this article will be made available by the authors on request.

Acknowledgments: English proofreading by D. Vaught is gratefully acknowledged.

Conflicts of Interest: The authors declare no conflicts of interest.

References

1. Schulze, D. *Powders and Bulk Solids. Behavior, Characterization, Storage and Flow*; Springer: Berlin, Germany, 2008.
2. Schulze, D.; Schwedes, J. Das Fließverhalten und die Silolagerung von REA-Gips. *Chem. Ing. Tech.* **1991**, *63*, 256–257. [CrossRef]
3. Lanzerstorfer, C. Characterization of the flowability of fly ashes from grate-fired combustion of forest residues. *Fuel Proc. Technol.* **2016**, *150*, 10–15. [CrossRef]
4. Lanzerstorfer, C. Properties of steelmaking dusts from dry dust separators. In Proceedings of the 27th METAL Conference, Brno, Czech Republic, 23–25 May 2018.
5. Prescott, J.K.; Ploof, D.A.; Carson, J.W. Developing a Better Understanding of Wall Friction. *Powder Handl. Process* **1999**, *11*, 27–36.
6. Iqbal, T.; Fitzpatrick, J.J. Effect of storage conditions on the wall friction characteristics of three food powders. *J. Food Eng.* **2006**, *72*, 273–280. [CrossRef]
7. Ooms, M.; Roberts, A.W. Significant Influence on Wall Friction in the Gravity Flow of Bulk Solids. *Bulk. Solids Handl.* **1985**, *5*, 1271–1277.

8. Heinrici, H.; Jacob, T. Die schüttgutmechanische Betrachtung der Siloauslegung und ihre Anwendung auf ein Silo unter Tage. *Schüttgut* **2005**, *11*, 430–435.
9. Lanzerstorfer, C. Influence of the surface roughness of grit-blasted steel on the wall friction angle of limestone powder. In Proceedings of the 14th Minisymposium Verfahrenstechnik, Linz, Austria, 4–5 April 2018.
10. Han, T. Comparison of wall friction measurements by Jenike shear tester and ring shear tester. *KONA Powder Part. J.* **2011**, *29*, 118–124. [CrossRef]
11. Bradley, M.S.A.; Pittman, A.N.; Bingley, M.; Farnish, R.J.; Pickering, J. Effects of wall material hardness on choice of wall materials for design of hoppers and silos for the discharge of hard bulk solids. *Tribol. Int.* **2000**, *33*, 845–853. [CrossRef]
12. Meng, Y.; Xu, J.; Jin, Z.; Prakash, B.; Hu, Y. A review of recent advances in tribology. *Friction* **2020**, *8*, 221–300. [CrossRef]
13. Hasebe, T.; Hotta, A.; Kodama, H.; Kamijo, A.; Takahashi, K.; Suzuki, T. Recent Advances in Diamond-Like Carbon Films in the Medical and Food Packing Fields. *New Diam. Front. C Tec.* **2007**, *17*, 263–279.
14. Héau, C. DLC Films in Mechanical and Manufacturing Industry. In *Tribology of Diamond-LIKE Carbon Films. Fundamentals and Applications*; Donnet, C., Erdemir, A., Eds.; Springer: New York, NY, USA, 2008; pp. 469–483.
15. van der Kolk, G.J. Wear Resistance of Amorphous DLC and Metal Containing DLC in Industrial Applications. In *Tribology of Diamond-LIKE Carbon Films. Fundamentals and Applications*; Donnet, C., Erdemir, A., Eds.; Springer: New York, NY, USA, 2008; pp. 484–493.
16. Iseki, T.; Mori, H.; Hasegawa, H.; Tachikawa, H.; Nakanishi, K. Structural analysis of Si-containing diamond-like carbon. *Diam. Relat. Mater.* **2006**, *15*, 1004–1010. [CrossRef]
17. Ferrari, A.C.; Robertson, J. Raman spectroscopy of amorphous, nanostructured, diamond-like carbon, and nanodiamond. *R. Soc.* **2004**, *362*, 2477–2512. [CrossRef] [PubMed]
18. Liu, Y.; Erdemir, A.; Meletis, E.I. A study of the wear mechanism of diamond-like carbon films. *Surf. Coat. Technol.* **1996**, *82*, 48–56. [CrossRef]
19. Liu, Y.; Erdemir, A.; Meletis, E.I. An investigation of the relationship between graphitization and frictional behavior of DLC coatings. *Surf. Coat. Technol.* **1996**, *82*, 564–568. [CrossRef]
20. Zhang, Y.Z.; Kovalev, A.; Meng, Y.G. Combined effect of boundary layer formation and surface smoothing on friction and wear rate of lubricated point contacts during normal running-in processes. *Friction* **2018**, *6*, 274–288. [CrossRef]
21. Jongwannasiri, C.; Yoshida, S.; Watanabe, S. Effects of Fluorine and Silicon Incorporation on Tribological Performance of Diamond-Like Carbon Films. *Mate Sci. Appl.* **2019**, *10*, 170–185. [CrossRef]
22. Donnet, C.; Erdemir, A. New horizon in the tribology of diamond like carbon films. *Surf. Eng.* **2008**, *24*, 399–401. [CrossRef]
23. Gangopadhyay, A.; Sinha, K.; Uy, D.; McWatt, D.G.; Zdrodowski, R.J.; Simko, S.J. Friction, Wear, and Surface Film Formation Characteristics of Diamond-Like Carbon Thin Coating in Valvetrain Application. *Tribol. Trans.* **2011**, *54*, 104–114. [CrossRef]
24. Björling, M.; Isaksson, P.; Marklund, P.; Larsson, R. The Influence of DLC Coating on EHL Friction Coefficient. *Tribol. Lett.* **2012**, *47*, 285–294. [CrossRef]
25. Vetter, J. 60 years of DLC coatings: Historical highlights and technical review of cathodic arc processes to synthesize various DLC types, and their evolution for industrial applications. *Surf. Coat. Technol.* **2014**, *257*, 213–240. [CrossRef]
26. Erdemir, A.; Martin, J.M. Superior wear resistance of diamond and DLC coatings. *Curr. Opin. Solid. State Mater. Sci.* **2018**, *22*, 243–254. [CrossRef]
27. Lanzerstorfer, C.; Forsich, C.; Heim, D. Reduction of wall friction of fine powders by use of wall surface coatings. *Coatings* **2021**, *11*, e427. [CrossRef]
28. Forsich, C.; Dipolt, C.; Heim, D.; Mueller, T.; Gebeshuber, A.; Holecek, R.; Lugmair, C. Potential of thick a-C:H:Si films as substitute for chromium plating. *Surf. Coat. Technol.* **2014**, *241*, 86–92. [CrossRef]
29. *ASTM D6773*; Standard Test Method for Bulk Solids Using Schulze Ring Shear Tester. ASTM International: West Conshohocken, PA, USA, 2008.
30. Ferrari, A.C.; Robertson, J. Interpretation of Raman spectra of disordered and amorphous carbon. *Phys. Rev. B* **2000**, *61*, 14095–14107. [CrossRef]
31. Casiraghi, C.; Ferrari, A.C.; Robertson, J. Raman spectroscopy of hydrogenated amorphous carbons. *Phys. Rev. B* **2005**, *72*, 085401. [CrossRef]
32. Robertson, J. Diamond-like amorphous carbon. *Mater. Sci. Eng. R.* **2002**, *37*, 129–281. [CrossRef]
33. Strijbos, S. Powder Wall Friction: The Effects of Orientation of Wall Grooves and Wall Lubricants. *Powder Technol.* **1977**, *18*, 209–214. [CrossRef]
34. Roberts, A.W.; Sollie, L.A.; De Silva, S.R. The interaction of bulk solid characteristics and surface parameters in surface or boundary friction measurements. *Tribol. Int.* **1993**, *26*, 335–343. [CrossRef]
35. Taylor, E.W. Correlation of the Mohs's scale of hardness with the Vickers's hardness numbers. *Mineral Mag. J. M. Soc.* **1949**, *28*, 718–721. [CrossRef]
36. Lanzerstorfer, C. Apparent density of compressible food powders under storage conditions. *J. Food Eng.* **2020**, *276*, e109897. [CrossRef]
37. Schulze, D. The characterization of bulk solids for silo design and flowability tests. *Aufbereitungstechnik* **1998**, *39*, 47–57.

38. Lanzerstorfer, C. Consideration of the stress dependence of the bulk density in silo storage of dusts from dry off-gas cleaning. *Adv. Powder Technol.* **2017**, *28*, 115–121. [CrossRef]
39. Rohilla, L.; Garg, V.; Mallick, S.S.; Setia, G. An experimental investigation on the effect of particle size into the flowability of fly ash. *Powder Technol.* **2018**, *330*, 164–173. [CrossRef]

materials

MDPI

Article

Development of Multifunctional Hybrid Coatings (Mechanically Resistant and Hydrophobic) Using Methyltrimethoxysilane–Diethoxydimethylsilane–Tetraethoxysilane Mixed Systems

Charlène Pellegrini, Sandrine Duluard, Marie Gressier, Viviane Turq, Florence Ansart and Marie-Joëlle Menu *

Centre Interuniversitaire de Recherche et d'Ingénierie des Matériaux (CIRIMAT), Université Toulouse 3 Paul Sabatier, Toulouse INP, CNRS, Université de Toulouse, 118 Route de Narbonne, CEDEX 9, 31062 Toulouse, France; sandrine.duluard@univ-tlse3.fr (S.D.); marie.gressier@univ-tlse3.fr (M.G.); viviane.turq@univ-tlse3.fr (V.T.); florence.ansart@univ-tlse3.fr (F.A.)

* Correspondence: marie-joelle.menu@univ-tlse3.fr

Abstract: For many industrial applications, the simultaneous presence in a material of different functional properties is necessary. The main interest lies in making a single material more versatile and durable, less fragile and more efficient. In this study, two concomitant properties in the same material were mainly studied: resistance to cracking and the increase in its hydrophobic properties. The chosen process was the sol-gel route due to its versatility and the ease of formulating materials from various precursors in order to obtain (multi)functional materials. In this paper, sol-gel coatings were prepared with tetraethoxysilane, methyltrimethoxysilane and diethoxydimethylsilane as precursors. Tetraethoxysilane was mainly used to improve the material's mechanical properties, especially hardness, and silicon oil was added to improve its hydrophobic behavior. The integration of silicon oil was monitored via ^{29}Si NMR. Microstructural characterizations were carried out to correlate the multi-scale properties with the microstructure of the derived films. Young's modulus and hardness were measured to highlight the effect of key formulation parameters on the mechanical strength of the coatings. The synergistic effect of these precursors is underlined as well as the beneficial effect of silicon oil (generated in situ or precondensed).

Keywords: sol-gel coating; hydrophobic; silicon oil; mechanical properties; mechanical resistance

1. Introduction

Organic–inorganic hybrid sol-gel coatings based on siloxane–oxide systems are very promising for many applications, such as anti-corrosion [1], anti-friction or anti-wear [2–5] coatings, for example, but also for the development of hydrophobic coatings [6–8]. Hybrid networks are obtained by mixing inorganic precursors such as tetraethoxysilane (TEOS), zirconium tetrapropoxide (ZrTPO) or titanium tetrapropoxide (TiTPO) and various organosilanes, including the well-known glycidoxypropyltrimethoxysilane (GPTMS), 3-methacryloxypropyltriethoxysilane (MAPTES) and methyltrimethoxysilane (MTMS). Diethoxydimethylsilane (DEDMS) is also well known for either obtaining hydrophobic protective coatings [6] or for processing metallic prostheses for tissue engineering applications [9,10]. Hybrid materials have given rise to numerous fundamental characterizations, which make it possible to link the chemical structure and microstructure to the working properties in working conditions. In our study, the hybrid material skeleton occurs mainly due to the MTMS network, and various quantities of TEOS or DEDMS are being added in order to improve the mechanical and/or hydrophobic properties.

Thus, Babonneau et al. studied TEOS/DEDMS and TEOS/PDMS-OH gels, where PDMS-OH is represented by OH-terminated poly(dimethyl)siloxane chains [11]. The structural evolution of this system has been followed via NMR from the molecular precursors to the final materials. Co-condensation reactions between PDMS-OH or precondensed DEDMS and TEOS have been highlighted via ^{29}Si NMR.

Deng et al. studied ORMOSIL TEOS/DEDMS compositions [12] and show that co-condensation between TEOS and DEDMS monomers occurs in acidic conditions. These result are in agreement with those of Babonneau. Richards et al. studied the antifouling potential of TEOS or DEDMS coatings [13]. They found that using TEOS as a precursor permits achieving interesting hydrophobic properties with a water contact angle > 100°. DEDMS-based coatings are also promising, with a water contact angle around 100°. Gavilan et al. studied the use of MTMS/TEOS, DEDMS/TEOS and PDMS-OH/TEOS sol-gel coatings on stainless steel for bone tissue engineering applications [9,10]. They concluded that the chemical structure obtained using the DEDMS precursor shows an interesting balance between protection and bioactivation and that PDMS/TEOS has the highest hydrophobic behavior.

Salazar-Hernandez et al. studied TEOS/PDMS-OH/colloidal silica sol-gel in stone consolidation [14]. It appeared that PDMS-OH improves elasticity thanks to siloxane chains chemically bonded to the inorganic silica skeleton (TEOS and colloidal silica). This phenomenon has been confirmed via ^{29}Si NMR. Malzbender et al. studied the mechanical properties of a MTMS/TEOS sol-gel coating filled with alumina or silica particles [15]. They measured a Young's modulus of 5 GPa and a hardness of 0.7 GPa for a coating without particle addition. Ballarre et al. [16] measured a Young's modulus of 6.5 GPa and a hardness of 0.92 GPa for a hybrid silica-based coating with TEOS and methyltrimethoxysilane (MTES) as precursors.

In this paper, the main aim of our study is to develop new MTMS-based coatings combining both hydrophobic and scratch-resistance properties. The matrix of the coating is based on an MTMS-based siloxane network associated, or not, with TEOS, which is known in the literature to improve coatings' mechanical properties [17]. DEDMS is introduced in order to improve hydrophobic properties through the following two different and original routes: either as a silicone oil after precondensation via hydrolysis of the DEDMS or directly as a silane precursor for in situ oil generation.

The integration of silicone oil is studied via NMR analysis coupled with SEM pictures of the surface. Its hydrophobic behavior is characterized by static and dynamic contact angle measurements. The mechanical properties are evaluated via nanoindentation.

These multi-scale characterizations are then correlated with the various formulations and finally compared in order to understand the effects of process parameters.

2. Materials and Methods

2.1. Materials

The first step involved preparing a sol from the following precursors. Methyltrimethoxysilane (MTMS, 98%), tetraethoxysilane (TEOS, 98%), diethoxydimethylsilane (DEDMS, 97%), glacial acetic acid (CH_3CO_2H, 99%) and 2-propanol (iPrOH, 99.5%) were purchased from Sigma-Aldrich (Saint-Quentin Fallavier, France), and 2-butoxyethanol ($C_4H_9OC_2H_4OH$, 99%) was obtained from Honeywell. All reagents were used without further purification. The roles and semi-developed formulas of the silanes used as precursors are indicated in Table 1.

All coatings were applied on sandblasted and dried aluminum substrates. Spray deposition was then performed with an air pressure of 4 bars with manual handled pulverization device using a high-quality SRI Pro Lite spray gun (Larius France, Estrablin, France). In order to compare all the samples in similar conditions, areal mass densities were controlled after deposition. An average value of 1.4 mg/cm^2 (+/−0.2) was measured; this deposited mass is sufficient to ensure a thickness higher than 5 microns. After deposition, a thermal treatment at 300 °C for 10 min was applied. To complete the study, xerogels were obtained from sols with the same heat treatment as for coatings.

Table 1. Role and semi-developed formula of silane precursors.

Precursors	Role	Structure
TEOS	Silica generation and siloxane network	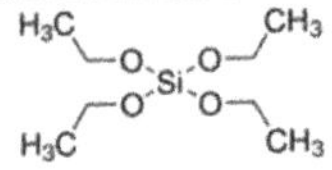
MTMS	Hybrid siloxane network precursor	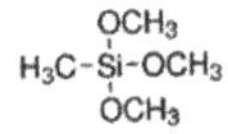
DEDMS	Hybrid siloxane network precursor and/or in situ oil generation	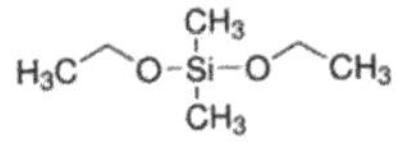

2.2. Polycondensation of Diethoxydimethylsilane, pDEDMS

The precondensation parameters of DEDMS to obtain a silicone oil were selected according to the publication of Kalinina et al. [18] who studied the condensation of DEDMS in active medium using several conditions (water and acetic acid contents, temperature, reaction time). A mixture of 1.3 g (0.070 mol) of water, 4.0 g (0.070 mol) of acetic acid and 2.0 g (0.013 mol) of DEDMS was stirred 4 h at room temperature. According to the literature, in these conditions, the reaction yielded 50 wt.% of linear polymer $HO(CH_3)_2SiO[Si(CH_3)_2O]nSi(CH_3)_2OH$ of 1600 Da and 50 wt.% of cyclic oligomers where $[(CH_3)_2SiO)]_4$ D4 and $[(CH_3)_2SiO)]_5$ D5 oligomers predominate in 37% and 11% proportions, respectively. The resulting silicone oil will be referred to as pDEDMS for precondensed DEDMS. The $^{29}Si\{^1H\}$ NMR spectrum of the silicone oil is presented in Figure S1 in the Supplementary Material. In this spectrum, three types of silicon atoms appear distinctly; they can be assigned according to published data on organosilicon linear polymers [19–21]: D^1 units in $(CH_3)_2(HO)Si^*$-O-$Si(CH_3)_2$- exhibit chemical shifts ranging from −10.5 (n = 0) to −12.5 (n > 5), while D^2 units in -O-$(CH_3)_2Si^*O)_n$ appear in the range from −20 to −22 ppm. The presence of small rings is confirmed by the signal at −18.1 ppm, which is characteristic of silicon atoms in D4 cyclic oligomers as also described by Kalinina et al. [18], Babonneau [22] and Suda [20]. The signal at −7 ppm, fine and of low intensity, does not correspond to any of the small cyclic oligomers (D3 at −9 ppm); thus, it is not assigned.

2.3. Preparation of Sols and Xerogels

Six formulations are studied in the following, and molar proportions are detailed in Table 2. The first one is the "reference" formulation, M100, which is based on hydrolysis and the condensation of MTMS in acidic medium (acetic acid) without any additive. In the other formulations, DEDMS is introduced for the ex situ or in situ generation of dimethylsiloxane oligomers as oil; in the last case, co-condensation of DEDMS with MTMS should be considered. This is investigated via the introduction of a small amount of precondensed DEDMS either in the reference formulation (M100-pDEDMS) or in a reference formulation with the substitution of 25% of MTMS by DEDMS (M75-D25-pDEDMS). The influence of TEOS in these formulations was evaluated by adding a large amount of TEOS in the formulations (nearly 1/1 vs. MTMS) (T-M100, T-M100-pDEDMS and T-M75-D25-pDEDMS). In these last cases, the hydrolysis rate was increased to favor the hydrolysis of all the silane functions. All formulations are acidified with acetic acid either added during the precondensation of the DEDMS or directly for the M100 and T-M100 formulations.

In all formulations, 2-propanol and 2-butoxyethanol are systematically introduced to obtain 10 wt.% for each in the formulations. Sols are prepared according to the protocol detailed below for the M100 and T-M75-D25-pDEDMS formulations; the preparation is similar for the other compositions.

For the M100 formulation, 2.7 g (0.045 mol) of acetic acid was added to 19 mL (1.05 mol) of water and stirred 2 min; then, 57.6 mL (0.40 mol) of MTMS was introduced and stirred

2 min more. A mixture of 13.2 mL of 2-propanol (0.17 mol) and 11.2 mL (0.087 mol) of 2-butoxyethanol was added, and the sol was stirred 2 h at room temperature. The resulting sol was aged during 24 h. The composition MTMS:H_2O:CH_3CO_2H:2-propanol:2-butoxyethanol is 55:19:3:10:10 in wt.% and 1:2.6:0.1:0.42:0.21 in molar ratios.

Table 2. Nomenclature and molar compositions of the different sols [1] and coatings.

	MTMS	H_2O	CH_3CO_2H	TEOS	DEDMS	p-DEDMS		
						DEDMS	H_2O	CH_3CO_2H
M100	1	2.6	0.1					
M100-pDEDMS	1	2.6				0.034	0.17	0.17
M75-D25-pDEDMS	1	4.0			0.3	0.051	0.26	0.26
T-M100	1	9.7	0.2	0.9				
T-M100-pDEDMS	1	9.7		0.9		0.09	0.42	0.42
T-M75-D25-pDEDMS	1	12.9		1.1	0.3	0.12	0.61	0.61

[1] The solvent is systematically composed of 2:1 ratio of 2-propanol and 2-butoxyethanol, respectively.

For the T-M75-D25-pDEDMS formulation, 2 g of p-DEDMS, 27.7 g (0.131 mol) of TEOS, and 26.4 g (1.47 mol) of water were stirred for 2 min. MTMS (15.8 g, 0.114 mol) and DEDMS (5.3 g, 0.036 mol) previously mixed were then introduced, and the mixture was stirred 2 min more. Finally, 2-propanol (8.8 g, 0.145 mol) and 2-butoxyethanol (5.3 g, 0.073 mol) complete the sol, which is stirred 2 h at room temperature and then left to mature for 24 h. The composition MTMS:H_2O:CH_3CO_2H:TEOS:DEDMS:pDEDMS:2-propanol:2-butoxyethanol is 55:19:3:10:10 in wt.% and 1:2.6:0.1:0.42:0.21 in molar ratios.

2.4. Characterization Techniques

The chemical structure of the coating constituted of siloxane network incorporating in situ synthesized silicon oil has been followed by ^{29}Si solid state NMR. The measurements were made on xerogels which underwent the same heat treatment as the coatings. ^{29}Si cross-polarization magic angle spinning nuclear magnetic resonance (^{29}Si CP MAS NMR) spectra were recorded using a Bruker Avance III 400WB (9.4 T) spectrometer (Bruker Biospin, Fällanden, Switzerland). Chemical shift references are tetramethylsilane (TMS), and a 4 mm zirconia rotor was used. MAS and CP MAS experiments were recorded. The rotation speed around the magic angle (MAS) was 8 kHz, and experiments were realized at room temperature. The observed silicon units were designed according to the usual notation employed in silicon chemistry: Q [SiO_4], T [$SiCO_3$], D [SiC_2O_2], and M [SiC_3O]. The number added as a superscript (Q^n, T^n or D^n) indicates the number of oxo bridging bonds to the corresponding Si site [23]. Data were processed using "MNova 11.0" software; deconvolution was completed using DmFit software (version # 20200306) [24].

SEM observations were performed with a FEG JEOL JSM 6700 F SEM (JEOL, Croissy sur seine, France) with a 5 kV operating voltage and back-scattered electrons observation mode. The contact angle has been measured with a drop shape analyzer KRUSS at room temperature. Static contact angles and hysteresis was measured with a method described in the literature [25], which is presented in Figure 1a,b. The measurement is made on a flat sample, the syringe holds the water drop, and the plate moves, causing the movement of the drop. The deformation of the drop and therefore the drop adhesion force instrument is quantified by $\Delta\theta$, the difference between the advancing θ_a and receding θ_r contact angles.

Roughness has been determined by confocal microscopy for all the samples using a "non-contact 3D optical profiler" S-neox (Sensofar) microscope (Sensofar metrology, Barcelona, Spain) used in confocal mode. The areal roughness parameter of arithmetic average of profile height deviation from the mean line, S_a parameter, lies in the range (0.5; 1.5 µm), indicating slightly rough coatings."

Nanoindentation tests were performed with an UltraNanoIndenter (CSM Instruments, Anton Paar, Les Ulis, France) at a maximal normal load of 2 mN, corresponding to a penetration depth of about 500 nm in the coating. The loading and unloading rates were

4 mN.min^{-1}. At least 3 indentation tests have been conducted for each sample. As the average thickness of all studied coatings lies above 5 µm, the ratio of the maximum penetration depth to coating thickness is smaller than 10%, thus ensuring that the substrate effect on the measured properties can be considered as negligible. A Berkovich indenter was used, and analysis was conducted using Oliver and Pharr's method [26]. A Poisson ratio of 0.3 was used for the calculation of the mechanical properties.

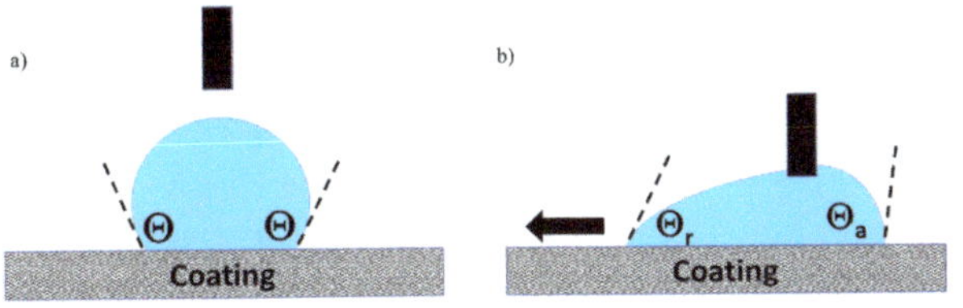

Figure 1. Static (**a**) and dynamic (**b**) contact angle method.

3. Results and Discussion

In this part, the various results corresponding to the coatings without TEOS (with or without DEDMS) are first of all presented (surface micrographs and ^{29}Si MAS and CP MAS NMR spectra).

After that, the results relative to the formulation including TEOS are then compared in terms of microstructural analyses and chemical analyses by ^{29}Si NMR spectra (chemical shifts and proportion).

3.1. MTMS Reference Formulation Coatings (M100)

The hydrolysis of MTMS in an acetic acid medium leads to a transparent sol after 24 h of aging. This sol is sprayed on 4000 series aluminum alloy substrates. The coatings obtained after drying are homogeneous and covering but reveal cracks, as observed in the micrographs in Figure 2 for M100. Cracks are probably formed during the rapid cooling after the curing step.

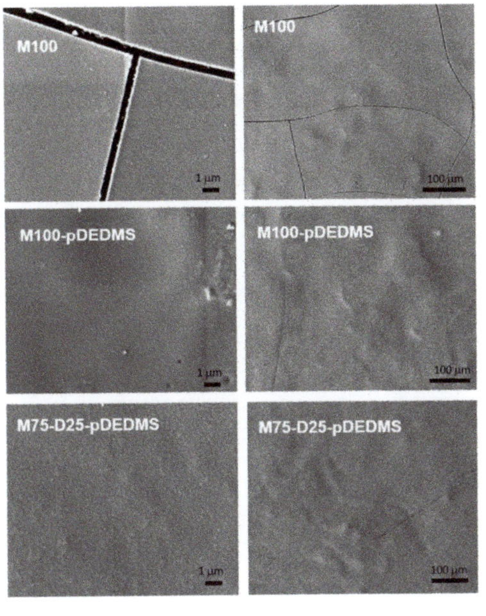

Figure 2. SEM images of coatings without TEOS: M100; M100-pDEDMS and M75-D25-pDEDMS (two magnifications).

In the ^{29}Si NMR study, the assignment of peaks in the different chemical shift ranges was determined based on works described on systems involving MTMS, TEOS and/or DEDMS [19,20,22,23,27]. The inorganic network of the M100 reference coating is only made up of siloxane bonds provided by MTMS. The two resonances observed in the ^{29}Si CPMAS NMR spectrum, Figure 3a, Table 3, at −58.2 and −67.1 ppm are assigned to the T^2 and T^3 silicon atoms, respectively, of the MTMS not completely condensed. The higher proportion of T^3 type silicon atoms is evident in the spectra and is confirmed by the deconvolution of the peaks, which gives a proportion of 19% T^2 silicon atom and 81% T^3 silicon atom. It is interesting to see that after a heat treatment of the coating of only 10 min at 300 °C, the condensation of the siloxane network is almost complete (93.7%, see Table 4), which is surely inherent to the structure of the MTMS precursor consisting of only the T silicon atom type.

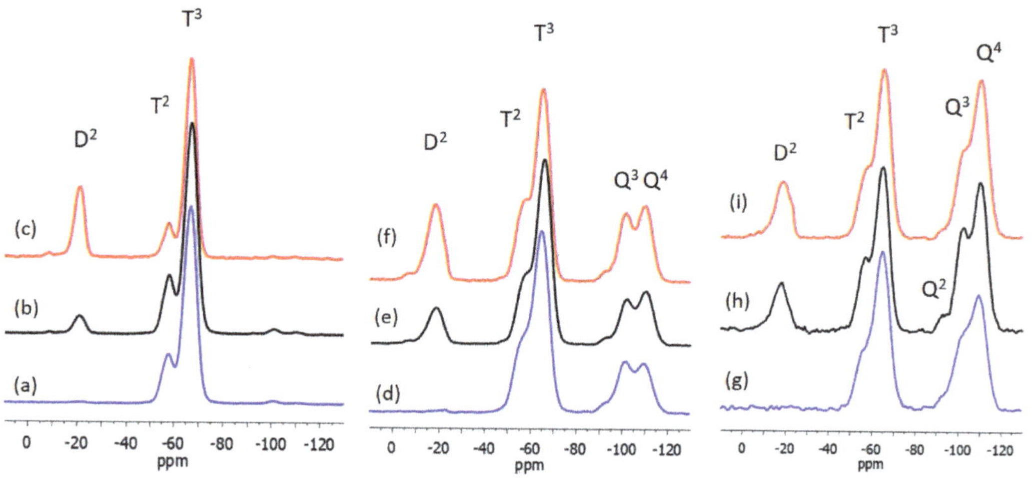

Figure 3. ^{29}Si CP MAS NMR of xerogels corresponding to (**a**) M100; (**b**) M100-pDEDMS; (**c**) M75-D25-pDEDMS; (**d**) T-M100; (**e**) T-M100-pDEDMS and (**f**) T-M75-D25-pDEDMS coatings and ^{29}Si MAS NMR of xerogels corresponding to (**g**) T-M100; (**h**) T-M100-pDEDMS and (**i**) T-M75-D25-pDEDMS coatings.

Table 3. Data extracted from deconvolution of ^{29}Si MAS NMR spectra of the different hybrid materials.

	δ in ppm (Ratio in %)								
	D^1	D^2	D^2	T^2	T^3	Q^2	Q^3	Q^3	Q^4
M100				−58.2 (19)	−67.1 (81)				
M100-pDEDMS		−19.4 (3)	−21.8 (2)	−57.9 (18.5)	−67.0 (76.4)				
M75-D25-pDEDMS		−21.4 (19.6)	−22.8 (1.3)	−58.3 (10.2)	−67.3 (68.9)				
T-M100				−56.3 (13.2)	−65.1 (39)		−98.6 (7.8)	−102.3 (12.2)	−110.1 (27.7)
T-M100-pDEDMS	−11.3 (1.3)	−18.5 (5.7)		−57.2 (14.5)	−65.6 (37.1)	−94.1 (1.1)		−101.9 (12.2)	−110.3 (28.1)
T-M75-D25-pDEDMS	−13.0 (3)	−19.1 (8.5)	−22.9 (0.9)	−57.0 (12.2)	−65.4 (29.4)	−92.9 (1.7)		−101.4 (13.6)	−110.2 (30.7)

Table 4. Degree of condensation of siloxane networks.

	Degree of Condensation (% ± 0.5%)		
	DEDMS	**MTMS**	**TEOS**
	$(D^1 + 2D^2)/2(D^1 + D^2)$	$(2T^2 + 3T^3)/3(T^2 + T^3)$	$(2Q^2 + 3Q^3 + 4Q^4)/4(Q^2 + Q^3 + Q^4)$
M100		93.7	
M100-pDEDMS	100	93.5	
M75-D25-pDEDMS	100	95.7	
T-M100		91.6	85.4
T-M100-pDEDMS	91	90.6	91.3
T-M75-D25-pDEDMS	100	90.2	90.8

3.2. DEDMS Incorporation in MTMS Based Coatings

DEDMS is introduced in the MTMS formulation either as a molecular precursor and/or as a precondensed DEDMS (pDEDMS) silicone oil. The introduction of only 2 wt.% of pDEDMS in the formulation M100-pDEDMS is enough to improve the properties of the coating, since the surface of the coating appears without any cracks, as it can be seen in Figure 2. In the ^{29}Si CPMAS NMR spectrum, as shown in Figure 3b, very few changes were observed since the T^2 and T^3 signals are always present at −57.9 and −67.0 ppm with almost the same proportion as in M100. The presence of pDEDMS slightly increases the T^2:T^3 proportion to 21:79 in favor of T^2, indicating a slight reduction in the condensation of the siloxane network. The D^2 signal due to the silicon atoms of the precondensed pDEDMS (SiC_2O_2) is visible around −21.0 ppm (two resonances at −19.4 and −21.8 ppm, Table 3) assigned, respectively, to D^2 units of short-chain linear polymers (n > 5), $HO(CH_3)_2SiO[Si^*(CH_3)_2O]_nSi(CH_3)_2OH$ [28]. Silicon atoms of the D^1 unit are expected to appear between −11 and −13 ppm, so an absence of resonance in this range indicates that the pDEDMS is linked to the siloxane network of the MTMS.

A new experiment was evaluated by introducing a large proportion of DEDMS as a precursor directly mixed with MTMS with the same amount of pDEDMS as in the M100-pDEDMS. This formulation corresponds to the M75-D25-pDEDMS coating whose molar proportions are detailed in Table 2. The coating is also covering and homogeneous, without cracks, revealing a relatively homogeneous nanostructure. This behavior could be due to the pDEDMS as already observed in the previous experiment. It is also expected that free DEDMS of this experiment would be co-condensed with the main organosilane MTMS, since it is introduced at the same time during the preparation of the sol leading to a higher flexibility of the network, and it also condensed itself for the in situ generation of a short chain of polydimethysiloxane oil, leading also to the same effect as observed for pDEDMS.

The ^{29}Si CP MAS NMR spectrum of the M75-D25-pDEDMS sample is presented in Figure 4c. Three types of silicon atoms clearly appear at −21.3, −58.4 and −67.3 ppm (see Table 3). In the low field range, the signal corresponds to the D^2 silicon atom of the short chain of the p-DEDMS. The presence of DEDMS in the bulk of the matrix is highlighted with the modification of the proportions of the T^2 and T^3 signals, respectively, at −58.4 and −67.3 ppm. The presence of DEDMS led to a slight increase in the proportion of T^3 and therefore to better condensation of the siloxane network; the condensation degree of MTMS increases from 93.5 to 95.7% (Table 4). Furthermore, this indicates that the DEDMS also condenses with the MTMS modifying the structure of the hybrid matrix, creating different T^3 units (T^3-D^2-T^3 or T^3-D^2-T^2).

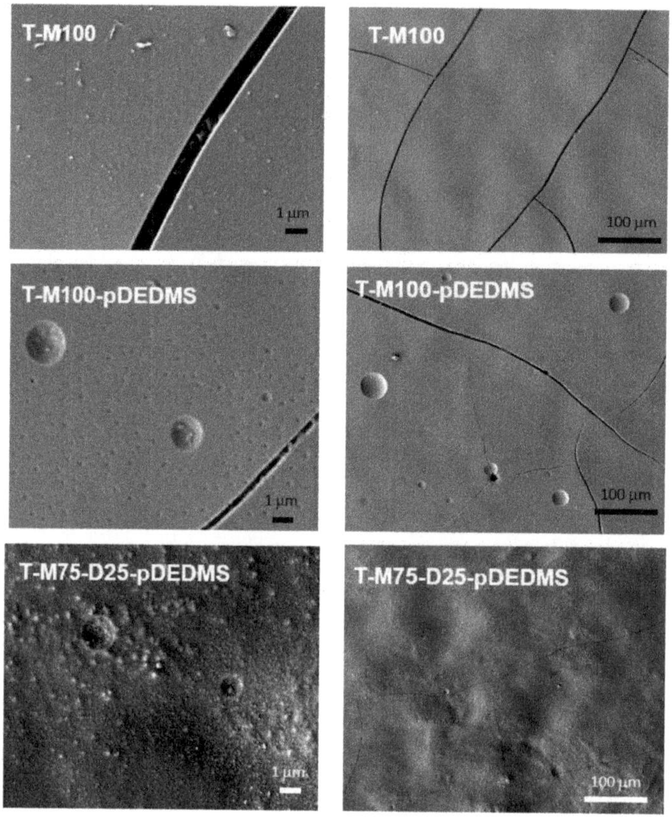

Figure 4. SEM images of coatings containing TEOS: T-M100, T-M100-pDEDMS and T-M75-D25-pDEDMS (2 magnifications).

3.3. *Effect of TEOS Addition on the MTMS and MTMS/DEDMS Formulations*

Based on the previous MTMS/DEDMS coatings, the introduction of TEOS into M100; M100-pDEDMS and M75-D25-pDEDMS formulations was investigated to evaluate the possibility of improving the mechanical properties of the coatings. The composition of these formulations is detailed in Table 2. TEOS was introduced in close to 1:1 molar amounts for MTMS:TEOS in all formulations. SEM images of the surface of the three corresponding coatings are presented in Figure 4 at two magnifications. In both cases, the coatings appeared covering and crack-free.

Round shape elements with a diameter of several microns appear on the coating, which is probably due to a heterogeneous incorporation of the DEDMS silicon oil from pDEDMS or from in situ condensed DEDMS at a submicronic scale.

In T-M100-pDEDMS coatings containing TEOS, the substitution of 25%mol of the MTMS by the DEDMS precursor, i.e., T-M75-D25-pDEDMS, makes the surfaces more heterogeneous (Figure 4).

The ^{29}Si CP MAS NMR spectra of the T-M100-pDEDMS and T-M75-D25-pDEDMS samples are presented in Figure 3e,f together with the corresponding MAS experiments in Figure 3h,i. Spectra of the T-M100 sample are also given for comparison (Figure 3d,g). The data, chemical shifts and proportions obtained by deconvolution are reported in Table 3. The degrees of condensation of the three different organosilanes are calculated using the proportions obtained by the deconvolution of the ^{29}Si MAS NMR spectra, and the values are summarized in Table 4.

The introduction of TEOS on the M100 formulation and coating increases the rigidity of the network as expected with this precursor capable of potentially creating four siloxane bonds. Even if the networks remain well condensed (>85% for TEOS and >90% for MTMS), the degree of condensation of the MTMS clearly decreases in the presence of TEOS from 93.7% to 91.6%, showing that MTMS and TEOS react well together. The co-condensation of MTMS and TEOS in T-M100 is also demonstrated by the low field shift of the T2 and T3 signals from −58.2 and −67.1 ppm, for xerogels without TEOS, to −56.3 and −65.1 ppm in T-M100.

In the ^{29}Si CP MAS NMR spectrum of T-M100-pDEDMS (Figure 4e), again, signals are observed in the three ranges of chemical shifts corresponding to the three types of silicon atoms of DEDMS, MTMS and TEOS precursors. These signals appear at –18.5 ppm (D^2), −57.2 (T^2), −65.6 (T^3), −94.1 (Q^2), −101.9 (Q^3), and −110.3 ppm (Q^4). These last two peaks are characteristic of silicon atoms of Q^3, $SiO_3(OH)$, and Q^4 units, SiO_4, which are usually observed for an incompletely condensed siloxane network obtained by the hydrolysis of TEOS. All condensation degrees were calculated from the deconvolution of the ^{29}Si MAS NMR experiments, which is especially important for the degree of condensation of the TEOS, which is minimized in the CP MAS experiments. In the absence of TEOS in the formulation, the MAS and CPMAS spectra are really very similar, which is why only the MAS spectra of T-M100, T-M100-pDEDMS and T-M75-D25-pDEDMS samples are presented, as shown in Figure 3g–i, respectively. The amount of Q^2, $SiO_2(OH)_2$ units is always very low (1.1%), and the degree of condensation of TEOS greatly increases from 85% in T-M100 to 91.3% in T-M100-pDEDMS. These changes in the NMR spectrum highlight the effect of pDEDMS, which seems to react with the siloxane network constituted by TEOS. The reaction of DEDMS with TEOS leads to a more condensed network more easily (fewer reactive sites required) than the MTMS-TEOS reaction. This can be seen in the slight decrease in the condensation degree of MTMS (91.6 to 90.6%). The T^2:T^3 ratio further evolves in favor of T^2 as already observed when DEDMS MTMS co-condensation occurs. In T-M100-pDEDMS, the T^2:T^3 ratio reaches 14.5:37.1.

When DEDMS is introduced in a larger amount, T-M75-D25-pDEDMS, the same effect is observed with condensation degrees of MTMS and TEOS (90.2 and 90.8%, respectively) higher than those of T-M100 for TEOS and lower for the condensation of the MTMS. In T-M75-D25-pDEDMS NMR spectra (Figure 3f,i), even if an additional resonance is observed in the D^2 units range at −22.9 ppm (0.9%) indicating the presence of a few longer linear polymer chains, the presence of D^2 units (Q-D^2-Q or D-D^2-Q) at −19 and −13 ppm highlights the co-condensation of DEDMS and TEOS in the siloxane network.

The hydrophobic and crack resistance properties were now investigated, and the results are reported in Figures 5 and 6.

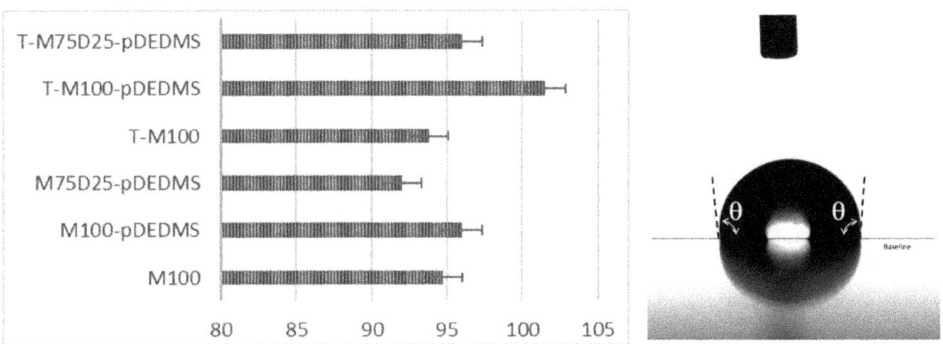

(a)

Figure 5. *Cont.*

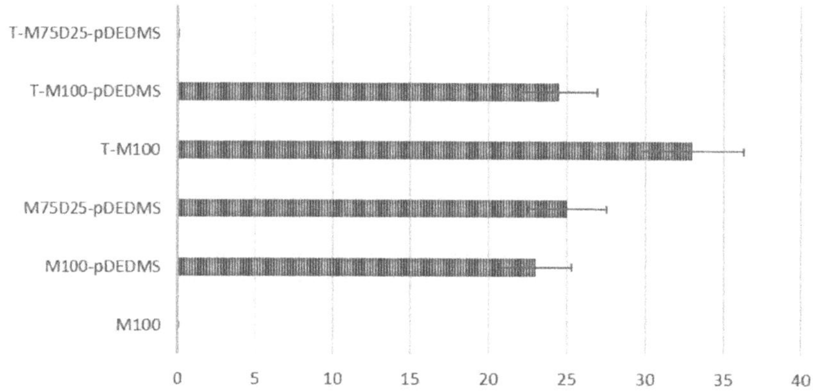

(b)

Figure 5. (**a**) Static contact angle θ (°) for the six coatings and typical image measured for M100-pDEDMS coating. (**b**) Dynamic contact angle hysteresis Δθ (°) for the six coatings.

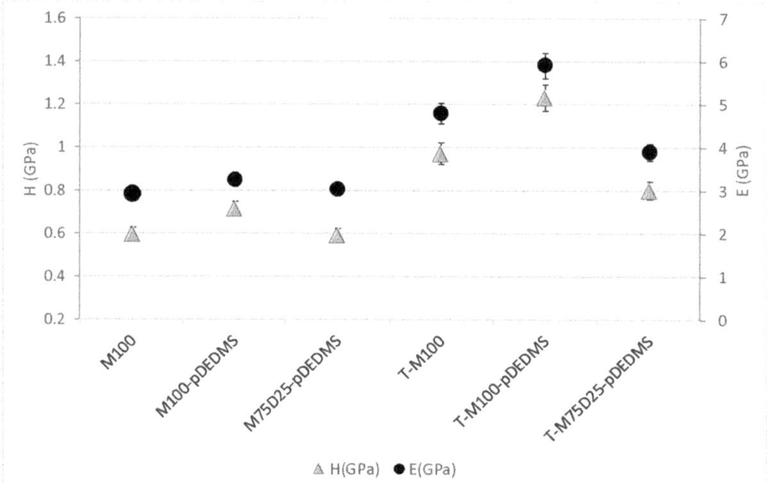

Figure 6. Hardness H and elastic modulus E for all coatings.

3.4. Hydrophobicity of DEDMS/MTMS Coatings

In this study, the incorporation of DEDMS was intended to improve the hydrophobic properties of the coatings. The presence of type D^2 silicon atoms (SiC_2O_2) reduces the number of residual Si-OH bonds in the siloxane network, which is observed in NMR with the evolution of the T^2:T^3 ratio in the presence of DEDMS or p-DEDMS, increasing the hydrophobic character. The contact angles in static mode are not very different in the series of six samples and remain around 95° except for the T-M100-pDEDMD sample for which it reaches 102°, characterizing hydrophobic coatings. Among these values, it is worth noting the contact angle value of the coating containing only MTMS. In fact, this very high value for this coating composed of 100% MTMS is among the highest contact angle values described for this type of coating. It can be compared to the value of 76° described by Juan-Diaz et al. [29]. This very hydrophobic character is explained in our case by the very high condensation rate measured by ^{29}Si NMR. Moreover, this hydrophobic character is preserved for all the studied coatings.

In addition to static contact angle measurements, dynamic contact angle measurements were performed to compare the hydrophobicity of the sol-gel coatings T-100, T-M100-pDEDMS and T-M75-D25-pDEDMS to previous measurements on free-TEOS coatings. Figure 5a,b present, respectively, the results obtained in statics and dynamics. Hysteresis was not measurable for the M100 formulation and for T-M75-D25 –pDEDMS. For the first coating, this is probably due to the presence of cracks leading to water penetration into the coating, while for the second, a higher sample roughness can be considered. The addition of silicone oil in the form of p-DEDMS has a great impact on the hydrophobic behavior, since a low contact angle hysteresis was measured for both formulations M100-pDEDMS and M75-D25-pDEDMS, the first being the lowest with a $\Delta\theta$ value less than 25° as compared to the M-100 reference formulation. A similar hysteresis angle between the two formulations M100-pDEDMS and M75-D25-pDEDMS indicates that the introduction of DEDMS as a precursor does not change the coating properties as drastically as the p-DEDMS does (both for hydrophobicity and cracking).

3.5. Mechanical Properties of DEDMS/MTMS Coatings

To evaluate the mechanical properties of these coatings, a quantitative analysis was carried out through nanoindentation tests. The hardness and elastic modulus of all the formulations are presented in Figure 6.

When TEOS precursor is added (Figure 6), it appears that the hardness and elastic modulus increase (40% increasing) for all coatings containing TEOS compared to formulations without TEOS. This effect can be linked to the increase in inorganic network fraction and to the formation of quaternary silicon-oxygen tetrahedra "Q" in the silicate structure, as seen in Table 3. This could lead to a stiffer network, and it is in good agreement with the literature. Bautista et al. [17] have indeed shown that a 20% addition of TEOS to a methacryloxypropyltrimethoxysilane (MAPTMS)-based sol-gel coating leads to an increase in both hardness (0.60 to 0.81 GPa) and elastic modulus (4.5 to 6.3 GPa).

The DEDMS precursor in substitution of the MTMS led to a decrease in the mechanical properties (both E and H) of the coating. This tendency is particularly noticeable in the TEOS formulation with a decrease of about 20%. This effect could be due to the fact that the partial substitution of the MTMS with DEDMS leads to either a lower degree of condensation (less stiff network) or a higher organic content due to the additional methyl group present in the DEDMS precursor and thus in the obtained film. As seen in Table 4, regarding whether the degree of condensation is similar or there is a light increase for M75-D25-pDEDMS and T-M75-D25-pDEDMS compared to M100-pDEDMS and M100-pDEDMS, data tend to support the latter hypothesis.

More surprisingly, precondensed DEDMS addition seems to have a slight increasing effect on the hardness and the elastic modulus of the obtained coatings. The introduction of precondensed DEDMS, as in M100-pDEDMS, increases the Young's modulus from 3 to 4 GPa and the hardness from 0.55 to 0.70 GPa as compared to the reference M100 coating. In that respect, the T-M100-pDEDMS presents the higher mechanical properties of all coatings.

4. Conclusions

Sol-gel coatings with a silicon oil addition have been carried out in the mixed MTMS-DEDMS-TEOS system. The aim was to modulate jointly or separately the mechanical properties and hydrophobicity of the coatings. First of all, the beneficial effect of silicon oil has been highlighted. In fact, there is a significative difference between in situ generated or precondensed DEDMS. In the first case, the effect is neutral but in the second one, there is a major effect both on the limitation of cracking and also on the hydrophobic properties. The formulation with TEOS, MTMS and only 2 wt.% of precondensed DEDMS silicon oil has shown the best hydrophobic behavior. The presence of TEOS in the coating also improves the mechanical properties by increasing the hardness and the Young modulus following the same evolution. To improve the comprehension of the mechanisms at a microscopic scale, the integration of silicon oil has been followed by NMR analysis. The addition of

DEDMS in the sol-gel network has a negative effect on the microstructure and does not improve the mechanical and hydrophobic properties. Relative to TEOS, the increase in the mechanical properties is proved. The presence of this precursor in the system acts on the decrease in MTMS condensation degree, but the beneficial effect of TEOS condensation results in a stiffening of the global system. A similar conclusion can be drawn on the effect of pDEDMS. Even in a very small proportion (2 wt.%), T-M100-pDEDMS compared to T-M100 presents a higher condensation degree of TEOS (91% vs. 85%, respectively), which explains that with pDEDMS, the stiffening increases and the mechanical properties H and E also increase. To evidence the correlation existing between the formulation (based on the three precursors) and the properties, a radar diagram was drawn, as shown in Figure 7. It reports the behavior of the 6 sol-gel systems with respect to the following properties: condensation ratio, hardness, Young's modulus, hydrophobicity (static contact angle) and resistance to cracking. In summary, these three precursors have synergistic effects and make it possible to modulate the functionalities of coatings.

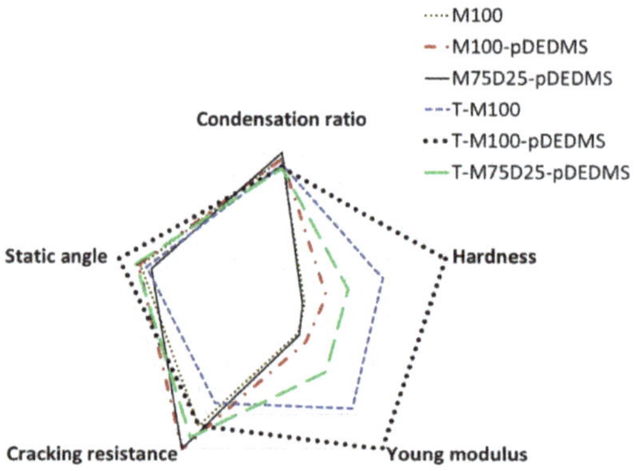

Figure 7. Correlation between the formulation and the functional properties.

Supplementary Materials: The following supporting information can be downloaded at https://www.mdpi.com/article/10.3390/ma17020368/s1, Figure S1: ^{29}Si{^{1}H} NMR spectra of precondensed DEDMS after 4 h in acetic acid medium. Figure S2: Young's modulus E and hardness H as a function of penetration depth, measured by sinus mode indentation (maximum normal force of 10 mN, loading and unloading rate of 20 mN·min^{-1}, sinus amplitude of 0.1 mN, sinus frequency of 12 Hz). Figure S3: Normal load as a function of penetration depth during static nanoindentation test (maximum normal load of 2 mN, loading and unloading rate of 4 mN·min^{-1}) measured for T-M100-pDEDMS sample.

Author Contributions: Conceptualization, F.A. and M.-J.M.; Investigation, C.P., S.D., M.G., V.T., F.A. and M.-J.M.; Project administration, M.-J.M.; Supervision, S.D., M.G., V.T., F.A. and M.-J.M.; Writing—original draft, C.P.; Writing—review and editing, S.D., M.G., V.T., F.A. and M.-J.M. All authors have read and agreed to the published version of the manuscript.

Funding: This research received no external funding.

Informed Consent Statement: Not applicable.

Data Availability Statement: Data are contained within the article and supplementary materials.

Acknowledgments: The authors thank Y. Coppel, research engineer at the Coordinating Chemistry Laboratory for the fruitful discussion on NMR experiments.

Conflicts of Interest: The authors declare no conflicts of interest.

References

1. Figueira, R.B.; Silva, C.J.R.; Pereira, E.V. Organic–inorganic hybrid sol-gel coatings for metal corrosion protection: A review of recent progress. *J. Coat. Technol. Res.* **2015**, *12*, 1–35. [CrossRef]
2. Zhang, W.; Liu, W.; Wang, C. Tribological investigations of sol-gel ceramic films. *Sci. China Ser. B Chem.* **2002**, *45*, 84–90. [CrossRef]
3. Celichowski, G.; Piwonski, I.; Cichomski, M.; Koralewski, K.; Plaza, S.; Olejniczak, W.; Grobelny, J. The influence of methyl group content on tribological properties of organo-silica thin films. *Tribol. Lett.* **2003**, *14*, 181–185. [CrossRef]
4. Rossi, S.; Valdrè, F.; Calovi, M. Validation of adhesion characterization methods for antistick coatings applied in cooking systems. *J. Coat. Technol. Res.* **2022**, *19*, 1287–1301. [CrossRef]
5. Atanacio, A.J.; Latella, B.A.; Barbé, C.J.; Swain, M.V. Mechanical properties and adhesion characteristics of hybrid sol-gel thin films. *Surf. Coat. Technol.* **2005**, *192*, 354–364. [CrossRef]
6. Morales-Florez, V.; Piñero, M.; Braza, V.; del Mar Mesa, M.; Esquivias, L.; de la Rosa-Fox, N. Absorption capacity, kinetics and mechanical behaviour in dry and wet states of hydrophobic DEDMS/TEOS-based silica aerogels. *J. Sol.-Gel. Sci. Technol.* **2017**, *81*, 600–610. [CrossRef]
7. Li, J.; Zhang, H.; Guo, C.; Hong, X.; Sun, L.; Ding, F. Superhydrophobic methylated silica sol for effective oil-water separation. *Materials* **2020**, *13*, 842. [CrossRef]
8. Wu, L.Y.L.; Tan, G.H.; Zeng, X.T.; Li, T.H.; Chen, Z. Synthesis and Characterization of Transparent Hydrophobic Sol-Gel Hard Coatings. *J. Sol.-Gel. Sci. Technol.* **2006**, *38*, 85–89. [CrossRef]
9. Romero-Gavilán, F.; Barros-Silva, S.; García-Cañadas, J.; Palla, B.; Izquierdo, R.; Gurruchaga, M.; Goñi, I.; Suay, J. Control of the degradation of silica sol-gel hybrid coatings for metal implants prepared by the triple combination of alkoxysilanes. *J. Non-Cryst. Solids* **2016**, *453*, 66–73. [CrossRef]
10. Romero-Gavilán, F. Sol-gel coatings made using methyl-modified alkoxysilanes: The balance between protection and bioactivation. *Prog. Org. Coat.* **2020**, *147*, 105770–105780. [CrossRef]
11. Babonneau, F. Hybrid siloxane-oxide materials via sol-gel processing: Structural characterization. *Polyhedron* **1994**, *13*, 1123–1130. [CrossRef]
12. Deng, Q.; Jarrett, W.; Moore, R.B.; Mauritz, K.A. Novel Nafion@/ORMOSIL Hybrids via in situ Sol-Gel Reactions: 2. Probe of ORMOSIL Phase Nanostructure by ^{29}Si Solid State NMR Spectroscopy. *J. Sol-Gel. Sci. Technol.* **1996**, *7*, 177–190. [CrossRef]
13. Richards, C.; Briciu-Burghina, C.; Jacobs, M.R.; Barrett, A.; Regan, F. Assessment of Antifouling Potential of Novel Transparent Sol Gel Coatings for Application in the Marine Environment. *Molecules* **2019**, *24*, 2983. [CrossRef] [PubMed]
14. Salazar-Hernández, C.; Alquiza, M.J.P.; Salgado, P.; Cervantes, J. TEOS–colloidal silica–PDMS-OH hybrid formulation used for stone consolidation. *Appl. Organomet. Chem.* **2010**, *24*, 481–488. [CrossRef]
15. Malzbender, J.; den Toonder, J.M.J.; Balkenende, A.R.; de With, G. Measuring mechanical properties of coatings: A methodology applied to nano-particle-filled sol-gel coatings on glass. *Mater Sci. Eng.* **2002**, *36*, 47–103. [CrossRef]
16. Ballarre, J.; López, D.A.; Cavalieri, A.L. Nano-indentation of hybrid silica coatings on surgical grade stainless steel. *Thin Solid Film.* **2008**, *516*, 1082–1087. [CrossRef]
17. Bautista, Y.; Gómez, M.P.; Ribes, C.; Sanz, V. Relation between the scratch resistance and the chemical structure of organic–inorganic hybrid coatings. *Prog. Org. Coat.* **2011**, *70*, 358–364. [CrossRef]
18. Kalinina, A.; Strizhiver, N.; Vasilenko, N.; Perov, N.; Demchenko, N.; Muzafarov, A. Polycondensation of Diethoxydimethylsilane in Active Medium. *Silicon* **2015**, *7*, 95–106. [CrossRef]
19. Harris, R.K.; Robins, M.L. ^{29}Si nuclear magnetic resonance studiers of oligomeric and polymeric siloxanes: 4. Chemical shift effects of end-groups. *Polymer* **1978**, *19*, 1123–1132. [CrossRef]
20. Suda, S.; Iwaida, M.; Yamashita, K.; Umegaki, T. The effect of sodium ion on hydrolysis and condensation of diethoxydimethylsilane. *J. Non-Cryst. Solids* **1996**, *197*, 65–72. [CrossRef]
21. Iwamoto, T.; Morita, K.; Mackenzie, J.D. Liquid state ^{29}Si NMR study on sol-gel reaction mechanisms of ormosils. *J. Non-Cryst. Solids* **1993**, *159*, 65–72. [CrossRef]
22. Babonneau, F.; Thorne, K.; Mackenzie, J.D. Dimethyldiethoxysilane/tetraéthoxysilane copolymers: Precursors for Si-O-C system. *Chem. Mater* **1989**, *1*, 554–558. [CrossRef]
23. Glaser, R.H.; Wilkes, G.L.; Bronnimann, C.E. Solid-State ^{29}Si NMR of TEOS-based multifunctional sol-gel materials. *J. Non-Cryst. Solids* **1989**, *113*, 73–87. [CrossRef]
24. Massiot, D.; Fayon, F.; Capron, M.; King, I.; Le Calvé, S.; Alonso, B.; Durand, J.-O.; Bujoli, B.; Gan, Z.; Hoatson, G. Modelling one- and two-dimensional solid-state NMR spectra. *Magn. Reason. Chem.* **2002**, *40*, 70–76. [CrossRef]
25. Butt, H.-J.; Liu, J.; Koynov, K.; Straub, B.; Hinduja, C.; Roismann, I.; Berger, R.; Li, X.; Vollmer, D.; Steffen, W.; et al. Contact angle hysteresis. *Curr. Opin. Colloid Interface Sci.* **2022**, *59*, 101574. [CrossRef]
26. Oliver, W.C.; Pharr, G.M. An improved technique for determining hardness and elastic modulus using load and displacement sensing indentation experiments. *J. Mater Res.* **1992**, *7*, 1564–1583. [CrossRef]
27. Olejniczak, Z.; Łęczka, M.; Cholewa-Kowalska, K.; Wojtach, K.; Rokita, M.; Mozgawa, W. ^{29}Si MAS NMR and FTIR study of inorganic–organic hybrid gels. *J. Mol. Struct.* **2005**, *744–747*, 465–471. [CrossRef]

28. Elvira, M.R.; Mazo, M.A.; Tamayo, A.; Rubio, F.; Rubio, J.; Oteo, J.L. Study and Characterization of Organically Modified Silica-Zirconia Anti-Graffiti Coatings Obtained by Sol-Gel. *J. Chem Chem. Eng.* **2013**, *7*, 120–131.
29. Juan-Diaz, M.J.; Martinez-Ibanez, M.; Hernandez-Escolano, M.; Cabedo, L.; Izquierdo, R.; Suay, J.F.; Gurruchaga, M.; Goni, I. Study of the Degradation of hybrid sol-gel coatings in aqueous medium. *Prog. Org. Coat.* **2014**, *77*, 1799–1806. [CrossRef]

MDPI AG
Grosspeteranlage 5
4052 Basel
Switzerland
Tel.: +41 61 683 77 34

Materials Editorial Office
E-mail: materials@mdpi.com
www.mdpi.com/journal/materials

www.ingramcontent.com/pod-product-compliance
Lightning Source LLC
Chambersburg PA
CBHW061616210926
51850CB00009B/2344
* 9 7 8 3 7 2 5 8 5 6 9 7 8 *